AN INTRODUCTION TO
DIFFERENTIAL
GEOMETRY

AN INTRODUCTION TO

DIFFERENTIAL GEOMETRY

BY

T. J. WILLMORE

OXFORD

AT THE CLARENDON PRESS

Oxford University Press, Ely House, London W. I

GLASGOW NEW YORK TORONTO MELBOURNE WELLINGTON
CAPE TOWN SALISBURY IBADAN NAIROBI LUSAKA ADDIS ABABA
BOMBAY CALCUTTA MADRAS KARACHI LAHORE DACCA
KUALA LUMPUR HONG KONG

FIRST PUBLISHED 1959

REPRINTED LITHOGRAPHICALLY IN GREAT BRITAIN
AT THE UNIVERSITY PRESS, OXFORD
FROM CORRECTED SHEETS OF THE FIRST EDITION
1961, 1964 (CORRECTED), 1966

PREFACE

THIS book is intended to be a University textbook, suitable for use in the second and third years of an Honours Mathematics course, or as an introduction to the subject at post-graduate level. It gives all that is likely to be required for an undergraduate course, and most of this material has in fact been taught to undergraduates. The book also gives a useful introduction to the methods of differential geometry or to tensor calculus for research students (e.g. in Physics or Engineering) who may wish to apply them.

It is intended to be an introduction to the essential ideas and methods of differential geometry rather than a comprehensive account of the subject. It does not, for example, consider all the special surfaces and other topics and applications to be found in the standard treatises. These could not possibly be properly included in one volume and are, I feel, more suitable for advanced or special study. For the prospective specialist this book would be a useful introduction but would not replace other standard texts. A student who is familiar with its contents would, however, have no difficulty in studying for himself any particular branch of the subject that he might require.

Part 1 is devoted to the classical theory of curves and surfaces, vector methods being used throughout. My view is that much of the classical theory should be taught before the tensor calculus since the problems and methods are special rather than general. Also, the teaching of tensor calculus is helped considerably by references to classical results as illustrations. The last chapter of Part 1 dealing with the global differential geometry of surfaces contains material which does not appear in any standard English text. Here the student is introduced at an elementary level to the essentially different viewpoint of differential geometry in the large. Although attention is confined to two-dimensional surfaces, many of the concepts involved can be easily extended to apply to n-dimensional differentiable manifolds, and in this way the student is introduced to some of the ideas and techniques which play a prominent role in current research in global differential geometry.

Part 2 introduces the idea of a tensor, first in algebra and then in calculus. It gives the basic theory of the absolute calculus and the fundamentals of Riemannian geometry. The final chapter gives a brief account of the application of tensor methods to yield results previously obtained in Part 1 and some new results in addition.

Several worked examples and exercises have been incorporated in the text, and, with the exception of Chapter IV, each chapter concludes with a set of exercises designed to test the understanding of the subject-matter of that chapter. A list of references is given at the end of each chapter. At the end of the book is a fairly large collection of miscellaneous exercises, most of which have appeared in examination papers of the University of Liverpool. I am grateful to the Senate of the University of Liverpool for permission to include these exercises.

In the preparation of the first three chapters of this book I have benefited very considerably from the valuable advice and criticism of Professor A. G. Walker, F.R.S. I am also very grateful to Dr. W. F. Newns and Dr. A. J. Ledger for their criticism of parts of the book, and to Dr. G. Horrocks and Dr. S. Robertson for their assistance in reading some of the proofs. Finally I wish to express my thanks to the staff of the Clarendon Press for the helpful cooperation which I have received while preparing the book for the press.

T. J. W.

Liverpool,
1958

CONTENTS

PART 1

THE THEORY OF CURVES AND SURFACES IN THREE-DIMENSIONAL EUCLIDEAN SPACE

Contents

PART 2

DIFFERENTIAL GEOMETRY OF
n-DIMENSIONAL SPACE

Contents

Part 1

THE THEORY OF CURVES AND SURFACES IN THREE-DIMENSIONAL EUCLIDEAN SPACE

I

THE THEORY OF SPACE CURVES

1. Introductory remarks about space curves

IN the theory of plane curves, a curve is usually specified either by means of a single equation or else by a parametric representation. For example, the circle centre $(0,0)$ and radius a is specified in Cartesian coordinates (x, y) by the single equation

$$x^2 + y^2 = a^2,$$

or else by the parametric representation

$$x = a \cos u, \qquad y = a \sin u, \qquad 0 \leqslant u \leqslant 2\pi.$$

In the theory of space curves similar alternatives are available. However, in three-dimensional Euclidean space E_3, a single equation generally represents a surface, and two equations are needed to specify a curve. Thus the curve appears as the intersection of the two surfaces represented by the two equations. Parametrically the curve may be specified in Cartesian coordinates by equations

$$x = X(u), \qquad y = Y(u), \qquad z = Z(u) \qquad (1.1)$$

where X, Y, Z are real-valued functions of the real parameter u which is restricted to some interval. Alternatively, in vector notation the curve is specified by a vector-valued function

$$\mathbf{r} = \mathbf{R}(u).$$

Suppose a curve is defined by equations $F(x, y, z) = 0$, $G(x, y, z) = 0$, and it is required to find parametric equations for the curve. If F and G have continuous first derivatives and if at least one of the Jacobian determinants

$$\frac{\partial(F, G)}{\partial(y, z)}, \qquad \frac{\partial(F, G)}{\partial(z, x)}, \qquad \frac{\partial(F, G)}{\partial(x, y)}$$

is not zero at a point (x_0, y_0, z_0) on the curve, it is known from the theory of implicit functions that the equations $F = 0$, $G = 0$ can be solved for two of the variables in terms of the third. For example, when the first Jacobian is non-zero, the variables y and z may be expressed as functions of x, say $y = Y(x)$, $z = Z(x)$, which leads to the parametrization $x = u$, $y = Y(u)$, $z = Z(u)$. However, this solution is valid only for a certain range of x and it will not in general give a parametrization of the whole curve.

Conversely, suppose a curve is given parametrically by equations (1.1) and it is required to find two equations which specify the curve. The straightforward method of solving the first equation to obtain $u = f(x)$ and substituting in the other two equations gives $y = Y(f(x))$, $z = Z(f(x))$, but this solution may be valid only over a restricted range. Other methods of elimination may produce new difficulties as may be seen from the following example of the cubic curve given parametrically by

$$x = u, \qquad y = u^2, \qquad z = u^3, \qquad -\infty < u < \infty.$$

Eliminate u to obtain the equations $xz = y^2$, $xy = z$. These equations represent two quadric surfaces which intersect not only along the given cubic but also along the x-axis.

A parametric representation of a curve specifies not only the curve but also a particular manner in which the curve is described. This is readily seen if the parameter u is interpreted as the time and the curve is considered as the locus of a moving point. The same curve may be parametrized in other ways, and some of the properties of a particular parametric representation may be peculiar to the parametrization and therefore not an intrinsic property of the curve. In this sense, a parametric representation of the curve specifies too much.

On the other hand, the specification of a curve by two equations gives too little information for the purpose of a differential geometer. Quite apart from the fact that, as has been seen, several different curves may be determined by the same pair of equations, there are other disadvantages in specifying a curve in this manner. For example, when considering the distance along a curve from a point P to a point Q, it is often necessary to specify the *sense* in which the curve has been described. To the differential geometer, a curve is not merely a set of points but it must have a sense of description. A parametric representation is not only a convenient way of giving a sense of description but it is also a useful tool for the further study of properties of the curve. A curve will therefore be specified by all its possible parametric representations which are equivalent in that they all describe the same curve with the same sense.

When a curve is regarded as a set of points, it is necessary to decide to what extent a set of points must be restricted before it can be regarded as a curve. Since we shall use parametric representations, it is similarly necessary for us to decide how general a manner of description is to be considered. Since our subject is *differential* geometry, we restrict the manner of description accordingly. As a result, the previous question becomes trivial, for any set of points which *can* be parametrized in the manner we require becomes a suitable object of study.

We are now in a position to make precise technical definitions.

2. Definitions

Functions of class m

Let I be a real interval and m a positive integer. A real-valued function f defined on I is said to be *of class m*, or to be a C^m-*function* if f has an mth derivative at every point† of I and if this derivative is continuous on I.

Briefly we can say that a C^m-function has a continuous mth derivative. When a function is infinitely differentiable we say it is of class ∞, or a C^∞-function, and when a function is analytic we say it is of class ω or a C^ω-function.

The extension of the concept of class to real-valued functions of several real variables is evident, and we leave the formulation of the precise definition to the reader. Briefly we can say that a C^m-function of several variables admits all continuous partial derivatives of the mth order.

A vector-valued function $\mathbf{R} = (X, Y, Z)$ defined on I is said to be of class m if it has an mth derivative at every point and if this derivative is continuous on I; or equivalently, if each of its components X, Y, Z is of class m. Such a function is frequently specified by the vector equation $\mathbf{R} = (X, Y, Z)$ or equivalently by the three

---

† If the left-hand end point of I belongs to I, f is required to have a right-hand derivative at this point. Similarly for the other end point.

equations for the Cartesian components

$$x = X(u), \qquad y = Y(u), \qquad z = Z(u).$$

If the derivative $d\mathbf{R}/du = \dot{\mathbf{r}}$ never vanishes on I—equivalently if $\dot{x}$, $\dot{y}$, $\dot{z}$ never vanish simultaneously—the function is said to be *regular*. A regular† vector-valued function of class m is called a *path* of class m.

Two paths $\mathbf{R}_1$, $\mathbf{R}_2$ of the same class m on I_1, I_2 are called *equivalent* if there exists a strictly increasing function ϕ of class m which maps I_1 onto I_2 and is such that $\mathbf{R}_1 = \mathbf{R}_2 \circ \phi$. The condition $\mathbf{R}_1 = \mathbf{R}_2 \circ \phi$ is equivalent to the three conditions

$$X_1(u) = X_2(\phi(u)), \qquad Y_1(u) = Y_2(\phi(u)), \qquad Z_1(u) = Z_2(\phi(u)).$$

This is easily seen to be a proper equivalence relation because it is reflexive, symmetric, and transitive. The first and third properties are evident while the symmetry follows from the fact that $\dot{\phi}$ is never zero and hence ϕ admits an inverse function of the same class m.

Any equivalence class of paths of class m determines a *curve of class m*. Any path $\mathbf{R}$ determines a unique curve and is called a *parametric representation of the curve*, the variable u being then called the *parameter*. The equations (1.1) are called *parametric equations* of the curve. The mapping ϕ which relates two equivalent paths is called a *change of parameter*. It produces a change in the manner of description of the curve whilst preserving the sense.

We now formally define a curve of class m in E_3 as a set of points in E_3 associated with an equivalence class of regular parametric representations of class m involving one parameter.

As an example of two equivalent representations, consider the *circular helix* given by

(i) $\mathbf{r} = (a\cos u, \ a\sin u, \ bu), \qquad 0 \leqslant u < \pi,$

(ii) $\mathbf{r} = \left(a\dfrac{1-v^2}{1+v^2}, \ \dfrac{2av}{1+v^2}, \ 2b\tan^{-1}v\right), \quad 0 \leqslant v < \infty.$

The change in parameter in this case is

$$v = \phi(u) = \tan\tfrac{1}{2}u.$$

It should be emphasized that not every property of a path is a property of the curve it represents, because some properties are

---

† A point where $\dot{\mathbf{r}} = \mathbf{0}$ is called a singular point. The study of singular points lies outside the scope of this book, and this restriction is made in order to exclude such points.

peculiar to the particular parameter chosen. The properties of the curve are those which are common to all parametric representations, i.e. properties which are invariant under a change of parameter. The verification of this invariance will frequently be left to the reader.

We note that when the function $\mathbf{R}(u)$ is linear, the equation $\mathbf{r} = \mathbf{R}(u)$ represents a straight line.

3. Arc length

The distance between two points $\mathbf{r}_1 = (x_1, y_1, z_1)$, $\mathbf{r}_2 = (x_2, y_2, z_2)$ in Euclidean space is the number

$$|\mathbf{r}_1 - \mathbf{r}_2| = \sqrt{(\mathbf{r}_1 - \mathbf{r}_2)^2} = \sqrt{\{(x_1 - x_2)^2 + (y_1 - y_2)^2 + (z_1 - z_2)^2\}}.$$

This distance in space will be used to define distance along a curve of class $m \geqslant 1$.

If we are given a path $\mathbf{r} = \mathbf{R}(u)$ and two numbers a, b $(a < b)$ in the range of the parameter, then the path $\mathbf{r} = \mathbf{R}(u)$ $(a \leqslant u \leqslant b)$ is an arc of the original path joining the points corresponding to a and b. To any subdivision Δ of the interval (a, b) by points

$$a = u_0 < u_1 < u_2 < \dots < u_n = b$$

there corresponds the length

$$L_\Delta = \sum_{i=1}^{n} |\mathbf{R}(u_i) - \mathbf{R}(u_{i-1})|$$

of the polygon 'inscribed' to the arc by joining successive points on it. Addition of further points of subdivision increases the length of the polygon (because two sides of a triangle are together greater than the third). It is therefore reasonable to *define* the length of the arc to be the upper bound of L_Δ taken over all possible subdivisions of (a, b). This upper bound is always finite, because for any Δ

$$L_\Delta = \sum_{i=1}^{n} \left| \int_{u_{i-1}}^{u_i} \dot{\mathbf{R}}(u)\, du \right| \leqslant \sum_{i=1}^{n} \int_{u_{i-1}}^{u_i} |\dot{\mathbf{R}}(u)|\, du = \int_a^b |\dot{\mathbf{R}}(u)|\, du,$$

$$(3.1)$$

and the right-hand member is finite and independent of Δ. (Note that we used the fact that $\mathbf{R}$ is at least C^1 in the first step. The inequality used in the second step follows as an easy consequence of the Schwarz inequality.)

We now show that this upper bound is actually equal to the right-hand term of (3.1), so that this term gives a formula for the arc length. The definition of arc length implies that if $a < c < b$, then the arc length from a to b is the sum of the arc lengths from a to c and from c to b. We denote by $s = S(u)$ the arc length from a to any point u. Then the arc length from u_0 to u is $S(u) - S(u_0)$. We have just seen that

$$S(u) - S(u_0) \leqslant \int_{u_0}^{u} |\dot{\mathbf{R}}(u)| \, du, \tag{3.2}$$

and it follows from the definition of arc length that

$$|\mathbf{R}(u) - \mathbf{R}(u_0)| \leqslant S(u) - S(u_0). \tag{3.3}$$

Hence

$$\left| \frac{\mathbf{R}(u) - \mathbf{R}(u_0)}{u - u_0} \right| \leqslant \frac{S(u) - S(u_0)}{u - u_0} \leqslant \frac{1}{u - u_0} \int_{u_0}^{u} |\dot{\mathbf{R}}(u)| \, du. \tag{3.4}$$

This formula is equally valid if $u < u_0$. As u tends to u_0 both extreme members of (3.4) tend to the same limit $|\dot{\mathbf{R}}(u_0)|$ and so the middle term tends to this limit, i.e. $\dot{S}(u_0)$ exists and has the value $|\dot{\mathbf{R}}(u_0)|$. Since this is true for any u_0 in the range I of the parameter, it follows that S *is a function of the same class as the curve*, and that

$$s = S(u) = \int_{a}^{u} |\dot{\mathbf{R}}(u)| \, du. \tag{3.5}$$

In terms of a Cartesian parametric representation, this formula becomes

$$s = \int_{a}^{u} \sqrt{(\dot{x}^2 + \dot{y}^2 + \dot{z}^2)} \, du. \tag{3.6}$$

Similarly, the equation $\dot{s} = |\dot{\mathbf{r}}|$ may be written

$$\dot{s}^2 = \dot{x}^2 + \dot{y}^2 + \dot{z}^2,$$

or, in terms of differentials,

$$ds^2 = dx^2 + dy^2 + dz^2.$$

Since $\dot{s}$ never vanishes, s can be used as a new parameter. The function S is the change of parameter from s to u. In order to change from u to s, S is inverted to obtain, say, $u = \phi(s)$; then the curve parametrized with respect to s is $\mathbf{r} = \mathbf{R}(\phi(s))$.

The verification that s is independent of the parametrization follows immediately from the rule for changing the variable of an integral. The above argument is similar to that used by W. F. Newns (1957) in obtaining the formula for the arc length of a plane curve.

EXAMPLE 3.1. Obtain the equations of the circular helix $\mathbf{r} = (a\cos u,\ a\sin u,\ bu)$, $-\infty < u < \infty$ where $a > 0$, referred to s as parameter, and show that the length of one complete turn of the helix is $2\pi c$, where $c = \sqrt{(a^2+b^2)}$.

From (3.6) $s = \int_0^u \sqrt{(a^2+b^2)}\,du = cu$, so the required equations are
$$\mathbf{r} = (a\cos(s/c),\ a\sin(s/c),\ bs/c).$$

The range of u corresponding to one complete turn of the helix is $u_0 \leqslant u \leqslant u_0+2\pi$, so the required length is $2\pi c$.

EXAMPLE 3.2. Find the length of the curve given as the intersection of the surfaces
$$\frac{x^2}{a^2} - \frac{y^2}{b^2} = 1, \qquad x = a\cosh(z/a),$$

from the point $(a, 0, 0)$ to the point (x, y, z).

Write the equation of the curve in the parametric form
$$x = a\cosh u, \qquad y = b\sinh u, \qquad z = au.$$

Then from (3.6) we get
$$s = \int_0^u \sqrt{(a^2\sinh^2 u + b^2\cosh^2 u + a^2)}\,du$$
$$= (a^2+b^2)^{\frac{1}{2}} y/b.$$

4. Tangent, normal, and binormal

From now on the same symbol $\mathbf{r}$ will be used to denote the position vector of a point on a curve and also as the function symbol of a path which represents the curve. With this convention, a curve may be represented by the equation $\mathbf{r} = \mathbf{r}(u)$. The object of the convention is to free the symbol $\mathbf{R}$, which will now be used to denote the position vector of a current point in space not necessarily lying on the curve.

Let γ be a curve of class $\geqslant 1$ and let P, Q be two neighbouring points on the curve. By 'neighbouring points' we do *not* mean that the points P, Q are near in space but that the points have neighbouring values of the parameter.

Let γ be represented by the equation $\mathbf{r} = \mathbf{r}(u)$, and let P and Q have parameters u_0 and u. Since γ has class $\geqslant 1$,
$$\mathbf{r}(u) = \mathbf{r}(u_0) + (u-u_0)\dot{\mathbf{r}}(u_0) + o\,(u-u_0) \qquad (4.1)$$

as $u \to u_0$. Hence

$$\lim_{u \to u_0} \frac{\mathbf{r}(u) - \mathbf{r}(u_0)}{|\mathbf{r}(u) - \mathbf{r}(u_0)|} = \frac{\dot{\mathbf{r}}(u_0)}{|\dot{\mathbf{r}}(u_0)|}, \tag{4.2}$$

i.e. the unit vector along the chord PQ tends to a unit vector at P as $Q \to P$. This is called the *unit tangent vector* to γ at P, and is denoted by $\mathbf{t}$. Using (3.4) we have

$$\mathbf{t} = \frac{\dot{\mathbf{r}}(u_0)}{|\dot{\mathbf{r}}(u_0)|} = \frac{\dot{\mathbf{r}}}{\dot{s}} = \frac{d\mathbf{r}}{ds}. \tag{4.3}$$

Observe that $\mathbf{t}$, like the curve, is *oriented* in that it points in the direction of increasing s. The line through P parallel to $\mathbf{t}$ is called the *tangent line* to γ at P. If R is any point on this line, the vector from the point of contact P to R is called a *tangent vector* to γ at P.

An alternative approach to the definition of tangent line is to note that there is a unique line which approximates to the curve to the first order near P—more precisely, there is a unique linear function $\mathbf{L}(u)$ such that

$$\mathbf{L}(u) = \mathbf{r}(u) + o\,(u - u_0) \quad \text{as } u \to u_0.$$

From (4.1) it can be seen that

$$\mathbf{L}(u) = \mathbf{r}(u_0) + (u - u_0)\dot{\mathbf{r}}(u_0).$$

The line determined by $\mathbf{L}(u)$ could then be defined as the tangent line to γ at P.

It is convenient to denote differentiation with respect to arc length by a prime; with this convention the unit tangent vector becomes

$$\mathbf{t} = \mathbf{r}'. \tag{4.4}$$

Osculating plane

Let γ be a curve of class $\geqslant 2$, and let P, Q be two neighbouring points on γ. Then the limiting position as $Q \to P$ of that plane which contains the tangent line at P and the point Q is called the *osculating plane* of γ at P.

Note that when γ is a straight line the osculating plane is indeterminate at each point, and we exclude this case in what follows. Assume that γ is parametrized with respect to arc length and that the parameters of P, Q are 0 and s respectively. The equation of the plane through the tangent line at P and the point Q is

$$[\mathbf{R} - \mathbf{r}(0),\ \mathbf{r}'(0),\ \mathbf{r}(s) - \mathbf{r}(0)] = 0. \tag{4.5}$$

Also we have

$$\mathbf{r}(s) - \mathbf{r}(0) = s\mathbf{r}'(0) + \tfrac{1}{2}s^2\mathbf{r}''(0) + o\,(s^2). \tag{4.6}$$

Using equations (4.5), (4.6) we find that the equation of the osculating plane is

$$[\mathbf{R} - \mathbf{r}(0),\ \mathbf{r}'(0),\ \mathbf{r}''(0)] = 0, \tag{4.7}$$

provided that the vectors $\mathbf{r}'(0)$, $\mathbf{r}''(0)$ are linearly independent. Since $\mathbf{t}^2 = 1$, i.e. $\mathbf{r}'^2 = 1$, differentiation gives $\mathbf{r}'.\mathbf{r}'' = 0$, and it follows that the vectors $\mathbf{r}'(0)$, $\mathbf{r}''(0)$ are linearly independent unless $\mathbf{r}''(0) = \mathbf{0}$. A point P where $\mathbf{r}'' = \mathbf{0}$ is called a *point of inflexion*, and the tangent line at P is called *inflexional*.

When P is not a point of inflexion, it follows from (4.7) that any vector lying in the osculating plane can be expressed as

$$\alpha\mathbf{r}' + \beta\mathbf{r}''$$

for some coefficients α, β. In particular, the vector $\mathbf{r}''$ lies in the osculating plane.

The question now arises, what happens at a point of inflexion ? *We now show that when a curve is analytic, we still obtain a definite osculating plane at a point of inflexion P unless the curve is a straight line.* To prove this we differentiate the relation $\mathbf{r}'.\mathbf{r}'' = 0$ to get $\mathbf{r}'.\mathbf{r}''' = 0$, since $\mathbf{r}'' = \mathbf{0}$ at P. Hence $\mathbf{r}'$ is linearly independent of $\mathbf{r}'''$ except when $\mathbf{r}''' = \mathbf{0}$. Repetition of this argument leads to the result that $\mathbf{r}'.\mathbf{r}^{(k)} = 0$, where $\mathbf{r}^{(k)}$ is the first non-zero derivative of $\mathbf{r}$ at P ($k \geqslant 2$). If $\mathbf{r}^{(k)} = \mathbf{0}$ for all $k \geqslant 2$, then since the curve is analytic we conclude that $\mathbf{t}$ is constant and the curve is a straight line. If $\mathbf{r}^{(k)} \neq \mathbf{0}$, then (4.6) gives

$$\mathbf{r}(s) - \mathbf{r}(0) = \frac{s^k}{k!}\mathbf{r}^{(k)}(0) + o\,(s^k) \tag{4.8}$$

as $s \to 0$, and the equation of the osculating plane is

$$[\mathbf{R} - \mathbf{r}(0),\ \mathbf{r}'(0),\ \mathbf{r}^{(k)}(0)] = 0. \tag{4.9}$$

Consider now the curve γ defined by

$$\mathbf{r}(u) = (u,\ e^{-1/u^2},\ 0), \quad u < 0,$$
$$\mathbf{r}(u) = (u,\ 0,\ e^{-1/u^2}), \quad u > 0,$$
$$\mathbf{r}(0) = (0, 0, 0).$$

It is readily verified that γ is a curve of class ∞ with the property that $\mathbf{r}^{(k)}(0) = \mathbf{0}$ for all $k \geqslant 2$.

The osculating plane at all points with parameter $u < 0$ is $Z = 0$, while the osculating plane at all points with parameter

$u > 0$ is $Y = 0$. The osculating plane at $u = 0$ is clearly indeterminate. Thus we see that *at a point of inflexion, even a curve of class ∞ need not possess an osculating plane.*

We have dealt with the equation of the osculating plane at P at some length because it enables us to illustrate the type of difficulties which occur at a very early stage, and to express our attitude towards these difficulties in the future. A differential geometer is interested primarily in properties of curves and surfaces which hold for a 'general' point, and tends to ignore what happens at 'special' points. An analyst, however, is interested in what happens at 'special' points, and is particularly interested in giving examples which contradict the general statement which the geometer wishes to assert. Usually a geometer is not interested in the precise class of a curve under discussion provided that it is sufficiently high to enable him to discuss relevant properties of the curve. However, an analyst cannot be uninterested in the problem of determining the smallest class of the curve or relevant functions which enter in the hypothesis of a theorem, and the differential geometer cannot afford to ignore these questions completely. For example, we have seen that the osculating plane may not exist at a point of inflexion on a C^∞-curve.

Throughout this book we shall be concerned essentially with properties at 'general' points, and we shall place considerably less emphasis on what happens at 'special' points. Our viewpoint is that in an introductory work of this kind the reader should not be side-tracked along paths whose interest is analytical rather than geometrical. Moreover, in many cases, the reader himself will be able to supply qualifications to the general argument which may be necessary to deal with 'special' cases.

EXAMPLE 4.1. Show that if a curve is given in terms of a general parameter u, then the equation of the osculating plane corresponding to (4.7) is
$$[\mathbf{R}-\mathbf{r},\ \dot{\mathbf{r}},\ \ddot{\mathbf{r}}] = 0.$$

We have $\mathbf{r}' = \dot{\mathbf{r}}/\dot{s}$, $\mathbf{r}'' = \ddot{\mathbf{r}}/\dot{s}^2 - \dot{\mathbf{r}}\ddot{s}/\dot{s}^3$. Substitution in (4.7) gives the required result.

EXAMPLE 4.2. Find the equation of the osculating plane at a general point on the cubic curve given by $\mathbf{r} = (u, u^2, u^3)$ and show that the osculating planes at any three points of the curve meet at a point lying in the plane determined by these three points.

The equation of the osculating plane is

$$\begin{vmatrix} X-u & Y-u^2 & Z-u^3 \\ 1 & 2u & 3u^2 \\ 0 & 2 & 6u \end{vmatrix} = 0,$$

which reduces to

$$3u^2X - 3uY + Z - u^3 = 0.$$

If u_1, u_2, u_3 are three distinct values of the parameter, the osculating planes at these points are linearly independent and the planes meet at a point (X_0, Y_0, Z_0). The parameters u_1, u_2, u_3 therefore satisfy the condition

$$u^3 - 3u^2X_0 + 3uY_0 - Z_0 = 0.$$

If $lX + mY + nZ + p = 0$ is the equation of the plane passing through the three points, then the parameters must also satisfy the condition
$$lu + mu^2 + nu^3 + p = 0.$$

Since this equation has three distinct roots we have $n \neq 0$; comparing coefficients in the two cubic equations gives

$$l = 3nY_0, \qquad m = -3nX_0, \qquad p = -nZ_0.$$

The equation of the plane is thus

$$3Y_0X - 3X_0Y + Z - Z_0 = 0,$$

and since this is satisfied by (X_0, Y_0, Z_0) the result follows.

The *normal plane* at a point P on a curve is that plane through P which is orthogonal to the tangent at P.

The *principal normal* at P is the line of intersection of the normal plane and the osculating plane at P. A unit vector along the principal normal is denoted by $\mathbf{n}$; its sense may be selected arbitrarily provided that it varies continuously along the curve.[†]

Curvature

The arc-rate at which the tangent changes direction as P moves along the curve is the *curvature* of the curve and is denoted by κ. Thus by definition $|\kappa| = |\mathbf{t}'|$ but the *sign* of κ is not determined. To find a suitable convention, we recall that $\mathbf{t}' = \mathbf{r}''$ lies in the osculating plane and is also normal to $\mathbf{t}$. It is therefore proportional to $\mathbf{n}$, i.e. $\mathbf{t}' = \pm\kappa\mathbf{n}$. We make the convention

$$\mathbf{t}' = \kappa\mathbf{n}. \tag{4.10}$$

[†] The C^∞-curve considered immediately after (4.9) shows that such a choice of $\mathbf{n}$ is not always possible.

If **n** is chosen to vary continuously with s, κ is determined in magnitude and sign by (4.10), but this value may change from positive to negative as P moves along the curve. It should be noted that the ambiguity of the sense of **n** and the sign of κ does not apply to the vector $\mathbf{t}' = \mathbf{r}''$ which is sometimes called the *curvature vector*. When the curve has class $\geqslant 2$, the curvature vector $\mathbf{r}'' = \kappa\mathbf{n}$ varies continuously along the curve, but this does not imply that **n** varies continuously along the curve (cf. footnote, p. 11).

A necessary and sufficient condition that a curve be a straight line is that $\kappa = 0$ at all points. Any straight line has equation of the form $\mathbf{r} = \mathbf{a}s + \mathbf{b}$, where **a** and **b** are constant vectors. Hence $\mathbf{t} = \mathbf{a}$ and $\mathbf{t}' = \mathbf{0}$, so that $\kappa = 0$ is necessary. Conversely if $\kappa = 0$ identically, then $\mathbf{r}'' = \mathbf{0}$ which gives on integration $\mathbf{r} = \mathbf{a}s + \mathbf{b}$, which is the equation of a straight line. Thus the condition is also sufficient.

The *binormal line* at P is the normal in a direction orthogonal to the osculating plane. The sense of the unit vector **b** along the binormal is chosen so that the triad **t**, **n**, **b** form a right-handed system of axes, i.e.

$$\mathbf{b} = \mathbf{t} \times \mathbf{n}. \tag{4.11}$$

Torsion

As P moves along a curve the arc-rate at which the osculating plane turns about the tangent is called the *torsion* of the curve and is denoted by τ. We now obtain the relation

$$\mathbf{b}' = -\tau\mathbf{n}. \tag{4.12}$$

Since $\mathbf{b}^2 = 1$ it follows that $\mathbf{b} \cdot \mathbf{b}' = 0$ and $\mathbf{b}'$ lies in the osculating plane. Also $\mathbf{b} \cdot \mathbf{t} = 0$ implies that $\mathbf{b}' \cdot \mathbf{t} + \mathbf{b} \cdot \mathbf{t}' = 0$; but as $\mathbf{b} \cdot \mathbf{t}' = \mathbf{b} \cdot (\kappa\mathbf{n}) = 0$ it follows that $\mathbf{b}'$ is orthogonal to **t**. But as $\mathbf{b}'$ lies in the osculating plane it must be parallel to **n**. Thus the equation $|\mathbf{b}'| = |\tau|$ follows, where $|\tau|$ is the absolute magnitude of the torsion. The negative sign in (4.12) is introduced since, as a convention, torsion is regarded as positive when the rotation of the osculating plane as s increases is in the direction of a right-handed screw travelling in the direction of **t**. Note that the torsion τ is determined both in magnitude and sign, whereas the curvature κ is determined only in magnitude.

*Let γ be a curve for which **b** varies differentiably with arc length. Then a necessary and sufficient condition that γ be a plane curve is that $\tau = 0$ at all points.* The necessity follows immediately since the osculating plane of a plane curve is just the plane containing the curve, and is therefore fixed. Conversely if $\tau = 0$, then **b** must be a constant vector.   In this case the identity $\mathbf{r}'.\mathbf{b} = 0$ implies $(\mathbf{r}.\mathbf{b})' = 0$, from which $\mathbf{r}.\mathbf{b} = \text{constant}$, showing that the curve is plane. Note that the C^∞-curve considered on p. 9 is not a plane curve but the torsion is zero where it exists.

EXAMPLE 4.3.   Prove that $[\mathbf{r}', \mathbf{r}'', \mathbf{r}'''] = \kappa^2\tau$. Equations (4.10) and (4.11) give $\mathbf{r}' \times \mathbf{r}'' = \kappa\mathbf{b}$. Differentiate and use (4.12) to get $\mathbf{r}' \times \mathbf{r}''' = \kappa'\mathbf{b} - \kappa\tau\mathbf{n}$. Scalar multiplication by $\mathbf{r}''$ then gives the required result.

EXAMPLE 4.4. Show that $[\dot{\mathbf{r}}, \ddot{\mathbf{r}}, \dddot{\mathbf{r}}] = 0$ is a necessary and sufficient condition that the curve be plane. Evidently

$$[\mathbf{r}', \mathbf{r}'', \mathbf{r}'''] = [\dot{\mathbf{r}}u', \ddot{\mathbf{r}}u'^2 + \dot{\mathbf{r}}u'', \dddot{\mathbf{r}}u'^3 + 3\ddot{\mathbf{r}}u'u'' + \dot{\mathbf{r}}u'''] = u'^6[\dot{\mathbf{r}}, \ddot{\mathbf{r}}, \dddot{\mathbf{r}}],$$

so, using example 4.3, $[\dot{\mathbf{r}}, \ddot{\mathbf{r}}, \dddot{\mathbf{r}}] = \kappa^2\tau(u')^{-6}$. If the left-hand member is zero, then either $\kappa = 0$ or $\tau = 0$. We prove that $\tau = 0$ *always* as follows. Suppose $\tau \neq 0$ at some point, then there is a neighbourhood of this point where $\tau \neq 0$. Hence $\kappa = 0$ in this neighbourhood and the arc is a straight line. Then $\tau = 0$ on this line, contrary to hypothesis, and hence $\tau = 0$ at all points and the curve is therefore plane. Conversely, if the curve is plane then $\tau = 0$ and so $[\dot{\mathbf{r}}, \ddot{\mathbf{r}}, \dddot{\mathbf{r}}] = 0$. The condition is thus necessary and sufficient.

EXAMPLE 4.5.   Calculate the curvature and torsion of the cubic curve given by $\mathbf{r} = (u, u^2, u^3)$.

Evidently $\qquad\qquad \dot{s}\mathbf{t} = \dot{\mathbf{r}} = (1, 2u, 3u^2).$ $\qquad\qquad$ (i)

Differentiate to get

$$\ddot{s}\mathbf{t} + \dot{s}^2\kappa\mathbf{n} = (0, 2, 6u). \qquad\qquad \text{(ii)}$$

Take the vector product of (ii) and (i) to get

$$\dot{s}^3\kappa\mathbf{b} = 2(3u^2, -3u, 1). \qquad\qquad \text{(iii)}$$

Differentiate to get

$$\mathbf{b}.\frac{d(\dot{s}^3\kappa)}{du} - \dot{s}^4\kappa\tau\mathbf{n} = 6(2u, -1, 0). \qquad\qquad \text{(iv)}$$

Take the scalar product of (ii) and (iv) to get

$$-\dot{s}^6\kappa^2\tau = -12. \qquad\qquad \text{(v)}$$

From (i), $\dot{s}^2 = (9u^4 + 4u^2 + 1)$. From (iii), $\dot{s}^6\kappa^2 = 4(9u^4 + 9u^2 + 1)$, from which

$$\kappa^2 = \frac{4(9u^4 + 9u^2 + 1)}{(9u^4 + 4u^2 + 1)^3}.$$

From (v), $\qquad\qquad\qquad \tau = \dfrac{3}{(9u^4 + 9u^2 + 1)}.$

This example shows the procedure which should be followed when solving problems of this type. It is not difficult to obtain general formulae from which the values of κ and τ for any given curve can be calculated by mere substitution, but it is advisable to use the method given and to treat each curve on its merits.

Serret–Frenet formulae

The relations

$$\mathbf{t}' = \kappa\mathbf{n}, \qquad \mathbf{n}' = \tau\mathbf{b}-\kappa\mathbf{t}, \qquad \mathbf{b}' = -\tau\mathbf{n}, \qquad (4.13)$$

are known as the Serret–Frenet formulae, and these underlie many investigations in the theory of curves and surfaces. The reader is well advised to commit them to memory.

The first and third relations have already been obtained. The second relation follows from differentiating the identity $\mathbf{n} = \mathbf{b}\times\mathbf{t}$ to get $\mathbf{n}' = -\tau\mathbf{n}\times\mathbf{t}+\mathbf{b}\times\kappa\mathbf{n} = \tau\mathbf{b}-\kappa\mathbf{t}$.

The behaviour of a curve in the neighbourhood of one of its points may be investigated by means of relations (4.13). At a point P on the curve let axes Ox, Oy, Oz be taken along $\mathbf{t}$, $\mathbf{n}$, and $\mathbf{b}$, and let X, Y, Z be the coordinates of a neighbouring point Q of the curve relative to these axes. If the curve is of class $\geqslant 4$ and if s denotes the small arc length PQ, then, using Taylor's theorem,

$$\mathbf{r}(s) = \mathbf{r}(0)+s\mathbf{r}'(0)+\frac{s^2}{2!}\mathbf{r}''(0)+\frac{s^3}{3!}\mathbf{r}'''(0)+\frac{s^4}{4!}\mathbf{r}^{(iv)}(0)+o\,(s^4) \text{ as } s \to 0.$$

When the derivatives of $\mathbf{r}$ are substituted from (4.13) this becomes

$$\mathbf{r}(s) = \mathbf{r}(0)+s\mathbf{t}+\frac{s^2}{2!}\kappa\mathbf{n}+\frac{s^3}{3!}(-\kappa^2\mathbf{t}+\kappa'\mathbf{n}+\kappa\tau\mathbf{b})+$$

$$+\frac{s^4}{4!}\{-3\kappa\kappa'\mathbf{t}+(\kappa''-\kappa\tau^2-\kappa^3)\mathbf{n}+(2\kappa'\tau+\kappa\tau')\mathbf{b}\}+o\,(s^4).$$

Hence

$$\left.\begin{aligned} X &= s-\frac{\kappa^2}{6}s^3-\frac{\kappa\kappa'}{8}s^4+o\,(s^4) \\[1ex] Y &= \frac{\kappa}{2}s^2+\frac{\kappa'}{6}s^3+\frac{\kappa''-\kappa\tau^2-\kappa^3}{24}s^4+o\,(s^4) \\[1ex] Z &= \frac{\kappa\tau}{6}s^3+\frac{2\kappa'\tau+\kappa\tau'}{24}s^4+o\,(s^4) \end{aligned}\right\}, \qquad (4.14)$$

the coefficients being evaluated at P. It follows that as a first-order approximation the chord PQ is along the tangent: its projection on the principal normal is a magnitude of the second order, and its projection on the binormal is of the third order.

From equation (4.14) two relations can be deduced which are analogous to Newton's formula for the curvature of plane curves. These are

$$\frac{2Y}{X^2} \sim \kappa \quad \text{as} \quad s \to 0$$

and

$$\frac{3Z}{XY} \sim \tau \quad \text{as} \quad s \to 0.$$

Moreover, it is easy to show that

$$(X^2+Y^2+Z^2)^{\frac{1}{2}} \sim s\left(1-\frac{\kappa^2 s^2}{24}\right),$$

from which it is seen that when $\kappa \neq 0$, the arc length PQ differs from the chord PQ by terms of the third order in s.

It is convenient to refer to the plane determined by the tangent and binormal at P as the *rectifying* plane.

EXAMPLE 4.6. Show that the projection of the curve near P on the osculating plane is approximately the curve $Z = 0, Y = \frac{1}{2}\kappa X^2$; its projection on the rectifying plane is approximately $Y = 0$, $Z = \frac{1}{6}\kappa\tau X^3$; and its projection on the normal plane is approximately $X = 0, Z^2 = \frac{2}{9}(\tau^2/\kappa)Y^3$.

These results follow immediately from equations (4.14) by retaining only the lowest powers of s in each equation and then eliminating s from the equations in pairs.

EXAMPLE 4.7. Show that the length of the common perpendicular d of the tangents at two near points distance s apart is approximately given by

$$d = \frac{\kappa\tau s^3}{12}.$$

Let P, Q have parameters 0 and s respectively. The unit tangent vectors at P and Q are $\mathbf{r}'(0)$, $\mathbf{r}'(s)$, so the unit vector of the common perpendicular is along $\mathbf{r}'(s) \times \mathbf{r}'(0)$. The projection of the vector $(\mathbf{r}(s)-\mathbf{r}(0))$ in this direction is equal to d, so

$$d = [\mathbf{r}(s)-\mathbf{r}(0), \mathbf{r}'(s), \mathbf{r}'(0)]/\, |\mathbf{r}'(s) \times \mathbf{r}'(0)|.$$

Use the relations

$$\mathbf{r}(s)-\mathbf{r}(0) = s\mathbf{t}+\tfrac{1}{2}s^2\kappa\mathbf{n}+\tfrac{1}{6}s^3(-\kappa^2\mathbf{t}+\kappa'\mathbf{n}+\kappa\tau\mathbf{b})+o\,(s^3) \text{ as } s \to 0,$$

$$\mathbf{r}'(s) = \mathbf{t}+s\kappa\mathbf{n}+\tfrac{1}{2}s^2(-\kappa^2\mathbf{t}+\kappa'\mathbf{n}+\kappa\tau\mathbf{b})+o\,(s^2) \text{ as } s \to 0,$$

and simplify to obtain the required formula.

5. Curvature and torsion of a curve given as the intersection of two surfaces

If a curve is given as the intersection of two surfaces

$$f(x, y, z) = 0, \qquad g(x, y, z) = 0,$$

and if a set of parametric equations for the curve cannot readily be obtained, then the curvature and torsion of the curve may be calculated by the following method.

Let the curve of intersection be represented by the equation $\mathbf{r} = \mathbf{r}(u)$, and let the two surfaces be given by $f(\mathbf{r}) = 0, g(\mathbf{r}) = 0$. Now the unit tangent vector to the curve is orthogonal to the normals of both surfaces. Thus, if $\nabla f = \left(\dfrac{\partial f}{\partial x}, \dfrac{\partial f}{\partial y}, \dfrac{\partial f}{\partial z}\right)$, it follows that $\mathbf{t}$ is parallel to $\nabla f \times \nabla g = \mathbf{h}$, say $\lambda \mathbf{r}' = \nabla f \times \nabla g$. Then

$$\lambda x' = h_1, \quad \lambda y' = h_2, \quad \lambda z' = h_3,$$

and

$$\lambda d/ds = \left(h_1 \frac{\partial}{\partial x} + h_2 \frac{\partial}{\partial y} + h_3 \frac{\partial}{\partial z}\right). \tag{5.1}$$

It is convenient to denote the operator (5.1) by Δ. Then

$$\Delta \mathbf{r} = \mathbf{h}. \tag{5.2}$$

From the definition of λ and $\mathbf{h}$ it follows that

$$\lambda \mathbf{t} = \mathbf{h}, \tag{5.3}$$

and so

$$\lambda^2 = \mathbf{h}^2. \tag{5.4}$$

Operating on (5.3) with Δ gives

$$\lambda^2 \kappa \mathbf{n} + \lambda \lambda' \mathbf{t} = \Delta \mathbf{h}. \tag{5.5}$$

Taking the vector product of (5.3) and (5.5) gives

$$\lambda^3 \kappa \mathbf{b} = \mathbf{h} \times \Delta \mathbf{h} = \mathbf{k}, \text{ say}, \tag{5.6}$$

from which

$$|\mathbf{k}| = \lambda^3 \kappa. \tag{5.7}$$

Operating on (5.6) with Δ gives

$$\lambda(\lambda^3 \kappa)' \mathbf{b} - \lambda^4 \kappa \tau \mathbf{n} = \Delta \mathbf{k}. \tag{5.8}$$

The scalar product of (5.5) and (5.8) now gives

$$-\lambda^6 \kappa^2 \tau = \Delta \mathbf{h} . \Delta \mathbf{k}. \tag{5.9}$$

From these equations κ and τ are calculated in the usual manner. It will be seen that the right-hand members of equations (5.2) to (5.9) are all readily expressible in terms of the given functions f and g.

Some procedure similar to that outlined above should be used when solving problems of this type; but sometimes it is advisable to modify slightly the definition of Δ in order to simplify the calculations, as is done in the example below. The method of this section should be used as a general guide and each separate example should be treated on its merits.

EXAMPLE 5.1. Obtain the curvature and torsion of the curve of intersection of the two quadric surfaces

$$ax^2+by^2+cz^2 = 1, \qquad a'x^2+b'y^2+c'z^2 = 1.$$

Write

$$f = \tfrac{1}{2}(ax^2+by^2+cz^2-1); \qquad g = \tfrac{1}{2}(a'x^2+b'y^2+c'z^2-1).$$

Then $\qquad \boldsymbol{\nabla}f = (ax, by, cz); \qquad \boldsymbol{\nabla}g = (a'x, b'y, c'z).$

Hence $\qquad\qquad \boldsymbol{\nabla}f \times \boldsymbol{\nabla}g = (Ayz, Bzx, Cxy)$

where $\qquad A = bc'-b'c, \qquad B = ca'-c'a, \qquad C = ab'-a'b.$

Then $\qquad\qquad \lambda\mathbf{t} = \lambda\mathbf{r}' = (A/x, B/y, C/z) \qquad\qquad$ (i)

from which $\qquad\qquad \lambda^2 = \sum (A/x)^2.$

Equation (5.1) gives

$$\lambda\frac{d}{ds} = \left(\frac{A}{x}\frac{\partial}{\partial x}+\frac{B}{y}\frac{\partial}{\partial y}+\frac{C}{z}\frac{\partial}{\partial z}\right). \qquad\text{(ii)}$$

Corresponding to (5.5) we have

$$\lambda^2\kappa\mathbf{n}+\lambda\lambda'\mathbf{t} = -(A^2/x^3, B^2/y^3, C^2/z^3). \qquad\text{(iii)}$$

The vector product of (i) and (iii) gives

$$\lambda^3\kappa\mathbf{b} = \left\{\frac{BC}{y^3z^3}(Bz^2-Cy^2), \frac{CA}{z^3x^3}(Cx^2-Az^2), \frac{AB}{x^3y^3}(Ay^2-Bx^2)\right\}.$$

Since $\qquad\qquad Bz^2-Cy^2 = (a'-a), \quad\text{etc.,}$

this simplifies to

$$\lambda^3\kappa\mathbf{b} = \frac{ABC}{x^3y^3z^3}\left\{\frac{x^3}{A}(a'-a), \frac{y^3}{B}(b'-b), \frac{z^3}{C}(c'-c)\right\}. \qquad\text{(iv)}$$

Write $\qquad \mu\mathbf{b} = \{x^3(a'-a)/A, y^3(b'-b)/B, z^3(c'-c)/C\}. \qquad\text{(v)}$

Then $\qquad\qquad\qquad \mu^2 = \sum x^6(a'-a)^2/A^2,$

and $\qquad\qquad\qquad \mu = \lambda^3\kappa x^3y^3z^3/ABC.$

Differentiate (v) to get

$$\mu'\lambda\mathbf{b}-\lambda\mu\tau\mathbf{n} = 3\{x(a'-a), y(b'-b), z(c'-c)\}. \qquad\text{(vi)}$$

Take the scalar product of (iii) and (vi) to get

$$\lambda^3 \mu \kappa \tau = 3 \sum A^2(a'-a)/x^2. \qquad \text{(vii)}$$

Hence

$$\kappa^2 = \frac{A^2 B^2 C^2}{x^6 y^6 z^6} \frac{\sum x^6(a'-a)^2/A^2}{\sum (A^2/x^2)^3}, \qquad \text{(viii)}$$

and

$$\tau = \frac{3x^3 y^3 z^3}{ABC} \frac{\sum A^2(a'-a)/x^2}{\sum x^6(a'-a)^2/A^2}. \qquad \text{(ix)}$$

Equations (viii) and (ix) give the curvature and torsion as required.

6. Contact between curves and surfaces

Let γ be a curve of sufficiently high class, given by the equation $\mathbf{r} = \{f(u), g(u), h(u)\}$; and let S be a surface given by $F(x, y, z) = 0$ where the function F has a sufficiently high class. Then the parameters of points of γ which also lie on S are zeros of the function $F(u) = F\{f(u), g(u), h(u)\}$. If u_0 is such a zero, then the function $F(u)$ may be expressed by Taylor's theorem in the form

$$F(u) = \epsilon F'(u_0) + \frac{\epsilon^2}{2!} F''(u_0) + \cdots + \frac{\epsilon^n}{n!} F^{(n)}(u_0) + O(\epsilon^{n+1}), \quad (6.1)$$

where $\epsilon = u - u_0$.

If $F'(u_0) \neq 0$, then u_0 is a simple zero of $F(u)$ and in this case γ and S have a *simple* intersection at $\mathbf{r}(u_0)$. If $F'(u_0) = 0$ but $F''(u_0) \neq 0$, then $F(u)$ is of the second order of ϵ; u_0 is a *double* zero of $F(u)$, and γ and S have *two-point contact*. If

$$F'(u_0) = F''(u_0) = 0$$

but $F'''(u_0) \neq 0$, then u_0 is a *triple* zero and γ and S have *three-point contact*. In general, if

$$F'(u_0) = F''(u_0) = \ldots = F^{(n-1)}(u_0) = 0$$

but $F^{(n)}(u_0) \neq 0$, then γ and S are said to have *n-point contact* at $\mathbf{r}(u_0)$. The reader should verify that these conditions remain invariant over a change of parameter, i.e. the property of having n-point contact with S is a property of the curve γ in the sense that any path which represents γ will have this property.

EXAMPLE 6.1. Show that the osculating plane at P has, in general, three-point contact with the curve at P.

With s as parameter measured from P ($s = 0$) we have

$$F(s) = [\mathbf{r}(s) - \mathbf{r}(0), \mathbf{r}'(0), \mathbf{r}''(0)],$$
$$= [s\mathbf{r}'(0) + \tfrac{1}{2}s^2\mathbf{r}''(0) + \tfrac{1}{6}s^3\mathbf{r}'''(0) + o(s^3), \mathbf{r}'(0), \mathbf{r}''(0)] \quad \text{as } s \to 0,$$
$$= -\tfrac{1}{6}s^3 \kappa^2 \tau + o(s^3) \quad \text{as } s \to 0.$$

It follows that, in general, the curve and plane have three-point contact. If, however, $\kappa = 0$ or $\tau = 0$ at P, then the plane has at least four-point contact with the curve.

In the theory of plane curves it is useful to consider not only the curvature of a curve at P but also the radius of curvature at P, defined to be the radius of that circle which has three-point contact with the curve at P. It is well known that the radius of curvature is the reciprocal of the curvature.

In the theory of space curves the radius of curvature of γ at P may similarly be defined as the radius of that circle which has three-point contact with the curve at P. We shall prove that, as with plane curves, the radius of curvature is the reciprocal of the curvature. However, in the case of space curves, we could also consider the sphere which has four-point contact with γ at P, and this gives rise to the so called radius of spherical curvature. We obtain formulae for the radius of curvature and radius of spherical curvature at a general point P of a curve γ, general in the sense that we shall assume that neither the curvature nor the torsion vanishes at P. We leave the reader to consider what modifications are required when these conditions are not satisfied.

Osculating circle. The osculating circle at a point P on a curve is the circle which has three-point contact with the curve at P. It evidently lies in the osculating plane at P, and its centre $\mathbf{c}$ is at some distance ρ along the principal normal at P, i.e. $\mathbf{c} - \mathbf{r} = \rho\mathbf{n}$. The osculating circle is thus the section of the sphere

$$(\mathbf{c} - \mathbf{R})^2 - \rho^2 = 0$$

by the osculating plane. Let the curve be parametrized with respect to arc length. If $F(s) \equiv (\mathbf{c} - \mathbf{r})^2 - \rho^2$, the conditions for three-point contact are $F = F' = F'' = 0$. These give rise to the relations $(\mathbf{c} - \mathbf{r}) \cdot \mathbf{t} = 0$ (which is otherwise obvious), and $(\mathbf{c} - \mathbf{r}) \cdot \kappa\mathbf{n} = 1$, from which $\rho = \kappa^{-1}$. Thus the radius of the osculating circle is $|\rho| = |\kappa^{-1}|$; ρ is called the *radius of curvature* of the curve at P. Note that ρ may be negative. The *centre of curvature* is the centre of the osculating circle, and its position vector is given by

$$\mathbf{c} = \mathbf{r} + \rho\mathbf{n}. \tag{6.2}$$

It is often convenient to introduce a parameter σ defined by the relation $\sigma = \tau^{-1}$, but although σ is called the *radius of torsion* it has no simple geometrical significance analogous to the radius of curvature.

Osculating sphere. The osculating sphere at a point P on a curve is the sphere which has four-point contact with the curve at P. If $\mathbf{C}$ is its centre and R its radius, the equation of the sphere is $(\mathbf{C}-\mathbf{R})^2 = R^2$. If $F(u) \equiv (\mathbf{C}-\mathbf{r})^2 - R^2$, the conditions for four-point contact are $F = F' = F'' = F''' = 0$, which give rise to the equations

$$(\mathbf{C}-\mathbf{r}).\mathbf{t} = 0, \qquad (\mathbf{C}-\mathbf{r}).\kappa\mathbf{n} = 1,$$

and

$$(\mathbf{C}-\mathbf{r}).(\kappa\tau\mathbf{b}+\kappa'\mathbf{n}) = 0.$$

It follows that

$$\mathbf{C}-\mathbf{r} = \rho\mathbf{n}+\sigma\rho'\mathbf{b}.$$

The centre of the osculating sphere is called the *centre of spherical curvature*, and its position vector is given by

$$\mathbf{C} = \mathbf{r}+\rho\mathbf{n}+\sigma\rho'\mathbf{b}. \tag{6.3}$$

The *radius of spherical curvature* is given by

$$R = (\rho^2+\sigma^2\rho'^2)^{\frac{1}{2}}. \tag{6.4}$$

Evidently if κ is constant then $R = \rho$ and the two centres of curvature coincide.

Locus of the centre of spherical curvature

As the point P traces out a curve C, the corresponding centre of spherical curvature traces out another curve C_1, whose curvature and torsion are simply related to the curvature and torsion of the original curve C. It will be convenient to distinguish entities belonging to the curve C_1 by a suffix, so that, for example, s_1 will denote arc length of the curve C_1.

Evidently $\mathbf{r}_1 = \mathbf{r}+\rho\mathbf{n}+\sigma\rho'\mathbf{b}$, and on differentiation

$$s_1'\mathbf{t}_1 = \mathbf{t}+\rho'\mathbf{n}+\rho(\tau\mathbf{b}-\kappa\mathbf{t})+(\sigma'\rho'+\sigma\rho'')\mathbf{b}-\rho'\mathbf{n}$$

$$= \left(\frac{\rho}{\sigma}+\sigma'\rho'+\sigma\rho''\right)\mathbf{b}.$$

Now, it should be remembered that C_1 is parametrized by s, and s_1 is an increasing function of s so that s_1' is non-negative. We write $\mathbf{t}_1 = e\mathbf{b}$, where $e = \pm 1$. Then

$$s_1' = e\left(\frac{\rho}{\sigma}+\sigma'\rho'+\sigma\rho''\right).$$

The equation $\mathbf{t}_1 = e\mathbf{b}$ gives on differentiation

$$\kappa_1 s_1'\,\mathbf{n}_1 = -e\tau\mathbf{n}.$$

We write $\mathbf{n}_1 = e_1\mathbf{n}$ where $e_1 = \pm 1$. Then $e_1\kappa_1 s_1' = -e\tau$.

The unit binormal $\mathbf{b}_1$ is parallel to $\mathbf{t}$ since

$$\mathbf{b}_1 = \mathbf{t}_1 \times \mathbf{n}_1 = e\mathbf{b} \times e_1 \mathbf{n} = -ee_1 \mathbf{t}.$$

The relation $\mathbf{b}_1 = -ee_1 \mathbf{t}$ gives on differentiation $\tau_1 s_1' \mathbf{n}_1 = ee_1 \kappa \mathbf{n}$, so that $\tau_1 s_1' = e\kappa$. It follows that $\kappa \kappa_1 = -e_1 \tau \tau_1$, i.e. the product of the curvatures at corresponding points is equal to the product of the torsions if $e_1 = -1$. Note that a point on C where

$$\rho\sigma^{-1} + \sigma'\rho' + \sigma\rho'' = 0$$

gives rise to a singular point on C_1.

EXERCISE 6.1. Prove that the radius of curvature of the locus of the centre of curvature of a curve is given by

$$\left[\left\{ \frac{\rho^2\sigma}{R^3} \frac{d}{ds}\left(\frac{\sigma\rho'}{\rho}\right) - \frac{1}{R} \right\}^2 + \frac{\rho'^2\sigma^4}{\rho^2 R^4} \right]^{-\frac{1}{2}}.$$

EXERCISE 6.2. If the radius of spherical curvature is constant, prove that the curve either lies on a sphere or has constant curvature.

EXERCISE 6.3. If a curve lies on a sphere show that ρ and σ are related by

$$\frac{d}{ds}(\sigma\rho') + \frac{\rho}{\sigma} = 0.$$

7. Tangent surface, involutes, and evolutes

In the previous section it was shown that a given space curve C determines in general two other space curves, viz. the locus of the centre of curvature and the locus of the centre of spherical curvature. In this section it is shown that a given space curve C determines two infinite systems of curves which are the involutes and evolutes of C. The theory of evolutes of space curves is essentially different from that of plane curves; a plane curve has a unique evolute while a space curve has infinitely many. Moreover, the evolute of a plane curve is often defined as the locus of its centre of curvature; but it will be seen that neither the locus of the centre of curvature nor the locus of the centre of spherical curvature are evolutes of a space curve. There is, however, a natural generalization to space curves of the concept of involute of a plane curve; and once an involute $\tilde{C}$ of a curve C has been defined, it is natural to define C to be an evolute of $\tilde{C}$.

The *tangent surface* of a curve C is the surface generated by lines tangent to C. Any point P on the tangent surface is determined by two parameters s and u; s is the arc length of C measured from some convenient base point on the curve to a point where the tangent passes through P, and u measures the distance of P along this tangent. The position vector of P can thus be written as

$$\mathbf{R}(s, u) = \mathbf{r}(s) + u\mathbf{t}(s). \tag{7.1}$$

Any additional relation between u and s of the form $u = \lambda(s)$ determines a curve which lies on the tangent surface of C, the class of the curve being the same as the class of λ or C, whichever is the smaller.

Involute. An *involute* of C is a curve which lies on the tangent surface of C and intersects the generators orthogonally.

It follows from (7.1) that since the involute lies on the tangent surface, the position vector of a point on the involute is

$$\mathbf{R} = \mathbf{r} + \lambda(s)\mathbf{t}. \tag{7.2}$$

A tangent vector to the involute at P is parallel to $\mathbf{R}'$, i.e. to $(1 + \lambda')\mathbf{t} + \lambda\kappa\mathbf{n}$. But this vector must be orthogonal to $\mathbf{t}$ and hence $1 + \lambda' = 0$; it follows that $\lambda = c - s$, where c is an arbitrary constant. Hence, the equation of the involute is

$$\mathbf{R} = \mathbf{r} + (c - s)\mathbf{t}. \tag{7.3}$$

Since c is an arbitrary constant, this equation represents an infinite system of involutes of C, a different curve arising from each different choice of the parameter c.

EXERCISE 7.1. Show that an involute of C is obtained by unwinding a string initially stretched along the curve, so that the string always remains taut.

Evolute. If $\tilde{C}$ is an involute of C, then C is, by definition, an *evolute* of $\tilde{C}$. Now we are given $\tilde{C}$ and the problem is to obtain a corresponding curve C. Let P be the point on C corresponding to the point Q on $\tilde{C}$. Then P must lie in the plane through Q normal to $\tilde{C}$; and if $\mathbf{R}$, $\mathbf{r}$ denote the position vectors of P, Q respectively, then

$$\mathbf{R} = \mathbf{r} + \lambda\mathbf{n} + \mu\mathbf{b}$$

for some functions λ, μ. Also, the vector

$$\mathbf{R}' = [(1 - \lambda\kappa)\mathbf{t} + (\lambda' - \mu\tau)\mathbf{n} + (\mu' + \lambda\tau)\mathbf{b}]$$

must be parallel to $\lambda\mathbf{n} + \mu\mathbf{b}$. It follows that

$$\lambda = \rho; \quad \text{and} \quad \frac{\lambda' - \mu\tau}{\lambda} = \frac{\mu' + \lambda\tau}{\mu},$$

from which
$$\tau = \frac{\mu\lambda' - \lambda\mu'}{\lambda^2 + \mu^2} = \frac{d}{ds}[\tan^{-1}\lambda/\mu].$$

Hence $\lambda = \mu\tan\left(\int\tau\,ds+c\right)$, where c is an arbitrary constant. Since $\lambda = \rho$, then $\mu = \rho\cot\left(\int\tau\,ds+c\right)$, and the equation of the evolute of $\tilde{C}$ is

$$\mathbf{R} = \mathbf{r}+\rho\mathbf{n}+\rho\cot\left(\int\tau\,ds+c\right)\mathbf{b}. \tag{7.4}$$

This equation represents an infinite system of evolutes of the given curve, one evolute arising from each choice of c.

It is evident from the form of (7.4) that the locus of the centre of curvature of a space curve is not an evolute. Neither can the locus of the centre of spherical curvature be an evolute, for the tangent vector at a point on this locus is parallel to $\mathbf{b}$ whilst the vector $\mathbf{R}'$ is parallel to $\rho\mathbf{n}+\mu\mathbf{b}$.

EXERCISE 7.2. Show that the involutes of a circular helix are plane curves.

EXERCISE 7.3. Show that the torsion of an involute of a curve is equal to

$$\frac{\rho(\sigma\rho'-\sigma'\rho)}{(\rho^2+\sigma^2)(c-s)}.$$

8. Intrinsic equations, fundamental existence theorem for space curves

In section 1 a curve was defined by expressing the coordinates of its points relative to three mutually orthogonal Cartesian axes. If the same curve be referred to a different set of Cartesian axes, then the defining equations are quite different, and it is by no means obvious that they refer to the same curve. For certain purposes it is convenient to describe a curve intrinsically, i.e. without reference to a particular set of Cartesian axes. This is achieved by using the *intrinsic equations* of the curve.

The intrinsic equations of a curve are of the form

$$\left.\begin{array}{l} \kappa = f(s) \\ \tau = g(s) \end{array}\right\} \tag{8.1}$$

which express the curvature and the torsion in terms of the arc length. Such equations cannot, of course, determine the position of the curve in space, and congruent curves will naturally have the same intrinsic equations. Conversely, as will now be proved, two curves with the same intrinsic equations are necessarily congruent.

More precisely, let C and C_1 be two curves defined in terms of their respective arc lengths s, and let points with the same values of s correspond. Then, if the curvature and torsion of C have the same values as the curvature and torsion at the corresponding points of C_1, the theorem states that C and C_1 are congruent.

Let C_1 be moved so that the two points on C and C_1 corresponding to $s = 0$ coincide, and suppose that C_1 is suitably oriented so that the two triads $(\mathbf{t}, \mathbf{n}, \mathbf{b})$, $(\mathbf{t}_1, \mathbf{n}_1, \mathbf{b}_1)$ coincide at $s = 0$. Then

$$\frac{d}{ds}(\mathbf{t}.\mathbf{t}_1) = \mathbf{t}.\kappa\mathbf{n}_1 + \kappa\mathbf{n}.\mathbf{t}_1, \tag{8.2}$$

$$\frac{d}{ds}(\mathbf{n}.\mathbf{n}_1) = \mathbf{n}.(\tau\mathbf{b}_1 - \kappa\mathbf{t}_1) + (\tau\mathbf{b} - \kappa\mathbf{t}).\mathbf{n}_1, \tag{8.3}$$

$$\frac{d}{ds}(\mathbf{b}.\mathbf{b}_1) = \mathbf{b}.(-\tau\mathbf{n}_1) + (-\tau\mathbf{n}).\mathbf{b}_1. \tag{8.4}$$

It follows by addition that

$$\frac{d}{ds}(\mathbf{t}.\mathbf{t}_1 + \mathbf{n}.\mathbf{n}_1 + \mathbf{b}.\mathbf{b}_1) = 0, \tag{8.5}$$

and so $\mathbf{t}.\mathbf{t}_1 + \mathbf{n}.\mathbf{n}_1 + \mathbf{b}.\mathbf{b}_1 = 3$ (the value when $s = 0$). But the sum of three cosines is equal to 3 only when each angle is zero, and so $\mathbf{t}_1 = \mathbf{t}$, $\mathbf{n}_1 = \mathbf{n}$, and $\mathbf{b}_1 = \mathbf{b}$ at all corresponding points. Thus $d(\mathbf{r} - \mathbf{r}_1)/ds = \mathbf{0}$ which gives $\mathbf{r} - \mathbf{r}_1$ a constant vector; but since $\mathbf{r} - \mathbf{r}_1 = \mathbf{0}$ when $s = 0$, it follows that $\mathbf{r} = \mathbf{r}_1$ identically. Thus C and C_1 are identical to within a Euclidean motion, and the theorem is proved.

EXERCISE 8.1. Show that the intrinsic equations of the curve given by $x = ae^u \cos u$, $y = ae^u \sin u$, $z = be^u$ are

$$\kappa = \frac{\sqrt{2}\,a}{(2a^2 + b^2)^{\frac{1}{2}}} \frac{1}{s}, \qquad \tau = \frac{b}{(2a^2 + b^2)^{\frac{1}{2}}} \frac{1}{s}.$$

A question which arises naturally is the existence of a curve having equations (8.1) as intrinsic equations when the functions f and g are prescribed. This is answered by the *Fundamental Existence Theorem for Space Curves* which states:

If $\kappa(s)$, $\tau(s)$ are continuous functions of the real variable s, where $s \geqslant 0$, then there exists a space curve for which κ is the curvature, τ is the torsion, and s is the arc length measured from some suitable base point. Such a curve is uniquely determined to within a Euclidean motion.

Consider the differential equations

$$\frac{d\alpha}{ds} = \kappa\beta, \qquad \frac{d\beta}{ds} = \tau\gamma - \kappa\alpha, \qquad \frac{d\gamma}{ds} = -\tau\beta. \tag{8.6}$$

It is proved in Appendix I. 1 that these equations admit a unique set of solutions which assume prescribed values $\alpha_0, \beta_0, \gamma_0$ when $s = 0$. In particular, there is a unique set $\alpha_1, \beta_1, \gamma_1$ which assume values 1, 0, 0 when $s = 0$; similarly, there is a unique set $\alpha_2, \beta_2, \gamma_2$ which assume initial values 0, 1, 0; and a unique set $\alpha_3, \beta_3, \gamma_3$ which assume initial values 0, 0, 1. We now prove that for all values of s, $\alpha_1^2 + \beta_1^2 + \gamma_1^2 = 1$. Evidently

$$d(\alpha_1^2 + \beta_1^2 + \gamma_1^2)/ds = 2(\alpha_1 \alpha_1' + \beta_1 \beta_1' + \gamma_1 \gamma_1')$$

and the right-hand member vanishes identically because of (8.6). Hence $(\alpha_1^2 + \beta_1^2 + \gamma_1^2) = \text{constant} = (\alpha_1^2 + \beta_1^2 + \gamma_1^2)_{s=0} = 1$, giving the required result. In the same way, it may be proved that $(\alpha_2^2 + \beta_2^2 + \gamma_2^2) = 1$, and $(\alpha_3^2 + \beta_3^2 + \gamma_3^2) = 1$. Similarly, it can be verified that the derivative of $(\alpha_1 \alpha_2 + \beta_1 \beta_2 + \gamma_1 \gamma_2)$ vanishes identically because of (8.6), and hence $(\alpha_1 \alpha_2 + \beta_1 \beta_2 + \gamma_1 \gamma_2) = 0$. In the same way $(\alpha_2 \alpha_3 + \beta_2 \beta_3 + \gamma_2 \gamma_3) = 0$ and $(\alpha_3 \alpha_1 + \beta_3 \beta_1 + \gamma_3 \gamma_1) = 0$.

These six relations can be summarized by saying that the matrix

$$A \equiv \begin{pmatrix} \alpha_1 & \beta_1 & \gamma_1 \\ \alpha_2 & \beta_2 & \gamma_2 \\ \alpha_3 & \beta_3 & \gamma_3 \end{pmatrix}$$

is orthogonal, i.e. $AA' = I$, where A' is the transpose of A. But the equation $AA' = I$ implies $A' = A^{-1}$ and so $A'A = I$, which is equivalent to the six relations

$$\alpha_1^2 + \alpha_2^2 + \alpha_3^2 = 1; \qquad \alpha_1 \beta_1 + \alpha_2 \beta_2 + \alpha_3 \beta_3 = 0;$$
$$\beta_1^2 + \beta_2^2 + \beta_3^2 = 1; \qquad \beta_1 \gamma_1 + \beta_2 \gamma_2 + \beta_3 \gamma_3 = 0;$$
$$\gamma_1^2 + \gamma_2^2 + \gamma_3^2 = 1; \qquad \gamma_1 \alpha_1 + \gamma_2 \alpha_2 + \gamma_3 \alpha_3 = 0.$$

It follows that there are three mutually orthogonal unit vectors $\mathbf{t} = (\alpha_1, \alpha_2, \alpha_3)$, $\mathbf{n} = (\beta_1, \beta_2, \beta_3)$, $\mathbf{b} = (\gamma_1, \gamma_2, \gamma_3)$ defined for each value of s. If $\mathbf{r} = \int_0^s \mathbf{t} \, ds$, then $\mathbf{r} = \mathbf{r}(s)$ is the position vector of a point on a curve which has $\mathbf{t}$ as tangent vector, $\mathbf{n}$ as principal normal, $\mathbf{b}$ as binormal, κ as curvature, τ as torsion, and s as arc length. This proves the existence of the required curve.

Of course the actual solution of equations (8.6) may present serious mathematical difficulties. If κ and τ are of class $\geqslant 3$, β and γ may be eliminated from (8.6), and this leads to a differential equation of the third order in α. It can be shown that the solution of this equation can be reduced to the solution of a Riccati equation of the first order (cf. Eisenhart (1909)).

9. Helices

A *cylindrical helix* is a space curve which lies on a cylinder and cuts the generators at a constant angle. Its tangent makes a constant angle α with a fixed line known as the axis of the helix. Helices more general than cylindrical helices have been considered by other authors, but in this book by helix we shall mean cylindrical helix.

A characteristic property of helices is that the ratio of the curvature to the torsion is constant at all points. To prove this, let $\mathbf{a}$ denote a unit vector in the direction of the axis; then

$$\mathbf{t}.\mathbf{a} = \cos\alpha.$$

By differentiating this relation, $\kappa\mathbf{n}.\mathbf{a} = 0$ from which (if the case $\kappa = 0$ is excluded) it follows that the vector $\mathbf{a}$ must lie in the rectifying plane. Writing

$$\mathbf{a} = \mathbf{t}\cos\alpha + \mathbf{b}\sin\alpha, \qquad (9.1)$$

and differentiating to get

$$\mathbf{a}' = (\kappa\cos\alpha - \tau\sin\alpha)\mathbf{n} = \mathbf{0},$$

it follows that $\qquad \kappa/\tau = \tan\alpha = \text{constant}. \qquad (9.2)$

Conversely, if $\kappa/\tau = \text{constant} = \tan\alpha$, then

$$(\kappa\cos\alpha - \tau\sin\alpha)\mathbf{n} = \mathbf{0},$$

i.e. $\qquad\qquad d(\mathbf{t}\cos\alpha + \mathbf{b}\sin\alpha)/ds = \mathbf{0}.$

It follows that $(\mathbf{t}\cos\alpha + \mathbf{b}\sin\alpha) = \mathbf{a}$, a constant unit vector. Hence $\mathbf{a}.\mathbf{t} = \cos\alpha$, and the curve is thus a helix. This proves that the property $\kappa/\tau = \text{constant}$ is characteristic of helices.

A *circular helix* is one which lies on the surface of a circular cylinder, the axis of the helix being that of the cylinder. If the z-axis is the axis of the helix, the parametric equation of the curve may be written as

$$x = a\cos u, \qquad y = a\sin u, \qquad z = bu, \qquad (9.3)$$

where $a > 0$. If $b > 0$, the helix is right-handed; if $b < 0$ it is left-handed. The *pitch* of the helix is equal to $2\pi b$, and represents the displacement along the axis corresponding to a complete turn round the axis. For a circular helix both κ and τ are constant, and in particular the ratio κ/τ is constant in agreement with the previous result.

For any general helix C there is a simple relation between its curvature and that of the plane curve C_1 obtained by projecting C

on a plane orthogonal to its axis. Using a suffix to denote entities belonging to C_1, then $\mathbf{r} = \mathbf{r}_1 + (\mathbf{a}.\mathbf{r})\mathbf{a}$, $\mathbf{t} = \mathbf{t}_1 ds_1/ds + (\mathbf{a}.\mathbf{t})\mathbf{a}$. Hence

$$ds_1/ds = \sin \alpha, \tag{9.4}$$

and

$$\mathbf{t} = \sin \alpha \, \mathbf{t}_1 + \cos \alpha \, \mathbf{a}. \tag{9.5}$$

By differentiating (9.5) and using (9.4) it follows that

$$\kappa \mathbf{n} = \sin \alpha \frac{ds_1}{ds} \frac{d\mathbf{t}_1}{ds_1} = \kappa_1 \sin^2 \alpha \, \mathbf{n}_1. \tag{9.6}$$

The normal of C_1 is thus parallel to the principal normal of the helix at the corresponding point, and the curvatures are related by

$$\kappa = \kappa_1 \sin^2 \alpha.$$

If the helix C has constant curvature, then C_1 is a plane curve with constant curvature and is thus a circle. It follows that a helix of constant curvature is necessarily a circular helix.

EXERCISE 9.1. Show that if C is a spherical helix, i.e. a helix lying on the surface of a sphere, then C_1 is an arc of an epicycloid.

EXERCISE 9.2. Show that if C is a helix lying on a paraboloid of revolution, then C_1 is the involute of a circle.

APPENDIX I. 1

Existence theorem on linear differential equations

The differential equations

$$\frac{du_i}{ds} = \sum_{k=1}^{p} c_{ik} u_k \quad (i = 1, 2, ..., p), \tag{1}$$

where c_{ik} are continuous functions of s in the interval $0 \leqslant s \leqslant r$, has a set of C^1-solutions which assume prescribed values u_i^0 when $s = 0$.

To prove this, let

$$\left.\begin{aligned}
u_i^1 &= u_i^0 + \int_0^s \sum_k c_{ik} u_k^0 \, ds \\
u_i^2 &= u_i^0 + \int_0^s \sum_k c_{ik} u_k^1 \, ds \\
&\cdot \quad \cdot \quad \cdot \quad \cdot \quad \cdot \quad \cdot \\
u_i^n &= u_i^0 + \int_0^s \sum_k c_{ik} u_k^{n-1} \, ds \\
&\cdot \quad \cdot \quad \cdot \quad \cdot \quad \cdot \quad \cdot
\end{aligned}\right\}. \tag{2}$$

Since c_{ik} are continuous functions they are bounded, i.e. there exists a constant C such that $|c_{ik}| < C/p$.

We assume that $|u_i^0| \leqslant K$. Then

$$|u_i^1 - u_i^0| < KCs.$$

Also

$$u_i^2 - u_i^1 = \int_0^s \sum_k c_{ik}(u_k^1 - u_k^0)\, ds$$

and so

$$|u_i^2 - u_i^1| < KC^2 \frac{s^2}{2}.$$

In a similar way it follows that

$$|u_i^n - u_i^{n-1}| < KC^n \frac{s^n}{n!} \leqslant KC^n \frac{r^n}{n!}.$$

Hence, by the Weierstrass test, the sequence $\{u_i^n(s)\}$ converges uniformly in the interval $0 \leqslant s \leqslant r$ and a continuous function $u_i(s)$ is defined by the equation

$$\lim_{n \to \infty} u_i^n(s) = u_i(s). \tag{3}$$

Also, from the uniformity of convergence, it follows from (2), as $n \to \infty$,

$$u_i = u_i^0 + \int_0^s \sum_k c_{ik} u_k\, ds. \tag{4}$$

Hence

$$\frac{du_i}{ds} = \sum_k c_{ik} u_k \quad \text{and} \quad u_i(0) = u_i^0.$$

This completes the proof of the existence theorem.

REFERENCES

EISENHART, L. P., *A Treatise on the Differential Geometry of Curves and Surfaces*, Ginn and Co. (1909), p. 26.

FLETT, T. M., *Edin. Math. Notes* (1957), pp. 1–9.

NEWNS, W. F., *Math. Gaz.* (1957), p. 213.

MISCELLANEOUS EXERCISES I

1. Determine the function $f(u)$ so that the curve given by
$$\mathbf{r} = (a\cos u, a\sin u, f(u))$$
shall be plane.

2. Find the curvature and the torsion of the curves given by
   (a) $\mathbf{r} = \{a(3u-u^3),\ 3au^2,\ a(3u+u^3)\}$.
   (b) $\mathbf{r} = \{a(u-\sin u),\ a(1-\cos u),\ bu\}$.
   (c) $\mathbf{r} = \{a(1+\cos u),\ a\sin u,\ 2a\sin \tfrac{1}{2}u\}$.

3. Find the coordinates of the centre of spherical curvature of the curve given by
$$\mathbf{r} = (a\cos u,\ a\sin u,\ a\cos 2u).$$

4. Prove that the curve given by $x = a\sin^2 u,\ y = a\sin u\cos u,\ z = a\cos u$ lies on a sphere.

5. Show that the principal normal to a curve is normal to the locus of centres of curvature at those points where the curvature is stationary.

6. Find the equation of the osculating plane of the curve given by
$$\mathbf{r} = \{a\sin u+b\cos u,\ a\cos u+b\sin u,\ c\sin 2u\}.$$
Find also the radius of spherical curvature at any point.

7. A pair of curves γ, γ_1 which have the same principal normals are called *Bertrand curves*. Prove that the tangents to γ and γ_1 are inclined at a constant angle; and show that, for each curve, there is a linear relation with constant coefficients between the curvature and torsion.

8. Show that the torsions at corresponding points P and P_1 of two Bertrand curves have the same sign, and that their product is constant. If C, C_1 are their centres of curvature, prove that the cross ratio (PCP_1C_1) is the same for all corresponding pairs of points.

9. The locus of a point whose position vector is the tangent vector $\mathbf{t}$ to a curve γ is called the *spherical indicatrix of the tangent to γ*. Prove that the tangent to the indicatrix is parallel to the principal normal at the corresponding point of γ. Show that the curvature κ_1 and the torsion τ_1 of the indicatrix are given by
$$\kappa_1^2 = (\kappa^2+\tau^2)/\kappa^2,$$
$$\tau_1 = \frac{(\kappa\tau'-\kappa'\tau)}{\kappa(\kappa^2+\tau^2)}.$$

10. The locus of a point whose position vector is the binormal $\mathbf{b}$ of a curve γ is called the *spherical indicatrix of the binormal to γ*. Prove that its curvature κ_2 and torsion τ_2 are given by
$$\kappa_2^2 = (\kappa^2+\tau^2)/\tau^2,$$
$$\tau_2 = \frac{(\tau\kappa'-\kappa\tau')}{\tau(\kappa^2+\tau^2)}.$$

11. Show that the spherical indicatrix of a curve is a circle if and only if the curve is a helix.

12. Prove that for any curve lying on the surface of a sphere,
$$\frac{d}{ds}(\sigma\rho')+\frac{\rho}{\sigma} = 0.$$

13. Prove that corresponding points on the spherical indicatrix of the tangent to γ and on the indicatrix of the binormal to γ have parallel tangent lines.

14. Find the equation of the tangent surface to the curve $\mathbf{r} = (u, u^2, u^3)$.

15. Show that a necessary and sufficient condition that a curve be a helix is that
$$[\mathbf{r^{iv}}, \mathbf{r'''}, \mathbf{r''}] = -\kappa^5 \frac{d}{ds}\left(\frac{\tau}{\kappa}\right) = 0.$$

16. Show that the locus C of the centre of curvature of a circular helix of curvature κ is a coaxial helix. Show that the locus of the centre of curvature of C is the original helix; and prove that the product of the torsions at corresponding points of the two helices is equal to κ^2.

17. Prove that all osculating planes to a circular helix which pass through a given point not lying on the helix have their points of contact in a plane. Show that the same property holds for any curve for which $x\,dy - y\,dx = c\,dz$, where c is a constant.

18. Show that the helices on a cone of revolution project on a plane perpendicular to the axis of the cone as logarithmic spirals.

19. Find the coordinates of the cylindrical helix whose intrinsic equations are $\kappa = \tau = 1/s$.

20. Show that the helix whose intrinsic equations are $\rho = \tau^{-1} = (s^2 + 4)/\surd 2$ lies upon a cylinder whose cross-section is a catenary.

21. Show that the locus of the centre of curvature of a curve is an evolute only when the curve is plane.

22. Find the involutes of a helix.

23. Find the involutes and evolutes of the twisted cubic given by $x = u$, $y = u^2$, $z = u^3$.

24. Show that the position vector $\mathbf{r}(s)$ of any space curve of class $\geqslant 4$ satisfies the differential equation
$$\mathbf{r^{iv}} - \left(\frac{2\kappa'}{\kappa} + \frac{\tau'}{\tau}\right)\mathbf{r'''} + \left(\kappa^2 + \tau^2 + \frac{\kappa'\tau'}{\kappa\tau} + \frac{2\kappa'^2 - \kappa\kappa''}{\kappa^2}\right)\mathbf{r''} + \kappa^2\left(\frac{\kappa'}{\kappa} - \frac{\tau'}{\tau}\right)\mathbf{r'} = 0.$$

THE METRIC: LOCAL INTRINSIC
PROPERTIES OF A SURFACE

1. Definition of a surface

A SURFACE often arises as the locus of a point P which satisfies some restriction, as a consequence of which the coordinates x, y, z of P satisfy a relation of the form

$$F(x, y, z) = 0. \qquad (1.1)$$

This is called the *implicit* or *constraint* equation of the surface, and is convenient for certain purposes, particularly for the study of algebraic surfaces as a whole. In the type of differential geometry considered in this and the following chapter, however, one focuses attention not so much on a surface as a whole but on small regions, and for these the implicit form of the equation has disadvantages. An explicit form, in which the coordinates of a point on the surface are expressed in terms of two parameters, is preferable and is always possible for regions which are not too large.

The *parametric* or *freedom* equations of a surface are of the form

$$x = f(u, v), \qquad y = g(u, v), \qquad z = h(u, v), \qquad (1.2)$$

where u and v are parameters which take real values and vary freely in some domain D. The functions f, g, and h are single valued and continuous, and are here assumed to possess continuous partial derivatives of the rth order. In this case the surface is said to be of *class r*. When the class of a surface is not stated explicitly it should be assumed that the functions f, g, and h possess as many derivatives as may be required, usually two or three. Parameters such as u and v are frequently called *curvilinear coordinates*; the point determined by the pair u, v is referred to as the point (u, v).

When the parametric equations of a surface are given it is a simple matter, at least in theory, to find the constraint equation. This is done by eliminating u and v between the three equations (1.2). For example, consider the surface given by the parametric equations

$$x = u + v, \qquad y = u - v, \qquad z = 4uv, \qquad (1.3)$$

where u and v take all real values. Eliminating u and v, the constraint equation is

$$x^2 - y^2 = z, \tag{1.4}$$

which represents the whole of a certain hyperbolic paraboloid.

The parametric equations are by no means unique; for example, the equations

$$x = u, \qquad y = v, \qquad z = u^2 - v^2, \tag{1.5}$$

where u and v take all real values, represent the same paraboloid.

Sometimes the constraint equation obtained by eliminating the parameters represents more than the given surface, so that parametric equations and constraint equations are not equivalent. This is illustrated by the following example.

Consider the surface given by the parametric equations

$$x = u \cosh v, \qquad y = u \sinh v, \qquad z = u^2, \tag{1.6}$$

where u and v take all real values. Eliminating the parameters, the constraint equation is (1.4), which represents the whole of the paraboloid. The parametric equations (1.6), however, represent only that part of the surface for which $z \geqslant 0$, since u takes only real values.

Two representations of the same surface, such as (1.3) and (1.5), are related by a *parameter transformation* of the form

$$u' = \phi(u, v), \qquad v' = \psi(u, v). \tag{1.7}$$

This transformation is said to be *proper* if ϕ and ψ are single valued and have non-vanishing Jacobian,

i.e. $$\frac{\partial(\phi, \psi)}{\partial(u, v)} \neq 0, \tag{1.8}$$

in some domain D. If D' is the domain of u', v' corresponding to D, condition (1.8) is necessary and sufficient that the transformation (1.7) can be inverted *near any point of D'*—the transformation is locally (1–1) but there may exist no inverse transformation defined on the whole of D.

Because of (1.2), the position vector $\mathbf{r} = (x, y, z)$ of a point on the surface is a function of u and v with the same continuity and differentiability properties as f, g, and h. Partial differentiation with respect to u and v will be denoted by suffixes 1 and 2 respectively, so that

$$\mathbf{r}_1 = \frac{\partial \mathbf{r}}{\partial u}, \qquad \mathbf{r}_2 = \frac{\partial \mathbf{r}}{\partial v}. \tag{1.9}$$

An *ordinary* point is defined as one for which $\mathbf{r}_1 \times \mathbf{r}_2 \neq 0$, i.e.

$$\operatorname{rank} \begin{pmatrix} x_1 & y_1 & z_1 \\ x_2 & y_2 & z_2 \end{pmatrix} = 2. \tag{1.10}$$

The significance of this condition is that u and v are determined uniquely by x, y, z in the neighbourhood of an ordinary point. This follows from the inversion theorem applied to any one of the non-vanishing 2×2 Jacobians in the above 2×3 matrix.

The property of being an ordinary point is unaltered by a proper parameter transformation. For from (1.7),

$$\mathbf{r}_1 \times \mathbf{r}_2 = \left(\frac{\partial \mathbf{r}}{\partial u'} \phi_1 + \frac{\partial \mathbf{r}}{\partial v'} \psi_1 \right) \times \left(\frac{\partial \mathbf{r}}{\partial u'} \phi_2 + \frac{\partial \mathbf{r}}{\partial v'} \psi_2 \right) = \frac{\partial(\phi, \psi)}{\partial(u, v)} \frac{\partial \mathbf{r}}{\partial u'} \times \frac{\partial \mathbf{r}}{\partial v'},$$
$$\tag{1.11}$$

and since the Jacobian is non-zero it follows that if $\mathbf{r}_1 \times \mathbf{r}_2 \neq 0$, then

$$\frac{\partial \mathbf{r}}{\partial u'} \times \frac{\partial \mathbf{r}}{\partial v'} \neq 0.$$

A point which is not an ordinary point is called a *singularity*. Some singularities are *essential*; they are due to particular geometrical features of the surface, and are independent of the choice of parametric representation. An example of such a singularity is the vertex of a cone. Other singularities are *artificial* and arise from the choice of a particular parametric representation. The simplest example of this is the origin of polar coordinates in the plane, for if $\mathbf{r} = (u \cos v, u \sin v, 0)$ then (1.10) is not satisfied when $u = 0$.

In the remainder of this book it will be understood that, unless otherwise stated, the domain D of u, v will be restricted so that every point (u, v) of the surface is ordinary.

Summarizing our preliminary discussion of surfaces, we can say that many of the problems which arose in Chapter I in defining a curve, reappear in connexion with surfaces. A surface could be defined by a constraint equation but this has disadvantages since it does not give a means of describing a *sense* round the surface. On the other hand, a particular parametrization gives too much information about one part of a surface and may give no information about the other parts of the surface, for it may be impossible to parametrize the whole surface at once without introducing artificial singularities.

A further difficulty arises because, as we have seen, proper parameter transformations are only *locally* (1–1) maps. It is much too restrictive to consider only those changes of parameter which are (1–1) over the domain D, while proper changes are only locally (1–1). If, following the procedure of Chapter I, we attempt a formal definition of a surface as an equivalence class of parametric representations, we are forced to consider only partial parametrizations. Thus we are led to consider a surface as a collection of parts of surface, each part being given by a parametric representation.

DEFINITION 1.1. *A representation R of a surface S of class r in E_3 is a set of points in E_3 covered by a system of overlapping parts $\{V_j\}$, each part V_j being given by parametric equations of class r. Each point lying in the overlap of two parts V_i, V_j is such that the change of parameters from those of one part to those of the other part is proper and of class r.*

Of course, it is possible to represent the same surface by a different system of overlapping parts $\{V'_j\}$, each part being given by parametric equations of class r, so that the conditions of definition 1.1 are satisfied. However, it is necessary to state precisely what we mean by asserting that another representation R' is as good as the first representation. This is achieved by

DEFINITION 1.2. *Two representations R, R' are said to be r-equivalent if the composite family of parts $\{V_j, V'_j\}$ satisfies the condition that at each point P lying in the overlap of any two parts, the change of parameters from those of one part to those of another is proper and of class r.*

This is easily seen to be an equivalence relation, and the various representations separate out into disjoint equivalence classes. We can now make a formal definition of a surface.

DEFINITION 1.3. *A surface S of class r in E_3 is an r-equivalence class of representations.*

In this chapter and in the following chapter we shall be mainly concerned with *local differential geometry*, i.e. with the study of those properties which hold in some neighbourhood of a point, the extent of the region in which these properties hold being unimportant. This study is contrasted with *global differential geometry* which examines the relations between local differential invariants, such as curvature, and global properties involving the surface as a whole, such as compactness. In order to deal with some of these global problems a more satisfactory definition of surface will be given in Chapter IV, but the present definition will be adequate for the purpose of this and the following chapter.

2. Curves on a surface

Let $\mathbf{r} = \mathbf{r}(u, v)$ be the equation of a surface of class r, defined on a domain D. Let $u = U(t)$, $v = V(t)$ be a curve of class s lying in D. Then $\mathbf{r} = \mathbf{r}(U(t), V(t))$ is a curve lying on the surface with class equal to the smaller of r and s. Thus the equations

$$u = U(t), \qquad v = V(t) \tag{2.1}$$

may be referred to as the (curvilinear) equations of the curve.

Curves of particular importance are obtained by keeping either u or v constant. Let v have the constant value c. Then as u varies the point $\mathbf{r} = \mathbf{r}(u, c)$ traces a curve, called the *parametric curve* $v = c$. There is one such curve for every value of c; together they form the system of parametric curves $v =$ constant. Similarly, keeping u constant and allowing v to vary, there is the system of parametric curves $u =$ constant.

Through every point of the surface there passes one and only one parametric curve of each system. For if P is the point (u_0, v_0), then u_0 and v_0 are uniquely determined by P, and there are just the two parametric curves $u = u_0$, $v = v_0$ through P. It follows that no two parametric curves of the same system intersect, and that the curves $u = u_0$, $v = v_0$ intersect once but not more than once if (u_0, v_0) belongs to D.

For a curve $v = c$, u serves as a parameter and determines a sense along the curve. Differentiating $\mathbf{r}(u, c)$ with respect to u, the tangent to $v = c$ in the sense of u increasing is in the direction $\mathbf{r}_1$. Similarly, the tangent to a curve $u =$ constant in the sense of v increasing is in the direction $\mathbf{r}_2$. Since $\mathbf{r}_1 \times \mathbf{r}_2 \neq \mathbf{0}$, parametric curves of different systems cannot touch each other.

The two parametric curves through a point P are orthogonal if $\mathbf{r}_1 . \mathbf{r}_2 = 0$ at P. If this condition is satisfied at every point, i.e. for all u, v in the domain D, the two systems of parametric curves are orthogonal.

For the general curve given by $u = u(t)$, $v = v(t)$, the tangent is in the direction

$$\frac{d\mathbf{r}}{dt} = \mathbf{r}_1 \frac{du}{dt} + \mathbf{r}_2 \frac{dv}{dt}.$$

Since $\mathbf{r}_1$ and $\mathbf{r}_2$ are non-zero and independent, the tangents to the curves (on the surface) through a point P lie in the plane which contains the two vectors $\mathbf{r}_1$ and $\mathbf{r}_2$ at P. This plane is the *tangent plane* at P.

The *normal* to the surface at P is the normal to the tangent plane at P and is therefore perpendicular to $\mathbf{r}_1$ and $\mathbf{r}_2$. The sense of the normal is fixed by the convention that if $\mathbf{N}$ is the unit normal vector, then $\mathbf{r}_1$, $\mathbf{r}_2$, and $\mathbf{N}$ in this order form a right-handed system. It follows that

$$\mathbf{N} = \frac{\mathbf{r}_1 \times \mathbf{r}_2}{H}, \qquad H = |\mathbf{r}_1 \times \mathbf{r}_2| \neq 0. \tag{2.2}$$

This convention for the sense of $\mathbf{N}$ depends upon the choice of parameters. If the parameters are transformed as in (1.7), and if $\mathbf{N}'$ is the new normal vector, then $\mathbf{N}'$ is in the direction

$$\partial \mathbf{r}/\partial u' \times \partial \mathbf{r}/\partial v'.$$

From (1.11), $\mathbf{N}$ and $\mathbf{N}'$ are therefore the same vector if

$$\partial(\phi, \psi)/\partial(u, v) > 0$$

and are opposite if $\partial(\phi, \psi)/\partial(u, v) < 0$. Since the Jacobian is continuous over D and does not vanish, it has the same sign throughout D, and a parameter transformation therefore either leaves every normal unchanged or reverses every normal.

The next two sections will describe two particular classes of surfaces.

3. Surfaces of revolution

The sphere. When the polar angles, i.e. the *co-latitude* u and the longitude v, are taken as parameters on a sphere of centre O and radius a, the position vector is

$$\mathbf{r} = a(\sin u \cos v, \ \sin u \sin v, \ \cos u). \tag{3.1}$$

The poles $u = 0$ and $u = \pi$ are (artificial) singularities, and the domain of u, v is $0 < u < \pi$, $0 \leqslant v < 2\pi$.

The parametric curves $v = $ constant are the *meridians* and $u = $ constant are the *parallels*. The two systems are clearly orthogonal, and it can be verified that $\mathbf{r}_1 . \mathbf{r}_2 = 0$ at every point. The normal $\mathbf{N}$ is directed outwards from the sphere.

The general surface of revolution. Taking the z-axis for the axis of revolution, let the generating curve in the xz-plane be given by the parametric equations

$$x = g(u), \qquad y = 0, \qquad z = f(u).$$

Then, if v is the angle of rotation about the z-axis, the position vector of the point (u, v) is

$$\mathbf{r} = (g(u)\cos v,\ g(u)\sin v,\ f(u)) \qquad (3.2)$$

and the domain of u, v is $0 \leqslant v < 2\pi$ together with the range of u.

As in the case of the sphere, $v = $ constant are the meridians given by the various positions of the generating curve, and $u = $ constant are the parallels, i.e. circles in planes parallel to the xy-plane. The vectors $\mathbf{r}_1$ and $\mathbf{r}_2$ are

$$\mathbf{r}_1 = (g'\cos v,\ g'\sin v,\ f'), \qquad \mathbf{r}_2 = (-g\sin v,\ g\cos v,\ 0),$$

so that $\mathbf{r}_1 . \mathbf{r}_2 = 0$ for all u, v, i.e. the parameters are orthogonal.

The normal $\mathbf{N}$ is found to be

$$\mathbf{N} = \frac{\mathbf{r}_1 \times \mathbf{r}_2}{H} = \frac{(-f'\cos v,\ -f'\sin v,\ g')}{(f'^2 + g'^2)^{\frac{1}{2}}},$$

using the fact that $g \neq 0$ at an ordinary point.

It is often convenient to take $g(u) = u$. The right circular cone of semi-vertical angle α, for example, is given by $g(u) = u$, $f(u) = u \cot \alpha$.

The anchor ring. This is obtained by rotating a circle of radius a about a line in its plane and at a distance b ($> a$) from its centre. It is therefore given by (3.2) with

$$g(u) = b + a\cos u, \qquad f(u) = a\sin u,$$

and the domain of u, v is $0 < u < 2\pi$, $0 < v < 2\pi$.

The meridians and parallels are all circles, and the normal $\mathbf{N}$ is directed inwards towards the centre of the meridian circle.

4. Helicoids

A helicoid is a surface generated by the screw motion of a curve about a fixed line, the *axis*. The various positions of the generating curve are thus obtained by first translating it through a distance λ parallel to the axis and then rotating it through an angle v about the axis, where λ/v has a constant value a. The constant $2\pi a$ is the *pitch* of the helicoid, being the distance translated in one complete revolution. It is positive or negative according as the helicoid is right or left handed (right or left screw), and is zero for a surface of revolution.

Right helicoid. This is the helicoid generated by a straight line which meets the axis at right angles. Taking the axis to be the z-axis, the position vector is

$$\mathbf{r} = (u\cos v,\ u\sin v,\ av) \tag{4.1}$$

where u is the distance from the axis, and v is the angle of rotation, the generator being assumed to be the x-axis when $v = 0$. Here u and v take all real values.

The curves $v = $ constant are the generators, and $u = $ constant are circular helices. Since $\mathbf{r}_1 . \mathbf{r}_2 = 0$, the helices are orthogonal to the generators.

The general helicoid. In the general case the sections of the surface by planes containing the axis are congruent plane curves, and the surface is generated by the screw motion of any one of these curves. There is no loss of generality, therefore, if the generating curve is assumed to be a plane curve given by equations of the form
$$x = g(u), \qquad y = 0, \qquad z = f(u).$$

The position vector of a point on the surface is then

$$\mathbf{r} = \big(g(u)\cos v,\ g(u)\sin v,\ f(u) + av\big). \tag{4.2}$$

The curves $v = $ constant are the various positions of the generating curve, and $u = $ constant are circular helices. It can be verified that $\mathbf{r}_1 . \mathbf{r}_2 = af'(u)$; the parametric curves are therefore orthogonal if either $f'(u) = 0$, in which case the surface is a right helicoid, or $a = 0$ which gives a surface of revolution.

EXAMPLE 4.1. A helicoid is generated by the screw motion of a straight line skew to the axis. Find the curve coplanar with the axis which generates the same helicoid.

If c is the shortest distance and α the angle between the axis and the given skew line, this line can be taken to be

$$x = c, \qquad y = u\sin\alpha, \qquad z = u\cos\alpha$$

where u is a parameter. Rotating through an angle v about the z-axis and translating a distance av parallel to this axis, the position vector of a point on the helicoid is found to be

$$\mathbf{r} = (c\cos v - u\sin\alpha\sin v,\ c\sin v + u\sin\alpha\cos v,\ u\cos\alpha + av).$$
$$\tag{4.3}$$

The required plane curve is the section of this surface by the plane $y = 0$ and is given by $u\sin\alpha = -c\tan v$. Substituting in $\mathbf{r}$, its equations are therefore

$$x = c \sec v, \qquad y = 0, \qquad z = av - c \cot \alpha \tan v,$$

where v is a parameter for the curve.

In the notation used above for the general helicoid, $g(u) = c \sec u$ and $f(u) = au - c \cot \alpha \tan u$. It can be verified that equation (4.3) is derived from (4.2) by the parameter transformation

$$u' = -c \operatorname{cosec} \alpha \tan u, \qquad v' = u + v.$$

5. Metric

On a given surface $\mathbf{r} = \mathbf{r}(u, v)$ consider the curve defined by $u = u(t)$, $v = v(t)$. Then $\mathbf{r}$ is a function of t along the curve, and the arc length s is related to the parameter t by

$$\left(\frac{ds}{dt}\right)^2 = \left(\frac{d\mathbf{r}}{dt}\right)^2 = \left(\mathbf{r}_1 \frac{du}{dt} + \mathbf{r}_2 \frac{dv}{dt}\right)^2 = E\left(\frac{du}{dt}\right)^2 + 2F \frac{du}{dt}\frac{dv}{dt} + G\left(\frac{dv}{dt}\right)^2,$$

where $\qquad E = \mathbf{r}_1^2, \qquad F = \mathbf{r}_1 . \mathbf{r}_2, \qquad G = \mathbf{r}_2^2. \qquad (5.1)$

This equation can conveniently be expressed in terms of differentials thus:

$$ds^2 = E\,du^2 + 2F\,du\,dv + G\,dv^2. \qquad (5.2)$$

The right-hand member of (5.2) does not involve the parameter t except in so far as u and v depend on t. Hence we regard this quadratic differential form as defined on the surface and we still denote it by ds^2. It must be remembered, however, that ds will no longer be a perfect differential, i.e. there is no function $\phi(u, v)$ such that $ds = d\phi$ identically. Geometrically ds can be interpreted as the 'infinitesimal distance' from the point (u, v) to the point $(u + du, v + dv)$. Because of (5.1) and the identity

$$(\mathbf{r}_1 \times \mathbf{r}_2)^2 = \mathbf{r}_1^2 \mathbf{r}_2^2 - (\mathbf{r}_1 . \mathbf{r}_2)^2,$$

the coefficients of (5.2) satisfy

$$E > 0, \quad G > 0, \quad H^2 = EG - F^2 > 0.$$

These inequalities show that the metric is a positive definite quadratic form in du, dv.

In (2.2), H was defined as $|\mathbf{r}_1 \times \mathbf{r}_2|$, so that H is the positive square root of $EG - F^2$.

EXAMPLE 5.1. For the paraboloid (1.5), $\mathbf{r}_1 = (1, 0, 2u)$ and $\mathbf{r}_2 = (0, 1, -2v)$, so that

$$E = \mathbf{r}_1^2 = 1 + 4u^2, \quad F = \mathbf{r}_1 . \mathbf{r}_2 = -4uv, \quad G = \mathbf{r}_2^2 = 1 + 4v^2;$$

$$H = (EG - F^2)^{\frac{1}{2}} = (1 + 4u^2 + 4v^2)^{\frac{1}{2}}.$$

The metric is chiefly used for the calculation of arc lengths on the surface, but the coefficients E, F, G also play important parts in other ways. For example, $F = \mathbf{r}_1 . \mathbf{r}_2 = 0$ has already appeared as the condition for the parametric curves to be orthogonal.

Angle between parametric curves. The parametric directions are given by $\mathbf{r}_1$ and $\mathbf{r}_2$. The angle ω $(0 < \omega < \pi)$ between them is therefore given by

$$\left.\begin{array}{ll} & \cos \omega = \dfrac{\mathbf{r}_1 . \mathbf{r}_2}{|\mathbf{r}_1||\mathbf{r}_2|} = \dfrac{F}{\surd(EG)} \\[3mm] \text{and} & \sin \omega = \dfrac{|\mathbf{r}_1 \times \mathbf{r}_2|}{|\mathbf{r}_1||\mathbf{r}_2|} = \dfrac{H}{\surd(EG)} \end{array}\right\}. \tag{5.3}$$

In general the angle between the parametric directions varies from point to point.

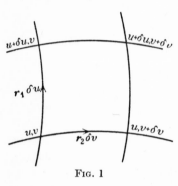

FIG. 1

Element of area. Consider the figure with vertices $(u,v), (u+\delta u, v)$, $(u+\delta u, v+\delta v)$, and $(u, v+\delta v)$ joined by parametric curves (Fig. 1). When δu and δv are small and positive, this figure is approximately a parallelogram with adjacent sides given by the vectors $\mathbf{r}_1 \delta u$, $\mathbf{r}_2 \delta v$, and area $|\mathbf{r}_1 \delta u \times \mathbf{r}_2 \delta v| = H \delta u \delta v$. This formula, with du, dv in place of δu, δv is taken to be[†] the element of area dS for the surface, so that

$$dS = H \, du \, dv.$$

EXAMPLE 5.2. For the anchor ring in section 3

$$E = a^2, \quad F = 0, \quad G = (b + a \cos u)^2, \quad H = a(b + a \cos u).$$

The whole anchor ring corresponds to the domain $0 \leqslant u \leqslant 2\pi$, $0 \leqslant v \leqslant 2\pi$. The area is therefore

$$\int_0^{2\pi} \int_0^{2\pi} a(b + a \cos u) \, du \, dv = 4\pi^2 ab,$$

a result which can, of course, be verified at once by applying Pappus's theorem on the area of a surface of revolution.

† See, for example, Goursat, i, § 131.

A transformation $u' = \phi(u,v)$, $v' = \psi(u,v)$ usually changes the fundamental coefficients of the metric. In terms of the primed parameters we have

$$E'du'^2 + 2F'du'dv' + G'dv'^2$$

$$= (\mathbf{r}_{1'}du' + \mathbf{r}_{2'}dv')^2$$

$$= \left\{\left(\mathbf{r}_1\frac{\partial u}{\partial u'} + \mathbf{r}_2\frac{\partial v}{\partial u'}\right)du' + \left(\mathbf{r}_1\frac{\partial u}{\partial v'} + \mathbf{r}_2\frac{\partial v}{\partial v'}\right)dv'\right\}^2$$

$$= (\mathbf{r}_1 du + \mathbf{r}_2 dv)^2$$

$$= E\,du^2 + 2F\,du\,dv + G\,dv^2.$$

The metric is thus invariant under a parameter transformation but the coefficients E, F, G are not invariant. In particular it should be noted that the coefficient E' considered as a function of u' and v' does not give the function $E(u,v)$ when the variables u', v' are replaced by $\phi(u,v)$, $\psi(u,v)$ respectively.

6. Direction coefficients

At a point P of a surface there are three independent vectors $\mathbf{N}$, $\mathbf{r}_1$, and $\mathbf{r}_2$. Every vector $\mathbf{a}$ at P can therefore be expressed in the form

$$\mathbf{a} = a_n\mathbf{N} + \lambda\mathbf{r}_1 + \mu\mathbf{r}_2,$$

where the scalars a_n, λ, and μ are defined uniquely by this relation. This gives $\mathbf{a}$ as the sum of two vectors, $a_n\mathbf{N}$ normal to the surface and $\lambda\mathbf{r}_1 + \mu\mathbf{r}_2$ in the tangent plane at P. The scalar a_n is called the *normal component* of $\mathbf{a}$, and is given by

$$a_n = \mathbf{a}.\mathbf{N}.$$

The vector $\mathbf{a}$ lies in the tangent plane if and only if $a_n = 0$.

The vector $\lambda\mathbf{r}_1 + \mu\mathbf{r}_2$ is called the *tangential part* of $\mathbf{a}$, and λ, μ are the *tangential components* of $\mathbf{a}$. These components depend only upon the tangential part of $\mathbf{a}$ and are both zero if and only if $\mathbf{a}$ is normal to the surface. The greater part of this chapter is concerned only with vectors which are tangent to the surface, i.e. have zero normal component. When there is no ambiguity we shall refer simply to their components, meaning their tangential components, and the vector (λ, μ) will be understood to mean the tangential vector with components (λ, μ).

If $\mathbf{a}$ is the vector (λ, μ), then

$$|\mathbf{a}| = |\lambda\mathbf{r}_1 + \mu\mathbf{r}_2| = (E\lambda^2 + 2F\lambda\mu + G\mu^2)^{\frac{1}{2}}. \tag{6.1}$$

This is the formula for the magnitude of a tangential vector in terms of its components; in general it depends upon the parameters of the point since it involves E, F, G.

A direction in the tangent plane at P is conveniently described by the components of the unit vector in this direction. These components are called *direction coefficients* and are written (l, m). They are analogous to direction cosines (l, m, n) in Cartesian geometry and satisfy an identity corresponding to $l^2 + m^2 + n^2 = 1$. Since the vector (l, m) has unit magnitude, the coefficients satisfy the identity

$$El^2 + 2Flm + Gm^2 = 1. \tag{6.2}$$

In the plane referred to rectangular Cartesian coordinates, a direction is determined by the angle ψ which it makes with the x-axis. The direction coefficients are then $\cos\psi$, $\sin\psi$, the metric is $dx^2 + dy^2$, and the identity (6.2) becomes simply $\cos^2\psi + \sin^2\psi = 1$.

In formulae for the angle between two (tangential) directions at the same point it is useful to introduce a convention. Angles in the tangent plane will be measured in the sense of rotation which carries the direction of $\mathbf{r}_1$ to that of $\mathbf{r}_2$ through an angle between 0 and π. This is also the positive sense of rotation about $\mathbf{N}$. By means of this convention a definite *orientation* is given to part of the surface.

If now (l, m) and (l', m') are coefficients of two directions at the same point, the corresponding unit vectors are

$$\mathbf{a} = l\mathbf{r}_1 + m\mathbf{r}_2, \qquad \mathbf{a}' = l'\mathbf{r}_1 + m'\mathbf{r}_2.$$

The angle between these directions, measured in the sense described above, is therefore given by $\cos\theta = \mathbf{a} . \mathbf{a}'$ and $\mathbf{N}\sin\theta = \mathbf{a} \times \mathbf{a}'$, i.e.

$$\left. \begin{array}{l} \cos\theta = Ell' + F(lm' + l'm) + Gmm' \\ \sin\theta = H(lm' - l'm) \end{array} \right\} . \tag{6.3}$$

From the definition of direction coefficients l, m it follows that the direction opposite (l, m) is $(-l, -m)$.

In (6.3) it must be remembered that l, m and l', m' are actual direction coefficients, i.e. they satisfy the identity (6.2). It is sometimes convenient to use *direction ratios*, numbers λ, μ proportional to the direction coefficients; they are the components of some vector in the given direction and the actual coefficients are obtained by dividing by the magnitude of the vector, thus

$$(l, m) = \frac{(\lambda, \mu)}{(E\lambda^2 + 2F\lambda\mu + G\mu^2)^{\frac{1}{2}}}.$$

The condition for orthogonal directions is $\cos\theta = 0$. In terms of direction ratios this becomes

$$E\lambda\lambda' + F(\lambda\mu' + \mu\lambda') + G\mu\mu' = 0. \tag{6.4}$$

The vectors $\mathbf{r}_1$ and $\mathbf{r}_2$ have components $(1, 0)$ and $(0, 1)$. These, then, are direction ratios for the parametric directions, the direction coefficients being $\left(\dfrac{1}{\sqrt{E}}, 0\right)$ and $\left(0, \dfrac{1}{\sqrt{G}}\right)$. Formulae (5.3) for the angle between the parametric directions can now be deduced from (6.3).

EXAMPLE 6.1. Find the coefficients of the direction which makes an angle $\frac{1}{2}\pi$ with the direction whose coefficients are (l, m).

If (l', m') are the required coefficients, then from (6.3)

$$l'(El + Fm) + m'(Fl + Gm) = 0,$$

i.e.
$$l' = -\alpha(Fl + Gm), \quad m' = \alpha(El + Fm)$$

for some α. Also, $H(lm' - l'm) = \sin\frac{1}{2}\pi = 1$, so that

$$\frac{1}{\alpha} = Hl(El + Fm) + Hm(Fl + Gm) = H,$$

since (l, m) are actual coefficients. Hence

$$l' = -\frac{1}{H}(Fl + Gm), \qquad m' = \frac{1}{H}(El + Fm).$$

It can be verified that l', m' satisfy the identity (6.2). This identity can be used to find α but gives only $\alpha^2 = 1/H^2$; the sine formula is needed to distinguish between the required direction and its opposite.

For the curve given by $u = u(t)$, $v = v(t)$, the position vector is $\mathbf{r} = \mathbf{r}(u, v) = \mathbf{r}(t)$, and $\dot{\mathbf{r}}$ is a tangent vector, a dot denoting differentiation with respect to t. Since

$$\dot{\mathbf{r}} = \dot{u}\mathbf{r}_1 + \dot{v}\mathbf{r}_2,$$

the components of $\dot{\mathbf{r}}$ are $(\dot{u}, \dot{v})$ which are therefore direction ratios for the tangent to the curve. The magnitude $|\dot{\mathbf{r}}|$ is $\dot{s}$ calculated from the metric. The unit tangent vector can therefore be written

$$\frac{d\mathbf{r}}{ds} = \frac{du}{ds}\mathbf{r}_1 + \frac{dv}{ds}\mathbf{r}_2,$$

and the direction coefficients are

$$l = \frac{du}{ds}, \qquad m = \frac{dv}{ds}.$$

These formulae show how a direction at a point is determined by differentials du, dv, corresponding to an infinitesimal displacement on the surface. Only the ratio $du:dv$ enters the formulae because the metric gives ds^2 as a quadratic in du, dv.

For example, in the direction of the curve $v = $ constant, $dv = 0$. Hence $m = 0$ and

$$l = \frac{du}{\sqrt{(E\,du^2)}} = \frac{1}{\sqrt{E}}.$$

For a curve given by an implicit equation $\phi(u, v) = 0$, $\phi_1\,du + \phi_2\,dv = 0$ along the curve, and direction ratios for the tangent are therefore $(-\phi_2, \phi_1)$ since these are proportional to du, dv (without regard to sense). Direction coefficients can now be found in the usual way.

7. Families of curves

By a *family* of curves on a surface we mean a system given by an implicit equation of the form

$$\phi(u, v) = c,$$

where ϕ is single valued and has continuous derivatives ϕ_1, ϕ_2 which do not vanish together, and c is a real parameter. There is just one curve of the family passing through every point (u, v) of the surface. The restriction on ϕ is desirable in order that a tangent shall exist at every point; from $\phi_1\,du + \phi_2\,dv = 0$ it follows that at any point (u, v) the tangent to the curve through the point has direction ratios $(-\phi_2, \phi_1)$.

The sense of the tangent vector $(-\phi_2, \phi_1)$ has some geometrical significance. It will be proved that for a variable direction at a point P, $|d\phi/ds|$ is maximum in a direction orthogonal to the curve $\phi = $ constant through P; and that the angle between the orthogonal direction in which ϕ is increasing and the vector $(-\phi_2, \phi_1)$ is $+\frac{1}{2}\pi$.

Let μ be the magnitude of the vector $(-\phi_2, \phi_1)$ and θ the angle between the direction with coefficients (l, m) and the vector $(-\phi_2, \phi_1)$. Then from the sine formula,

$$\mu\sin\theta = H(l\phi_1 + m\phi_2) = H\left(\frac{du}{ds}\phi_1 + \frac{dv}{ds}\phi_2\right) = H\frac{d\phi}{ds},$$

since $l = du/ds$ and $m = dv/ds$ for differentials du, dv in the direction (l, m). H and μ are independent of (l, m), so that $|d\phi/ds|$ has its maximum value μ/H when $\theta = \pm\frac{1}{2}\pi$, i.e. in a direction orthogonal to ϕ = constant. Since $H > 0$ and $\mu > 0$, the orthogonal direction for which $d\phi/ds > 0$ is such that $\theta = +\frac{1}{2}\pi$.

The curves of the family ϕ = constant are the solutions of the differential equation $\phi_1 du + \phi_2 dv = 0$. Conversely, a first order differential equation of the form

$$P(u,v)\,du + Q(u,v)\,dv = 0, \qquad (7.1)$$

where P and Q are C^1-functions which do not vanish simultaneously, always defines a family of curves. (See, for example, Goursat, ii, chapter xix.) It is well known that for an equation of this kind, functions $\lambda(u,v)$ ($\neq 0$) and $\phi(u,v)$ can be found so that $\lambda P = \phi_1$ and $\lambda Q = \phi_2$; the solutions of the equations are therefore the curves ϕ = constant.

For the curves defined by (7.1), the tangent vector at the point (u,v) is given by direction ratios $(-Q, P)$, since these are proportional to (du, dv).

Orthogonal trajectories

For a given family of curves there always exists a second family, the orthogonal trajectories, such that at every point the two curves, one from each family, are orthogonal.

Let the given family be defined by $P\,du + Q\,dv = 0$. The tangent at the point (u,v) is in the direction $(-Q, P)$; and, if du, dv are now taken to be differentials in an orthogonal direction, then

$$E(-Q)\,du + F(-Q\,dv + P\,du) + GP\,dv = 0$$

giving

$$(FP-EQ)\,du + (GP-FQ)\,dv = 0. \qquad (7.2)$$

The coefficients of du and dv are continuous and do not vanish together since $EG \neq F^2$ and P and Q do not vanish together. This is therefore the differential equation of a family of curves having the required property of being orthogonal to the given curves.

If the integral of (7.2) is $\psi(u,v)$ = constant, then $FP-EQ = \mu\psi_1$ and $GP-FQ = \mu\psi_2$ for some $\mu \neq 0$. If also the given family is ϕ = constant, so that $P = \lambda\phi_1$ and $Q = \lambda\phi_2$ ($\lambda \neq 0$), then

$$\frac{\partial(\phi,\psi)}{\partial(u,v)} = \frac{1}{\lambda\mu}\begin{vmatrix} P & Q \\ FP-EQ & GP-FQ \end{vmatrix} = \frac{1}{\lambda\mu}(EQ^2 - 2FPQ + GP^2).$$

This shows that $\partial(\phi,\psi)/\partial(u,v) \neq 0$ because the quadratic in the bracket is positive definite and P and Q do not vanish together. Thus the proper transformation $u' = \phi(u,v)$, $v' = \psi(u,v)$ transforms the given family of curves and the orthogonal family into the two families of parametric curves. Hence, *parameters can always be chosen so that the curves of a given family and their orthogonal trajectories become parametric curves.*

EXAMPLE 7.1. On the paraboloid $x^2-y^2 = z$, find the orthogonal trajectories of the sections by the planes $z = $ constant.

Taking $\mathbf{r} = (u, v, u^2-v^2)$, the given curves are $u^2-v^2 = $ constant. Along the given curves, $u\,du-v\,dv = 0$. (Note that since $P\ (= u)$ and $Q\ (= -v)$ vanish simultaneously at the origin, this point must be excluded.) The direction of the tangent at the point (u, v) is therefore that of (v, u). If (du, dv) is now orthogonal to this direction,
$$Ev\,du+F(v\,dv+u\,du)+Gu\,dv = 0;$$
and since $E = 1+4u^2$, $F = -4uv$, $G = 1+4v^2$, this equation reduces to
$$v\,du+u\,dv = 0.$$

The orthogonal trajectories are therefore given by $uv = $ constant. They are the sections of the paraboloid by the hyperbolic cylinders $xy = $ constant.

EXAMPLE 7.2. A helicoid is generated by the screw motion of a straight line which meets the axis at an angle α. Find the orthogonal trajectories of the generators. Find also the metric of the surface referred to the generators and their orthogonal trajectories as parametric curves.

The surface is given by (4.2) with $g = u\sin\alpha$ and $f = u\cos\alpha$, i.e.
$$\mathbf{r} = (u\sin\alpha\cos v,\ u\sin\alpha\sin v,\ u\cos\alpha+av),$$
where u and v take all real values. Hence $E = 1$, $F = a\cos\alpha$, $G = a^2+u^2\sin^2\alpha$.

The generators are given by $v = $ constant and have direction ratios $(1, 0)$. The direction (du, dv) is orthogonal to $(1, 0)$ if $E\,du+F\,dv = 0$, i.e. $du+a\cos\alpha\,dv = 0$. The orthogonal trajectories of the generators are therefore given by
$$u+av\cos\alpha = \text{constant}.$$

To examine these trajectories, note that $u = 0$ for some value of v on every curve, so that every trajectory meets the axis of the helicoid. For a particular curve there is no loss of generality in taking its intersection with the axis to be the origin. Then

$u = -av\cos\alpha$, and the curve is given by

$$\mathbf{r} = a\sin\alpha(-v\cos\alpha\cos v, \; -v\cos\alpha\sin v, \; v\sin\alpha),$$

with v as parameter. It is the intersection of the cone

$$x^2+y^2 = z^2\cot^2\alpha$$

and the cylinder whose cross-section (by the xy-plane) is the spiral $r = -\tfrac{1}{2}a\theta\sin 2\alpha$.

A transformation which takes the generators and their orthogonal trajectories into parametric curves is

$$u' = u+av\cos\alpha, \qquad v' = v.$$

The metric is

$$ds^2 = du^2+2a\cos\alpha\,du\,dv+(a^2+u^2\sin^2\alpha)\,dv^2$$
$$= du'^2+\sin^2\alpha\{a^2+(u'-av'\cos\alpha)^2\}\,dv'^2,$$

and the new coefficients are

$$E' = 1, \quad F' = 0, \quad G' = \sin^2\alpha\{a^2+(u'-av'\cos\alpha)^2\}.$$

Double family of curves

If P, Q, and R are continuous functions of u and v which do not vanish together, the quadratic differential equation

$$P\,du^2+2Q\,du\,dv+R\,dv^2 = 0 \tag{7.3}$$

has for solutions two families of curves provided that $Q^2-PR > 0$. The differential equations for the separate families are found by solving the quadratic for the ratio $du:dv$.

The direction coefficients for the two tangents at a point are (l,m) and (l',m') where l/m and l'/m' are the roots of the quadratic in du/dv. Hence

$$\frac{ll'}{R} = \frac{mm'}{P} = \frac{lm'+ml'}{-2Q}.$$

The two families are therefore orthogonal if and only if

$$ER-2FQ+GP = 0. \tag{7.4}$$

If $P = R = 0$, equation (7.3) becomes $du\,dv = 0$, giving the two families of parametric curves. The condition for orthogonality is now $F = 0$.

EXERCISE 7.1. Prove that, if θ is the angle at the point (u,v) between the two directions given by (7.3), then

$$\tan\theta = \frac{2H(Q^2-PR)^{\frac{1}{2}}}{ER-2FQ+GP}.$$

8. Isometric correspondence

In this chapter we shall consider examples of classes of surfaces with the property that surfaces in the same class are specially related to one another. The fundamental notion common to all these examples is that of correspondence of points between two surfaces, and the two surfaces are regarded as equivalent if this correspondence or mapping preserves some geometrical structure on the surfaces. An example is an *isometric* correspondence between points P on a surface S and points P' on S' such that, as P traces out an arc on S then P' traces out an arc of equal length on S'. Other examples will be discussed in sections 19 and 20, but it will be convenient to mention here certain properties common to all types of correspondence to be considered.

In this chapter we are concerned only with local properties of a surface, and in discussing correspondences between surfaces S, S' we shall consider only correspondence between *parts* of the surfaces. Each part will be assumed to carry a parametric system so that if the point (u', v') on S' corresponds to the point (u, v) on S, then u', v' are single-valued functions of u and v, say

$$u' = \phi(u,v), \qquad v' = \psi(u,v). \tag{8.1}$$

If S and S' are of class r and r' respectively we assume that ϕ and ψ are functions of class $\min(r, r')$, with Jacobian

$$\partial(\phi, \psi)/\partial(u, v) \neq 0$$

in the domain of u, v. We also assume that the mapping is (1–1) throughout this domain. Effectively we have restricted the maps between a part of S and a part of S' to be *differentiable homeomorphisms of sufficiently high class, regular at each point of the domain of u, v.* In what follows we shall use the word 'surface' to mean 'part of a surface'. In this chapter the emphasis is on local properties of parts of surfaces, and provided this is borne in mind no confusion should arise.

Consider a curve C of class $\geqslant 1$ passing through P and lying on S, given parametrically by equations $u = u(t)$, $v = v(t)$. If S' is related to S by (8.1), then C will map into a curve C' on S', passing through P', with parametric equations

$$u' = \phi\{u(t), v(t)\},$$
$$v' = \psi\{u(t), v(t)\}.$$

The direction of the tangent to C at P will map into a definite direction at P', namely that of the tangent to C', given by direction ratios $(\dot{u}', \dot{v}')$ where

$$\dot{u}' = \frac{\partial\phi}{\partial u}\dot{u} + \frac{\partial\phi}{\partial v}\dot{v},$$

$$\dot{v}' = \frac{\partial\psi}{\partial u}\dot{u} + \frac{\partial\psi}{\partial v}\dot{v}.$$

Solving these equations for $\dot{u}$, $\dot{v}$ we find

$$\dot{u} = \left(\dot{u}'\frac{\partial\psi}{\partial v} - \dot{v}'\frac{\partial\phi}{\partial v}\right)\!\bigg/ J,$$

$$\dot{v} = \left(\dot{v}'\frac{\partial\phi}{\partial u} - \dot{u}'\frac{\partial\psi}{\partial u}\right)\!\bigg/ J,$$

where $J = \partial(\phi,\psi)/\partial(u,v)$ and, since $J \neq 0$, it follows that to a given direction at P' will correspond a definite direction at P.

We now prove that it is possible to make a proper parameter transformation in S' (or S), *so that corresponding points P, P' carry identical parameter values*. This follows immediately because the functions ϕ, ψ of (8.1) satisfy the conditions for a proper parameter transformation, and after transforming the parameters of S' in this way the correspondence $S \to S'$ gives $(u,v) \to (u,v)$ as required.

The previous remarks apply to all correspondences between parts of surfaces considered in this chapter. So far a correspondence has been restricted to be (1–1), of sufficiently high class, and regular ($J \neq 0$) at each point. In the rest of this section we shall assume that correspondences are restricted in this manner, and we impose additional restrictions by requiring certain geometrical properties to be preserved under the correspondence.

Two surfaces S, S' are said to be *isometric* or *applicable* if there is a correspondence between the points of S and S' such that corresponding arcs of curves have the same length. The correspondence is called an *isometry*.

An example is a region S (not too large) of a plane and a region S' of a cylinder. The plane can be considered as being fitted onto the cylinder so that S coincides with S', and since no part of S is cut or stretched in this process the length of an arc in S remains unaltered. Geometrically, S is continuously deformed in space until it coincides with S' so that continuity and arc lengths in S are preserved. Points of S and S' which ultimately coincide are corresponding points of the isometry. This gives a general picture of the relation between two isometric surfaces and illustrates the fact that S and S' need not be congruent in order to be isometric.

The application of a plane to a circular cylinder also illustrates the idea of *local* isometry. If the whole plane S is wrapped round the cylinder S', infinitely many points of S correspond to the same point of S', so that the correspondence $S \to S'$ is not one–one but many–one. The plane and cylinder are not isometric 'in the large'; they are, however, *locally isometric* because every point of the plane has a neighbourhood which is isometric with a region of the cylinder.

For an isometry the length of any arc in S must be equal to the length of the corresponding arc in S'. It follows that $ds = ds'$ where ds and ds' are corresponding linear elements, and this must be true for all u, v, du, dv, and the corresponding u', v', du', dv' given by (8.1). The metric of S therefore transforms into the metric of S' under the transformation (8.1), and we have:

If surfaces S and S' are isometric, there exists a correspondence (8.1) between their parameters, where ϕ and ψ are single valued and have non-vanishing Jacobian, such that the metric of S transforms into the metric of S'.

EXAMPLE 8.1. To find a surface of revolution which is isometric with a region of the right helicoid.

From section 3, a surface of revolution is given by

$$\mathbf{r} = \big(g(u)\cos v,\ g(u)\sin v,\ f(u)\big)$$

for some functions f and g, and its metric is

$$(g_1^2 + f_1^2)\,du^2 + g^2\,dv^2,$$

where $f_1 = df/du$, etc. The right helicoid of pitch $2\pi a$ is given by $\mathbf{r} = (u'\cos v',\ u'\sin v',\ av')$ and its metric is

$$du'^2 + (u'^2 + a^2)\,dv'^2.$$

The problem is to find a transformation $(u, v) \to (u', v')$ which makes these two metrics identical.

Taking $v' = v$ and $u' = \phi(u)$ for simplicity, then $du' = \phi_1\,du$ and the metrics are identical if

$$g^2 = \phi^2 + a^2, \qquad g_1^2 + f_1^2 = \phi_1^2.$$

These are two equations for three functions f, g, ϕ. If ϕ is eliminated there remains a differential equation for f as a function of g. Or more simply, putting $\phi(u) = a \sinh u$ and $g(u) = a \cosh u$ to satisfy the first equation, the second gives $f_1^2 = a^2$ so that we can take $f(u) = au$. Hence, the right helicoid is isometric with the surface obtained by revolving the curve $x = a \cosh u, y = 0, z = au$ about the z-axis.

The generating curve is the catenary $x = a \cosh(z/a)$, with parameter a and directrix the z-axis, and the surface of revolution is a *catenoid*.

The correspondence $u' = a \sinh u, v' = v$ shows that the generators $v' = $ constant on the helicoid correspond to the meridians $v = $ constant on the catenoid, and the helices $u' = $ constant correspond to the parallels $u = $ constant.

On the helicoid u' and v' can take all values, but on the catenoid $0 \leqslant v < 2\pi$. The correspondence is therefore an isometry only for that region of the helicoid for which $0 \leqslant v' < 2\pi$, i.e. one period (given by one rotation in the screw motion). *Hence, one period of a right helicoid of pitch $2\pi a$ corresponds isometrically to the whole catenoid of parameter a.*

Without the limitation to one period of the helicoid the correspondence would be locally isometric.

The reader may be able to imagine a period of the helicoid, between two parallel generators, deformed without further cutting or stretching until it fits on to the catenoid, with the two generators coinciding along a meridian.

EXERCISE 8.1. By taking $v' = pv$ in the above problem, where $p > 1$, prove that the region of the right helicoid given by

$$|u'| < a/\sqrt{(p^2-1)}, \qquad 0 < v' < 2p\pi,$$

corresponds isometrically to the surface obtained by revolving that part of the curve

$$x = ap \cosh u, \qquad y = 0, \qquad z = a \int_0^u (\cosh^2 t - p^2 \sinh^2 t)^{\frac{1}{2}}\,dt$$

given by $|u| < \cosh^{-1} \dfrac{p}{\sqrt{(p^2-1)}}$ about the z-axis.

EXERCISE 8.2. Discuss the problem similar to that of Exercise 8.1 but with $0 < p < 1$.

EXERCISE 8.3. A surface is such that E, F, and G are functions of u only. Prove that every point has a neighbourhood which can be mapped isometrically on a region of a surface of revolution.

[HINT. Find a transformation $(u, v) \to (u, \bar{v})$ giving new co-efficients $\bar{E}(u)$, $\bar{F} = 0$, $\bar{G}(u)$. Then proceed as in the previous worked example.]

It will be seen later that not every surface has a region which can be mapped isometrically on a region of a surface of revolution.

Some surfaces have the property that they can be mapped iso-metrically on themselves (the trivial identity mapping is excepted). Obvious examples are the plane, circular cylinder, sphere, and helicoid since each of these admits a rigid motion in which the surface slides over itself. An example which is not given simply by rigid motion is as follows.

EXERCISE 8.4. A surface of revolution is defined by the equations

$$x = \cos u \cos v, \quad y = \cos u \sin v, \quad z = -\sin u + \log \tan(\tfrac{1}{4}\pi + \tfrac{1}{2}u),$$

where $0 < u < \tfrac{1}{2}\pi$ and $0 < v < 2\pi$. Show that the metric is

$$\tan^2 u \, du^2 + \cos^2 u \, dv^2,$$

and prove that the region $0 < u < \tfrac{1}{2}\pi$, $0 < v < \pi$ is mapped isometrically on the region $\tfrac{1}{3}\pi < u' < \tfrac{1}{2}\pi$, $0 < v' < 2\pi$ by the correspondence

$$u' = \cos^{-1}(\tfrac{1}{2}\cos u), \qquad v' = 2v.$$

EXERCISE 8.5. Find the metric of the surface in Exercise 8.4 referred to parameters U, v where $U = \log \sec u$, and find the domain of U, v for the whole surface. Prove that, when $n > 1$, the correspondence $U' = U + \log n$, $v' = nv$ maps a region of the surface isometrically on the whole surface.

9. Intrinsic properties

Let E, F, G be any real single-valued continuous functions of u and v satisfying $E > 0$ and $EG - F^2 > 0$ in some domain D of u, v. Then it will be seen later, in Chapter VII, section 19, *that every point of D has a neighbourhood D' (in D) in which $E \, du^2 + 2F \, du \, dv + G \, dv^2$ is the metric of a surface referred to u and v as parameters*. This is the first fundamental existence theorem, and shows that there is no hidden identity relating E, F, and G. It asserts the existence of a vector function $\mathbf{r}(u, v)$ satisfying the partial differential equations $\mathbf{r}_1^2 = E$, $\mathbf{r}_1 . \mathbf{r}_2 = F$, $\mathbf{r}_2^2 = G$, in some domain D'.

The surface having a given metric is certainly not unique, however, even apart from rigid displacements in space. Any two isometric surfaces, for example, have the same metric when corresponding points are assigned the same parameters, although, as seen in section 8, they need not be congruent. In fact, the class of surfaces having a given metric is the class of surfaces isometric with any one member.

It follows that any formula or property of a surface which is deducible from the metric alone, without recourse to the vector function $\mathbf{r}(u, v)$, automatically applies to the whole class of isometric surfaces. Properties of this kind will be described as *intrinsic*. In studying surfaces in general we make a point of distinguishing between those features which are intrinsic and those which are not; the latter are described as *non-intrinsic* and they usually involve normal components of vectors associated with the surface.

If a formula, equation, or theorem is intrinsic, it should be possible to derive it by an intrinsic argument, without introducing normal properties. This is done in the present book as far as is practicable. It is not merely an academic exercise, since it paves the way for Riemannian geometry, which is mainly intrinsic. It is realized, of course, that the quadratic differential form of metric is itself deduced from $\mathbf{r}(u, v)$; some such justification is needed for this particular form of ds^2, which might otherwise be taken to be, say, the square root of a quartic differential form, or any other homogeneous form of degree 2.

As soon as the quadratic form for ds^2 is adopted it is possible to proceed intrinsically, though this is not the way we have so far proceeded since we chose not to deny ourselves the advantage of referring to the basis vectors $\mathbf{r}_1$ and $\mathbf{r}_2$. A vector in the tangent plane may be defined by its components (λ, μ), and is intrinsic; all such vectors at a point form a *vector space* as defined in algebra, with a *norm* (magnitude) defined so that the norm of (du, dv) is the linear element ds given by the metric. Geometrically, the vector $(\epsilon\lambda, \epsilon\mu) = \epsilon(\lambda, \mu)$, where ϵ is small, can be regarded as representing the (small) displacement from the point (u, v) to the point

$$(u + \epsilon\lambda,\ v + \epsilon\mu).$$

The angle between two vectors (λ, μ) and (λ', μ') at a point (u, v) can now be defined by the Euclidean cosine formula applied to the small triangle with vertices (u, v), $(u+\epsilon\lambda, v+\epsilon\mu)$, and $(u+\epsilon'\lambda', v+\epsilon'\mu')$, where ϵ and ϵ' are small. It can be verified that this definition of angle is consistent with that previously given in section 6. We now have all that is needed at a point for the intrinsic study of a surface, viz. linear and area elements, vector components, vector magnitude, direction coefficients, and angle formulae.

10. Geodesics

On any surface there are special intrinsic curves, called *geodesics*, which are analogous to straight lines in Euclidean space because they are curves of shortest distance. The problem is, given any two points A and B on the surface, to find, out of all the arcs joining A and B, those which give the least arc length. This problem, treated properly, is difficult and beyond the scope of this book. For example, it is by no means clear that a solution exists, for although the lengths of the various arcs AB certainly have a non-zero greatest lower bound, it does not follow that there is an arc of this length. However, the problem does lead to a definite answer in the form of differential equations for the functions $u = u(t)$, $v = v(t)$ defining the curve. Every curve given by these equations is called a geodesic, whether it is a curve of shortest distance or not, and geodesics may be regarded as curves of stationary rather than strictly shortest distance on the surface.

We shall now derive the geodesic differential equations mentioned above by formulating a more restricted problem.

Let A, B be any two points, and consider the arcs which join A and B and are given by equations of the form $u = u(t)$, $v = v(t)$ where $u(t)$ and $v(t)$ are of class 2. Without loss of generality it can be assumed that for every arc α, $t = 0$ at A and $t = 1$ at B, so that α is given by $0 \leqslant t \leqslant 1$. Then the length of α is

$$s(\alpha) = \int_0^1 (E\dot{u}^2 + 2F\dot{u}\dot{v} + G\dot{v}^2)^{\frac{1}{2}}\, dt, \tag{10.1}$$

where $u(t)$ and $v(t)$ are substituted for u and v in E, F, and G.

Suppose now that an arc α' is obtained by deforming α slightly, keeping its end points A and B fixed. Then α' is given by equations of the form

$$u = u'(t) = u(t) + \epsilon\lambda(t), \qquad v = v'(t) = v(t) + \epsilon\mu(t),$$

where ϵ is small, and λ and μ are arbitrary functions of t of class 2 in $0 \leqslant t \leqslant 1$ and satisfying $\lambda = \mu = 0$ at $t = 0$ and $t = 1$. The length of α' is $s(\alpha')$ given by (10.1) with u', v' in place of u, v. The *variation* in $s(\alpha)$ is $s(\alpha') - s(\alpha)$ and is in general of order ϵ. If, however, α is such that the variation in $s(\alpha)$ is at most of order ϵ^2 for all small variations in α (i.e. for all $\lambda(t)$ and $\mu(t)$), then $s(\alpha)$ is said to be *stationary* and α is a *geodesic*.

The geodesics given in this way are clearly intrinsic and independent of any particular parametric representation of the surface.

To find the equations for geodesics, we follow the usual procedure as in the calculus of variations. Writing $f = \sqrt{(2T)}$ where

$$T(u, v, \dot{u}, \dot{v}) = \tfrac{1}{2}(E\dot{u}^2 + 2F\dot{u}\dot{v} + G\dot{v}^2),$$

then

$$s(\alpha') - s(\alpha) = \int_0^1 \{f(u + \epsilon\lambda, v + \epsilon\mu, \dot{u} + \epsilon\dot{\lambda}, \dot{v} + \epsilon\dot{\mu}) - f(u, v, \dot{u}, \dot{v})\}\, dt$$

$$= \epsilon \int_0^1 \left(\lambda \frac{\partial f}{\partial u} + \mu \frac{\partial f}{\partial v} + \dot{\lambda} \frac{\partial f}{\partial \dot{u}} + \dot{\mu} \frac{\partial f}{\partial \dot{v}}\right) dt + O(\epsilon^2).$$

Integrating by parts,

$$\int_0^1 \dot{\lambda} \frac{\partial f}{\partial \dot{u}}\, dt = \left[\lambda \frac{\partial f}{\partial \dot{u}}\right]_0^1 - \int_0^1 \lambda \frac{d}{dt}\left(\frac{\partial f}{\partial \dot{u}}\right) dt,$$

and the first term on the right is zero because $\lambda = 0$ at $t = 0$ and $t = 1$. Similarly,

$$\int_0^1 \dot{\mu} \frac{\partial f}{\partial \dot{v}}\, dt = -\int_0^1 \mu \frac{d}{dt}\left(\frac{\partial f}{\partial \dot{v}}\right) dt$$

and

$$s(\alpha') - s(\alpha) = \epsilon \int_0^1 (\lambda L + \mu M)\, dt + O(\epsilon^2),$$

where

$$L = \frac{\partial f}{\partial u} - \frac{d}{dt}\left(\frac{\partial f}{\partial \dot{u}}\right), \qquad M = \frac{\partial f}{\partial v} - \frac{d}{dt}\left(\frac{\partial f}{\partial \dot{v}}\right). \tag{10.2}$$

From the definition, therefore, $s(\alpha)$ is stationary and α is a geodesic if and only if $u(t)$ and $v(t)$ are such that

$$\int_0^1 (\lambda L + \mu M)\, dt = 0 \tag{10.3}$$

for all *admissible* λ, μ, i.e. functions of class 2 in $0 \leqslant t \leqslant 1$ which satisfy $\lambda = \mu = 0$ at $t = 0$ and $t = 1$.

It will now be proved that this condition implies $L = M = 0$.

LEMMA. *If $g(t)$ is continuous for $0 < t < 1$ and if*

$$\int_0^1 \nu(t) g(t)\, dt = 0$$

for all admissible functions $\nu(t)$ as defined above, then $g(t) = 0$.

Suppose there is a t_0 between 0 and 1 such that $g(t_0) \neq 0$, say $g(t_0) > 0$. Then, since g is continuous, $g(t) > 0$ in some interval (a, b) where $0 < a < t_0 < b < 1$. Now we define $\nu(t)$ as follows: $\nu(t) = 0$ for $0 \leqslant t < a$ and for $b < t \leqslant 1$, and $\nu(t) = (t-a)^3(b-t)^3$ for $a \leqslant t \leqslant b$. Then $\nu(t)$ is admissible, and

$$\int_0^1 \nu(t) g(t)\, dt = \int_a^b \nu(t) g(t)\, dt > 0,$$

since $g > 0$ and $\nu > 0$ for $a < t < b$. The supposition that there is a t_0 such that $g(t_0) \neq 0$ is therefore false, and the lemma is proved.

The functions L and M in (10.2) are continuous because E, F, G are assumed to be of class 1 and $u(t)$, $v(t)$ of class 2. The lemma can therefore be applied to (10.3), first with $\mu = 0$ and λ, L in place of ν, g and then with $\lambda = 0$ and μ, M in place of ν, g. It follows that (10.3) is satisfied for all admissible functions λ, μ if and only if $L = M = 0$. These, then, are differential equations for $u(t)$ and $v(t)$. They do not involve the points A and B explicitly and are therefore the same for all geodesics on the surface.

Substituting $f = \sqrt{(2T)}$, then

$$L = \frac{1}{\sqrt{(2T)}} \frac{\partial T}{\partial u} - \frac{d}{dt}\left(\frac{1}{\sqrt{(2T)}} \frac{\partial T}{\partial \dot{u}}\right)$$

$$= \frac{1}{\sqrt{(2T)}}\left\{\frac{\partial T}{\partial u} - \frac{d}{dt}\left(\frac{\partial T}{\partial \dot{u}}\right)\right\} + \frac{1}{(2T)^{\frac{3}{2}}} \frac{dT}{dt} \frac{\partial T}{\partial \dot{u}},$$

with a similar expression for M. The geodesic equations are therefore

$$\left.\begin{aligned}
U &\equiv \frac{d}{dt}\left(\frac{\partial T}{\partial \dot{u}}\right) - \frac{\partial T}{\partial u} = \frac{1}{2T} \frac{dT}{dt} \frac{\partial T}{\partial \dot{u}} \\
V &\equiv \frac{d}{dt}\left(\frac{\partial T}{\partial \dot{v}}\right) - \frac{\partial T}{\partial v} = \frac{1}{2T} \frac{dT}{dt} \frac{\partial T}{\partial \dot{v}}
\end{aligned}\right\}, \qquad (10.4)$$

where $\qquad T(u, v, \dot{u}, \dot{v}) = \tfrac{1}{2}(E\dot{u}^2 + 2F\dot{u}\dot{v} + G\dot{v}^2)$,

and the left-hand members of the equations are denoted by U and V for convenience.

The expressions U and V so defined are important in relation to any curve, whether it is a geodesic or not. They satisfy the identity

$$\dot u U + \dot v V = \frac{dT}{dt},\tag{10.5}$$

because

$$\dot u U + \dot v V = \frac{d}{dt}\!\left(\dot u\frac{\partial T}{\partial \dot u}+\dot v\frac{\partial T}{\partial \dot v}\right)-\ddot u\frac{\partial T}{\partial \dot u}-\ddot v\frac{\partial T}{\partial \dot v}-\dot u\frac{\partial T}{\partial u}-\dot v\frac{\partial T}{\partial v}$$

$$=\frac{d}{dt}(2T)-\frac{dT}{dt}=\frac{dT}{dt},$$

remembering that T is a function of u, v, $\dot u$, $\dot v$ homogeneous of degree 2 in $\dot u$, $\dot v$.

Since also the expressions on the right in (10.4) satisfy the same identity, i.e.

$$\dot u\!\left(\frac{1}{2T}\frac{dT}{dt}\frac{\partial T}{\partial \dot u}\right)+\dot v\!\left(\frac{1}{2T}\frac{dT}{dt}\frac{\partial T}{\partial \dot v}\right)=\frac{1}{2T}\frac{dT}{dt}\!\left(\dot u\frac{\partial T}{\partial \dot u}+\dot v\frac{\partial T}{\partial \dot v}\right)=\frac{dT}{dt},$$

it follows that the two equations in (10.4) are not independent; they are therefore equivalent to only one equation for the two unknown functions $u(t)$ and $v(t)$. This is to be expected because the parameter t has not been defined in any special way; the reader should verify formally that any transformation $t' = \phi(t)$, where ϕ is of class 2, would leave the differential equations unaltered. It is convenient to regard a curve as defined by two functions $u = u(t)$, $v = v(t)$, but strictly speaking there is only one function of one variable involved, as in the equation $v = f(u)$.

Eliminating dT/dt between the two equations (10.4), we obtain

$$U\frac{\partial T}{\partial \dot v}-V\frac{\partial T}{\partial \dot u}=0.\tag{10.6}$$

This, then, is necessary for a geodesic. To prove that it is also sufficient, suppose that it is satisfied by functions $u(t)$ and $v(t)$, whose first derivatives do not vanish simultaneously at any point. Then $\partial T/\partial \dot u$ and $\partial T/\partial \dot v$ cannot vanish together since this would imply $E\dot u + F\dot v = 0 = F\dot u + G\dot v$, and therefore $\dot u = \dot v = 0$. Hence,

$$U = \theta\frac{\partial T}{\partial \dot u},\qquad V = \theta\frac{\partial T}{\partial \dot v}$$

for some θ, and from the identity (10.5),

$$\frac{dT}{dt} = \theta\!\left(\dot u\frac{\partial T}{\partial \dot u}+\dot v\frac{\partial T}{\partial \dot v}\right) = 2T\theta,$$

i.e. $\theta = (1/2T)(dT/dt)$. The functions $u(t)$ and $v(t)$ therefore satisfy equations (10.4).

EXAMPLE 10.1. Prove that the curves of the family $v^3/u^2 = $ constant are geodesics on a surface with metric

$$v^2\,du^2 - 2uv\,dudv + 2u^2\,dv^2 \quad (u > 0, v > 0).$$

Consider $v^3/u^2 = c\ (> 0)$ and put this into a convenient parametric form $u = ct^3$, $v = ct^2$. Then $\dot u = 3ct^2$, $\dot v = 2ct$ and

$$\frac{\partial T}{\partial u} = -v\dot u\dot v + 2u\dot v^2 = 2c^3t^5, \qquad \frac{\partial T}{\partial v} = v\dot u^2 - u\dot u\dot v = 3c^3t^6,$$

$$\frac{\partial T}{\partial \dot u} = v^2\dot u - uv\dot v = c^3t^6, \qquad \frac{\partial T}{\partial \dot v} = -uv\dot u + 2u^2\dot v = c^3t^7,$$

$$U = \frac{d}{dt}(c^3t^6) - 2c^3t^5 = 4c^3t^5, \qquad V = \frac{d}{dt}(c^3t^7) - 3c^3t^6 = 4c^3t^6.$$

Hence $V\dfrac{\partial T}{\partial \dot u} - U\dfrac{\partial T}{\partial \dot v} = 0$, i.e. the curve is a geodesic for every value of c.

EXAMPLE 10.2. Prove that, on the general surface, a necessary and sufficient condition that the curve $v = c$ be a geodesic is

$$EE_2 + FE_1 - 2EF_1 = 0 \tag{10.7}$$

when $v = c$, for all values of u.

On the curve $v = c$, u can be taken as parameter, i.e. the curve is $u = t$, $v = c$. Then $\dot u = 1$, $\dot v = 0$, and on substituting these values (*after* calculating the partial derivatives of T),

$$\frac{\partial T}{\partial u} = \tfrac12 E_1, \qquad \frac{\partial T}{\partial \dot u} = E, \qquad U = \frac{dE}{dt} - \tfrac12 E_1 = \tfrac12 E_1,$$

$$\frac{\partial T}{\partial v} = \tfrac12 E_2, \qquad \frac{\partial T}{\partial \dot v} = F, \qquad V = \frac{dF}{dt} - \tfrac12 E_2 = F_1 - \tfrac12 E_2.$$

The curve is therefore a geodesic if

$$E(F_1 - \tfrac12 E_2) - F(\tfrac12 E_1) = 0$$

when $v = c$. This is condition (10.7) which is therefore necessary. Conversely when (10.7) is satisfied so is (10.6) and the curve $v = c$ is a geodesic.

If (10.7) is satisfied for all values of u and v, the parametric curves $v = $ constant are all geodesics.

Similarly, the curve $u = c$ is a geodesic if and only if

$$GG_1 + FG_2 - 2GF_2 = 0 \qquad (10.8)$$

when $u = c$.

In the neighbourhood of a point of a geodesic at which $\dot{u} \neq 0$, u can be taken as the parameter, as in Example 10.2 above. Then $\dot{u} = 1$,

$$\frac{\partial T}{\partial \dot{u}} = E + F\dot{v}, \qquad \frac{d}{du}\left(\frac{\partial T}{\partial \dot{u}}\right) = E_1 + (E_2 + F_1)\dot{v} + F_2\dot{v}^2 + F\ddot{v},$$

and
$$U = F\ddot{v} + (F_2 - \tfrac{1}{2}G_1)\dot{v}^2 + E_2\dot{v} + \tfrac{1}{2}E_1.$$

Also
$$\frac{\partial T}{\partial \dot{v}} = F + G\dot{v},$$

$$V = G\ddot{v} + \tfrac{1}{2}G_2\dot{v}^2 + G_1\dot{v} + F_1 - \tfrac{1}{2}E_2.$$

Hence

$$\frac{\partial T}{\partial \dot{u}} V - \frac{\partial T}{\partial \dot{v}} U = H^2(\ddot{v} + P\dot{v}^3 + Q\dot{v}^2 + R\dot{v} + S),$$

where $H^2P = \tfrac{1}{2}(GG_1 + FG_2 - 2GF_2)$, etc. The curve $v = v(u)$ is therefore a geodesic if v satisfies a second-order differential equation of the form

$$\ddot{v} + P\dot{v}^3 + Q\dot{v}^2 + R\dot{v} + S = 0,$$

where P, Q, R, and S are functions of u and v determined by E, F, G, and their first derivatives.

This gives some idea of the complicated nature of the geodesic equation in general. A form which is more convenient for theoretical investigations will be given in the next section.

11. Canonical geodesic equations

In equations (10.4) the parameter t is arbitrary and can conveniently be taken to be the arc length s of the curve measured from some fixed point on it. (This could not be done earlier because in the variational problem the limits of the independent variable were required to be fixed.)

When there is no ambiguity a prime will denote differentiation with respect to s. Then with s as parameter, $\dot{u}$, $\dot{v}$ are replaced by u', v' and

$$T = \tfrac{1}{2}(Eu'^2 + 2Fu'v' + Gv'^2). \qquad (11.1)$$

Along the curve, u' and v' satisfy the identity for direction coefficients. Hence $T = \tfrac{1}{2}$, $dT/ds = 0$, and equations (10.4) become the *canonical equations* for geodesics:

$$U \equiv \frac{d}{ds}\left(\frac{\partial T}{\partial u'}\right) - \frac{\partial T}{\partial u} = 0 \left.\begin{array}{c} \\ \\ \\ \\ \end{array}\right\} . \qquad (11.2)$$

$$V \equiv \frac{d}{ds}\left(\frac{\partial T}{\partial v'}\right) - \frac{\partial T}{\partial v} = 0$$

It must be remembered that in these equations the partial derivatives of T are calculated from (11.1) *before* values for u' and v' are substituted; T is not equal to $\frac{1}{2}$ identically for all u, v, u', v', but only along the curve.

The identity (10.5) now becomes

$$u'U + v'V = 0$$

confirming that equations (11.2) are not independent. For a curve other than a parametric curve, $u' \neq 0$, $v' \neq 0$, and the conditions $U = 0$ and $V = 0$ are equivalent, either being sufficient for a geodesic. For a parametric curve $u = $ constant, $u' = 0$, $v' \neq 0$, and $V = 0$ for all s, so that the equation is satisfied automatically; the condition for a geodesic is therefore $U = 0$. Similarly, $V = 0$ is the sufficient condition for a curve $v = $ constant to be a geodesic.

EXAMPLE 11.1. To find the geodesics on a surface of revolution. Let the surface be given as in section 3. Then

$$T = \tfrac{1}{2}\{(f_1^2 + g_1^2)u'^2 + g^2 v'^2\},$$

where $f_1 = df/du$, etc., and since $\partial T/\partial v = 0$ the canonical equation $V = 0$ can be integrated immediately to give

$$g^2 v' = \alpha,$$

where α is an arbitrary constant which can be assumed non-negative, taking the positive sense along the curve to be that in which v increases. If $\alpha = 0$, then v is constant and every meridian is a geodesic. Assume now that α is positive. Then the first order differential equation can be written

$$g^4 dv^2 = \alpha^2 ds^2 = \alpha^2\{(f_1^2 + g_1^2)du^2 + g^2 dv^2\},$$

giving $\qquad \alpha\sqrt{(f_1^2 + g_1^2)}\,du \pm g\sqrt{(g^2 - \alpha^2)}\,dv = 0,$

the $\pm$ being included although α is arbitrary because dv/du may change sign along the same geodesic. If $g^2 \neq \alpha^2$, by integration the geodesics are given by an equation of the form

$$v = \alpha\phi(u, \alpha) + \beta,$$

where α, β are arbitrary constants.

If $g^2 = \alpha^2$, then $u = $ constant. However, for curves $u = $ constant the equation $V = 0$ is automatically satisfied. To see whether $u = c$ is a geodesic it is necessary to apply the condition $U = 0$. Since now $u' = 0$ and $v' = g^{-1}$ from the identity for direction coefficients,

$$\frac{\partial T}{\partial u'} = 0, \qquad \frac{\partial T}{\partial u} = g_1/g, \qquad U = -g_1/g.$$

The curve $u = c$ is therefore a geodesic if and only if $g_1(c) = 0$. Since g is the radius of the parallel $u = c$ on the surface of revolution, a parallel is a geodesic if its radius is stationary.

The method used in the above example can be applied to give the following result which will be left to the reader to verify. If E, F, and G are functions of only one parameter, u say, the geodesics can all be found by quadratures. This applies not only to the general surface of revolution but also to the general helicoid (section 4). The geodesics are given by the equation

$$v = \int \left\{ -\frac{F}{G} \pm \frac{\alpha H}{G(G-\alpha^2)^{\frac{1}{2}}} \right\} du + \beta$$

where α and β are arbitrary constants; and also by the equation $u = c$ where c is any root of the equation $G_1 = 0$. From (10.7) it follows that if F^2/E is constant, then every curve $v = $ constant is a geodesic.

EXERCISE 11.1. Prove that for a helicoid of non-zero pitch the sections by planes containing the axis are geodesics if and only if these sections are straight lines.

EXAMPLE 11.2. On a right helicoid of pitch $2\pi a$, a geodesic makes an angle α with a generator at a point distant c from the axis $(0 < \alpha < \frac{1}{2}\pi, c > 0)$. Prove that the geodesic meets the axis if $c \tan \alpha < a$, but that if $c \tan \alpha > a$, its least distance from the axis is $(c^2 \sin^2\alpha - a^2 \cos^2\alpha)^{\frac{1}{2}}$. Find the equation of the geodesic in the case $c \tan \alpha = a$.

From the equations given in section 4 the metric of the right helicoid is $du^2 + (u^2 + a^2) dv^2$. As in the above examples, a first integral of the geodesic equations is

$$\frac{dv}{du} = \frac{\pm k}{\{(u^2 + a^2)(u^2 + a^2 - k^2)\}^{\frac{1}{2}}},$$

where k is an arbitrary positive constant. Further integration in general requires elliptic functions.

The given point is $(c, 0)$ for a suitable choice of axes, and α is the angle between the directions $(1, 0)$ and (u', v') at this point, i.e. $\tan \alpha = Hv'/u' = k(c^2+a^2-k^2)^{-\frac{1}{2}}$. This gives $k = (c^2+a^2)^{\frac{1}{2}} \sin \alpha$. There are two geodesics satisfying the given initial conditions, but it will be sufficient to consider the one for which $dv/du < 0$ initially.

From the form of dv/du it appears that there are three cases.

(i) $k^2 > a^2$, i.e. $c \tan \alpha > a$. Since $dv/du < 0$ initially, u decreases as v increases until $u = (k^2-a^2)^{\frac{1}{2}} = (c^2 \sin^2\alpha - a^2 \cos^2\alpha)^{\frac{1}{2}}$. As v continues to increase, the sign of dv/du changes and u increases indefinitely. The least distance from the axis is therefore

$$(c^2 \sin^2\alpha - a^2 \cos^2\alpha)^{\frac{1}{2}}.$$

(ii) $k^2 < a^2$, i.e. $c \tan \alpha < a$. In this case $dv/du < 0$ for all v, and u decreases indefinitely as v increases. There is a point on the curve at which $u = 0$, i.e. the curve meets the axis.

(iii) $k^2 = a^2$, i.e. $c \tan \alpha = a$. In this special case

$$\frac{dv}{du} = \frac{-a}{u(u^2+a^2)^{\frac{1}{2}}}$$

and $v = -\beta + \sinh^{-1}(a/u)$ where $\beta = +\sinh^{-1}(a/c)$, since $v = 0$ when $u = c$. The geodesic is therefore given by

$$u \sinh(v+\beta) = a, \qquad \beta = \sinh^{-1}(a/c).$$

As v increases, the curve approaches the axis without reaching it. In the opposite sense, $u \to \infty$ as $v \to -\beta$, showing that the generator $v = -\beta$ is an asymptote.

12. Normal property of geodesics

The geodesic equations can be expressed in terms of $\mathbf{r}(u, v)$ by means of the following identities which hold for any functions $u(t)$, $v(t)$ of a general parameter t;

$$\left. \begin{array}{ll} \dfrac{\partial T}{\partial \dot{u}} = \dot{\mathbf{r}} . \mathbf{r}_1 & \dfrac{\partial T}{\partial \dot{v}} = \dot{\mathbf{r}} . \mathbf{r}_2 \\[2mm] U(t) = \ddot{\mathbf{r}} . \mathbf{r}_1, & V(t) = \ddot{\mathbf{r}} . \mathbf{r}_2 \end{array} \right\} , \qquad (12.1)$$

where, as before, $T = \frac{1}{2}(E\dot{u}^2 + 2F\dot{u}\dot{v} + G\dot{v}^2)$.

To prove these, consider the relations

$$T = \frac{1}{2}\dot{\mathbf{r}}^2, \qquad \dot{\mathbf{r}} = \mathbf{r}_1 \dot{u} + \mathbf{r}_2 \dot{v}.$$

Then
$$\frac{\partial T}{\partial \dot{u}} = \dot{\mathbf{r}}.\frac{\partial \dot{\mathbf{r}}}{\partial \dot{u}} = \dot{\mathbf{r}}.\mathbf{r}_1,$$

$$\frac{\partial T}{\partial u} = \dot{\mathbf{r}}.\frac{\partial \dot{\mathbf{r}}}{\partial u} = \dot{\mathbf{r}}.(\mathbf{r}_{11}\dot{u}+\mathbf{r}_{21}\dot{v}) = \dot{\mathbf{r}}.\frac{d}{dt}(\mathbf{r}_1),$$

$$U(t) = \frac{d}{dt}(\dot{\mathbf{r}}.\mathbf{r}_1)-\dot{\mathbf{r}}.\frac{d}{dt}(\mathbf{r}_1) = \ddot{\mathbf{r}}.\mathbf{r}_1,$$

and similarly for $\partial T/\partial \dot{v}$ and $V(t)$.

With s as parameter the geodesic equations are $U(s) = 0$; $V(s) = 0$. They can therefore be written

$$\mathbf{r}''.\mathbf{r}_1 = 0, \qquad \mathbf{r}''.\mathbf{r}_2 = 0, \tag{12.2}$$

showing that, at every point P of the geodesic, $\mathbf{r}''$ is perpendicular to the tangent plane at P. This condition is sufficient as well as necessary. Hence:

A characteristic property of a geodesic is that at every point its principal normal is normal to the surface. Every curve having this property is a geodesic.

In terms of a general parameter t, equation (10.6) can be written

$$(\dot{\mathbf{r}}.\mathbf{r}_1)(\ddot{\mathbf{r}}.\mathbf{r}_2)-(\dot{\mathbf{r}}.\mathbf{r}_2)(\ddot{\mathbf{r}}.\mathbf{r}_1) = 0,$$

i.e. $$(\dot{\mathbf{r}}\times\ddot{\mathbf{r}}).(\mathbf{r}_1\times\mathbf{r}_2) = 0. \tag{12.3}$$

This says that the binormal of the curve is perpendicular to the normal to the surface, from which it follows that the principal normal is normal to the surface.

An equivalent statement of the above normal property is that at every point of a geodesic the rectifying plane is tangent to the surface.

The above property often makes it possible to intuit that a curve is a geodesic. For example, every great circle of a sphere and every meridian of a surface of revolution clearly have the normal property of geodesics. Again, it is now clear that the only parallels of a surface of revolution which are geodesics are those whose radii have stationary lengths.

EXAMPLE 12.1. A particle is constrained to move on a smooth surface under no force except the normal reaction. Prove that its path is a geodesic.

The acceleration is in the direction $\ddot{\mathbf{r}}$ which is therefore in the direction of the force, i.e. normal to the surface. Since $\dot{\mathbf{r}}$ is tangent to the surface, $\dot{\mathbf{r}} \cdot \ddot{\mathbf{r}} = 0$ and $\dot{s} = |\dot{\mathbf{r}}| = $ constant, showing that the speed is constant. It follows that $\mathbf{r}''$ is in the direction $\ddot{\mathbf{r}}$, i.e. is normal to the surface, and the curve is therefore a geodesic.

This problem can also be solved by using the Lagrange equation of dynamics, taking u and v as generalized coordinates; this leads to equations (11.2) for the path.

EXERCISE 12.1. Prove that every helix on a cylinder is a geodesic.

The normal property is sometimes taken as the definition of a geodesic. It has the advantage of simplicity but obscures the intrinsic character of geodesics and could not apply to Riemannian geometry which is similar to the intrinsic geometry of a surface but with any number of dimensions. Also, the normal property strictly fails in the case of a straight line on a surface, for then the principal normal is indeterminate. Such a line is clearly a geodesic according to the intrinsic definition.

It is instructive to see how the differential equations for geodesics arise out of equations (12.2), and to see how certain *Christoffel symbols* arise at this stage, although they will arise in different contexts later in the book.

Differentiating $\mathbf{r}' = \mathbf{r}_1 u' + \mathbf{r}_2 v'$, we find

$$\mathbf{r}'' = \mathbf{r}_1 u'' + \mathbf{r}_2 v'' + \mathbf{r}_{11} u'^2 + 2\mathbf{r}_{12} u'v' + \mathbf{r}_{22} v'^2.$$

The geodesic equations $\mathbf{r}'' \cdot \mathbf{r}_1 = 0$ and $\mathbf{r}'' \cdot \mathbf{r}_2 = 0$ thus become

$$\left. \begin{array}{l} Eu'' + Fv'' + \Gamma_{111} u'^2 + 2\Gamma_{112} u'v' + \Gamma_{122} v'^2 = 0 \\ Fu'' + Gv'' + \Gamma_{211} u'^2 + 2\Gamma_{212} u'v' + \Gamma_{222} v'^2 = 0 \end{array} \right\}, \qquad (12.4)$$

where
$$\Gamma_{ijk} = \mathbf{r}_i \cdot \mathbf{r}_{jk} \quad (i, j, k = 1, 2). \qquad (12.5)$$

The coefficients Γ_{ijk} are called *Christoffel symbols of the first kind* and can be expressed in terms of first derivatives of the fundamental coefficients. It can easily be verified that

$$\tfrac{1}{2}\{(\mathbf{r}_i \cdot \mathbf{r}_j)_k + (\mathbf{r}_i \cdot \mathbf{r}_k)_j - (\mathbf{r}_j \cdot \mathbf{r}_k)_i\} = \mathbf{r}_i \cdot \mathbf{r}_{jk} = \Gamma_{ijk}. \qquad (12.6)$$

Thus
$$\left. \begin{array}{l} \Gamma_{111} = \tfrac{1}{2}E_1, \quad \Gamma_{112} = \Gamma_{121} = \tfrac{1}{2}E_2, \quad \Gamma_{122} = F_2 - \tfrac{1}{2}G_1 \\ \Gamma_{211} = F_1 - \tfrac{1}{2}E_2, \quad \Gamma_{212} = \Gamma_{221} = \tfrac{1}{2}G_1, \quad \Gamma_{222} = \tfrac{1}{2}G_2 \end{array} \right\}. \qquad (12.7)$$

Since $EG - F^2 \neq 0$, equations (12.4) can be solved for u'' and v''. The resulting equations, which are equivalent to (12.4), are written

$$\left. \begin{array}{l} u'' + \Gamma_{11}^1 u'^2 + 2\Gamma_{12}^1 u'v' + \Gamma_{22}^1 v'^2 = 0 \\ v'' + \Gamma_{11}^2 u'^2 + 2\Gamma_{12}^2 u'v' + \Gamma_{22}^2 v'^2 = 0 \end{array} \right\}, \qquad (12.8)$$

where the coefficients Γ^i_{jk}, called the *Christoffel symbols of the second kind*, are given by

$$\Gamma^1_{jk} = H^{-2}(G\Gamma_{1jk} - F\Gamma_{2jk}), \quad \Gamma^2_{jk} = H^{-2}(E\Gamma_{2jk} - F\Gamma_{1jk}). \quad (12.9)$$

EXERCISE 12.2. By using equations (12.7), show that (12.4) and (11.2) are the same equations in different notation.

13. Existence theorems

With s as parameter the geodesic equations can be written in the form (cf. (12.8))

$$u'' = f(u, v, u', v'), \qquad v'' = g(u, v, u', v'), \qquad (13.1)$$

where f and g are quadratic forms in u', v' with single-valued continuous functions of u and v as coefficients. These are simultaneous second order differential equations for u and v as functions of s, and from the theory of such equations,† if f and g are of class $\geqslant 1$, a solution exists and is determined uniquely by arbitrary initial values of u, v, u', and v'. Hence:

A geodesic can be found to pass through any given point and have any given direction at that point. The geodesic is determined uniquely by these initial conditions.

From the above existence theorem it is to be expected that if a point Q is sufficiently near any given point P, then it is possible to find a direction at P such that the geodesic through P in this direction also passes through Q. The following theorem can in fact be proved, assuming merely that the surface is of class 3.

Every point P of the surface has a neighbourhood N with the property that every point of N can be joined to P by a unique geodesic arc which lies wholly in N.

This does not, of course, state that if Q is a point of N then the geodesic arc PQ which lies in N is the only geodesic joining P and Q; there may be other geodesic arcs PQ but they leave N. Examples of this will be given later in this section.

The proof of this theorem involves the study of differential equations of the form (13.1) and will be given in Appendix II.1.

This theorem gives all that we can say at present about the existence of geodesics joining two given points; it says that Q can be joined to P if it is sufficiently near P. Nothing more than that can be said as long as the region of the surface being considered

† See, for example, L. Bieberbach (1926), pp. 115–16.

is arbitrary. Later, however, when a *complete* surface has been defined, it will appear that any two points can be joined by at least one geodesic.

A region R is *convex* if any two points of it can be joined by a geodesic arc lying wholly in R, and is *simple* if there is not more than one such geodesic arc. In the Euclidean plane a convex region is necessarily simple but this is not so for a surface in general. The surface of a sphere, for example, is convex but not simple. An existence theorem due to J. H. C. Whitehead (1932) states that

Every point P of a surface has a neighbourhood which is convex and simple.

The difference between this and the previous theorem is that it is no longer just one particular point which is joined to the others of the neighbourhood; every point is joined uniquely to every other point. Whitehead's theorem is in fact much deeper than the previous theorem and its proof is beyond the scope of this book. It will not be used in the sequel.

A particular and interesting form of Whitehead's theorem is concerned with a geodesic disk of given centre P and radius r, defined as the set of points Q such that there is a geodesic arc PQ of length not greater than r. Whitehead proved that *for every point P there is an $\epsilon > 0$ such that every geodesic disk of centre P and radius $r \leqslant \epsilon$ is convex and simple.*

EXERCISE 13.1. On a circular cylinder of radius a, find the least upper bound for the radius of a simple convex geodesic disk, and prove that a geodesic disk of greater radius is convex but not simple.

This section will be concluded with examples of the multiplicity of geodesics joining two points. They are mostly constructed by using the intrinsic property of geodesics, that if surfaces S and S' are isometric, then the curve on S' which corresponds to a geodesic on S is a geodesic on S'. In fact the correspondence need only be locally isometric since a curve is a geodesic if every small arc is a geodesic arc.

Consider, for example, the mapping of a plane on a circular cylinder obtained by wrapping the plane round the cylinder. A geodesic on the plane is a straight line, and this corresponds to a helix (or meridian or circular section) on the cylinder. The helix is therefore a geodesic on the cylinder. Conversely, every helix on the cylinder corresponds to a straight line (or strictly to a family of parallel straight lines) on the plane; thus every helix is a geodesic.

It follows at once that any two points P, Q of the cylinder, not on the same parallel, are joined by infinitely many geodesic arcs because there are infinitely many helices joining the two points. When the cylinder is unrolled into a plane there are infinitely many images of Q, and the geodesics PQ correspond to the straight lines joining all the images of Q to any one image of P. There is a geodesic arc PQ making any desired number of turns round the cylinder, in either sense.

A similar result holds for the anchor ring. Joining two points P, Q not on the same meridian there are infinitely many geodesic arcs; an arc can be found to make any number of turns, in either sense, of the kind made by the meridian circles, and at the same time any number of turns, in either sense, of the kind made by the parallel circles. This cannot be proved by the simple method used for the cylinder because the anchor ring is not locally isometric to a plane. The geodesic equations can, however, be integrated by the method of section 11.

An example of a surface on which the number of geodesics joining two given points may be more than one but is strictly limited is a right circular half-cone. Here again the different geodesic arcs PQ are obtained by taking different numbers of turns round the cone in either sense.

A local isometry can be set up by rolling the cone over the plane. The surface of the cone corresponds isometrically to a sector of the plane which is reproduced according to the number of revolutions of the cone, in either sense. The images of Q are points $Q_1, Q_2,...$ in one sense and $Q_{-1}, Q_{-2},...$ in the other. If P' is the first image of P (in the sector between Q_1 and Q_{-1}), then the geodesic arcs PQ which pass round the cone in one sense correspond to the straight lines $P'Q_1, P'Q_2,...$, and those which pass round the cone in the opposite sense correspond to the lines $P'Q_{-1}, P'Q_{-2},...$. In either sense the number is limited; the lines $P'Q_r$ must all be on one side of $P'V$ and the lines $P'Q_{-r}$ must all be on the other side of $P'V$. In Fig. 2 there are three geodesic arcs in one sense and two in the other.

Clearly, the smaller the solid angle of the cone the greater the number of geodesics. Fig. 3 illustrates the case when there is only one geodesic arc PQ.

It is interesting to note that, on the cone, there may be non-trivial geodesic arcs joining a point to itself; in the above argument P can coincide with Q.

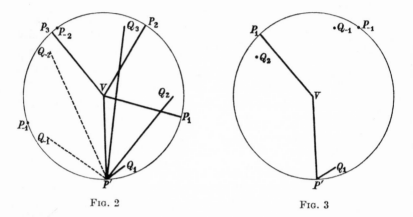

FIG. 2   FIG. 3

With the cone as a guide it is not difficult to construct other surfaces on which there may be any finite number of geodesics joining two points, or joining a point to itself. An example is the paraboloid of revolution, on which the geodesic equation can be integrated by the method of section 11. Again, on the paraboloid

$$\frac{x^2}{a^2} + \frac{y^2}{b^2} = \frac{2z}{c},$$

or on one sheet of the hyperboloid

$$\frac{x^2}{a^2} + \frac{y^2}{b^2} - \frac{z^2}{c^2} = -1,$$

the larger c is in comparison with a and b the more geodesics there are joining two points.

EXERCISE 13.2. Prove that, on a right circular cone of semi-vertical angle α, every point can be joined to itself by a geodesic arc if $\alpha < \frac{1}{6}\pi$. If this condition is satisfied prove that the number of geodesic arcs joining a point to itself is the greatest integer less than $(2\sin\alpha)^{-1}$. Prove also that this is the number of times a geodesic other than a generator intersects itself.

EXERCISE 13.3. Prove that, on a paraboloid of revolution, every geodesic other than a meridian intersects itself infinitely often.

14. Geodesic parallels

Suppose a family of geodesics is given, and that a parameter system is chosen so that the geodesics of the family are the curves v = constant and their orthogonal trajectories are the curves u = constant. Then $F = 0$ and condition (10.7) for the curves v = constant to be geodesics becomes $E_2 = 0$. The metric is therefore of the form

$$ds^2 = E(u)\,du^2 + G(u,v)\,dv^2. \qquad (14.1)$$

Consider the distance between any two of the orthogonal trajectories, say $u = u_1$ and $u = u_2$, measured along the geodesic $v = c$. Along $v = c$, $dv = 0$, and $ds = \sqrt{E}\,du$, so that the distance is

$$\int_{u_1}^{u_2} \sqrt{E(u)}\,du,$$

a number independent of c. The distance is thus the same along whichever geodesic v = constant it is measured. Because of this, the orthogonal trajectories are called *geodesic parallels*.

In the plane, a family of geodesics is a family of straight lines enveloping some curve C, and the geodesic parallels are the involutes of C. In particular, when the geodesics are concurrent straight lines, the parallels are concentric circles.

In the above metric the parameter u can be specialized by taking it to be the distance from some fixed parallel to the parallel determined by u, the distance being measured along any geodesic $v = c$. Then $ds = du$ when $dv = 0$, i.e. $E = 1$. Hence: *for any given family of geodesics, a parameter system can be chosen so that the metric takes the form $du^2 + G\,dv^2$. The given geodesics are the parametric curves v = constant and their orthogonal trajectories are u = constant, u being the distance measured along a geodesic from some fixed parallel.*

The transformation $u \to u' : du' = \sqrt{E}\,du$ also gives the simplified metric from (14.1).

EXERCISE 14.1. If a surface admits two orthogonal families of geodesics, it is isometric with the plane.

Geodesic polars

A particularly useful system of geodesics and parallels is found by taking the geodesics which pass through a given point O. By the second existence theorem there is a neighbourhood of O in which, when the point O itself is excluded, the geodesics constitute a family. Parameters u, v can therefore be chosen as above. In particular u can be taken to be the distance measured from O along the geodesics and v can be taken to be the angle measured at O between a fixed geodesic $v = 0$ and the one determined by v. In this way u and v correspond to polar coordinates r and θ in the plane. The metric is therefore

$$du^2 + G\,dv^2,$$

where G is such that, when u is small, the metric approximates to the plane polar form with u, v in place of r, θ, i.e. to $du^2 + u^2\,dv^2$. Hence $G \sim u^2$, i.e.

$$\lim_{u \to 0} \frac{\sqrt{G}}{u} = 1.$$

In geodesic polar parameters the parallels $u = $ constant are geodesic circles.

15. Geodesic curvature

For any curve on a surface the *curvature vector* at a point P is $\mathbf{r}'' = \kappa\mathbf{n}$, where κ is the curvature and $\mathbf{n}$ is the principal normal. This can be written

$$\mathbf{r}'' = \kappa_n\mathbf{N} + \lambda\mathbf{r}_1 + \mu\mathbf{r}_2, \tag{15.1}$$

where κ_n is the normal component of $\mathbf{r}''$, called the *normal curvature*. The vector $\lambda\mathbf{r}_1 + \mu\mathbf{r}_2$, with components (λ, μ), is zero for a geodesic because then $\mathbf{r}''$ is normal to the surface. This suggests that for any curve the vector (λ, μ) is intrinsic so that its magnitude measures in some sense the deviation of the curve from a geodesic. The vector (λ, μ) is, in fact, intrinsic, for from (15.1), taking scalar products with $\mathbf{r}_1$ and $\mathbf{r}_2$ and using the identities (12.1),

$$E\lambda + F\mu = \mathbf{r}''.\mathbf{r}_1 = U, \qquad F\lambda + G\mu = \mathbf{r}''.\mathbf{r}_2 = V, \tag{15.2}$$

where U and V are calculated with s as parameter. Thus λ and μ are given by the intrinsic formulae

$$\lambda = H^{-2}(GU - FV), \qquad \mu = H^{-2}(EV - FU). \tag{15.3}$$

The vector (λ, μ) is called the *geodesic curvature vector* of the curve under consideration. In the notation introduced at the end of section 12 the components λ, μ are given by

$$\begin{aligned}
\lambda &= u'' + \Gamma^1_{11}u'^2 + 2\Gamma^1_{12}u'v' + \Gamma^1_{22}v'^2, \\
\mu &= v'' + \Gamma^2_{11}u'^2 + 2\Gamma^2_{12}u'v' + \Gamma^2_{22}v'^2.
\end{aligned} \tag{15.4}$$

The geodesic curvature vector of any curve is orthogonal to the curve. This follows at once from (15.1) since the tangent vector $\mathbf{r}'$ is orthogonal to $\mathbf{r}''$ and to $\mathbf{N}$ and therefore also to $\lambda\mathbf{r}_1 + \mu\mathbf{r}_2$, which is the geodesic curvature vector. It can also be proved intrinsically;

the orthogonality condition for the vectors (u', v') and (λ, μ) can be written
$$u'(E\lambda + F\mu) + v'(F\lambda + G\mu) = 0$$
which from (15.2) becomes the identity $u'U + v'V = 0$.

EXERCISE 15.1. Prove that the components λ, μ of the geodesic curvature vector are given by the following formulae, with s as parameter.

$$\lambda = \frac{1}{H^2}\frac{U}{v'}\frac{\partial T}{\partial v'} = -\frac{1}{H^2}\frac{V}{u'}\frac{\partial T}{\partial v'}, \qquad \mu = \frac{1}{H^2}\frac{V}{u'}\frac{\partial T}{\partial u'} = -\frac{1}{H^2}\frac{U}{v'}\frac{\partial T}{\partial u'}.$$

The *geodesic curvature*, κ_g, of any curve is defined as the magnitude of the geodesic curvature vector with a sign attached, positive or negative according as the angle between the tangent and the geodesic curvature vector is $+\frac{1}{2}\pi$ or $-\frac{1}{2}\pi$. The geodesic curvature is therefore intrinsic. From the sine formula for the angle between the vectors (u', v') and (λ, μ) it follows that

$$\kappa_g = H(u'\mu - v'\lambda). \tag{15.5}$$

The geodesic curvature of a geodesic is zero. Conversely, a curve with zero geodesic curvature at every point has zero geodesic curvature vector and is therefore a geodesic.

Since the unit tangent vector $\mathbf{r}'$ is orthogonal to $\mathbf{N}$, the unit vector which lies in the tangent plane and makes an angle $+\frac{1}{2}\pi$ with $\mathbf{r}'$ is $\mathbf{N} \times \mathbf{r}'$. The geodesic curvature vector is therefore $\kappa_g \mathbf{N} \times \mathbf{r}'$, and (15.1) can be written

$$\mathbf{r}'' = \kappa_n \mathbf{N} + \kappa_g \mathbf{N} \times \mathbf{r}'. \tag{15.6}$$

Taking the scalar product with the unit vector $\mathbf{N} \times \mathbf{r}'$, we have

$$\kappa_g = [\mathbf{N}, \mathbf{r}', \mathbf{r}'']. \tag{15.7}$$

In this formula for κ_g it is a simple matter to pass from s to a general parameter t. Since $\mathbf{r}' = \dot{\mathbf{r}}/\dot{s}$ and $\mathbf{r}' \times \mathbf{r}'' = \dot{\mathbf{r}} \times \ddot{\mathbf{r}}/\dot{s}^3$ the formula becomes

$$\kappa_g = \dot{s}^{-3}[\mathbf{N}, \dot{\mathbf{r}}, \ddot{\mathbf{r}}]. \tag{15.8}$$

Substituting $\mathbf{N} = H^{-1}\mathbf{r}_1 \times \mathbf{r}_2$, we have

$$\kappa_g = H^{-1}\dot{s}^{-3}(\mathbf{r}_1 \times \mathbf{r}_2).(\dot{\mathbf{r}} \times \ddot{\mathbf{r}})$$
$$= H^{-1}\dot{s}^{-3}\{(\mathbf{r}_1.\dot{\mathbf{r}})(\mathbf{r}_2.\ddot{\mathbf{r}}) - (\mathbf{r}_2.\dot{\mathbf{r}})(\mathbf{r}_1.\ddot{\mathbf{r}})\},$$

and because of the identities (12.1) this can be written

$$\kappa_g = \frac{1}{H\dot{s}^3}\left(\frac{\partial T}{\partial \dot{u}}V(t) - \frac{\partial T}{\partial \dot{v}}U(t)\right). \tag{15.9}$$

EXAMPLE 15.1. To find the geodesic curvature of the parametric curve $v = c$. Taking u as parameter, then $\dot{u} = 1$, $\dot{v} = 0$, and

$$\frac{\partial T}{\partial \dot{u}} = E, \qquad \frac{\partial T}{\partial \dot{v}} = F, \qquad U = \tfrac{1}{2}E_1, \qquad V = F_1 - \tfrac{1}{2}E_2.$$

Also, $\dot{s} = E^{\frac{1}{2}}$. Hence the required curvature is given by

$$\kappa_g = \tfrac{1}{2}H^{-1}E^{-\frac{3}{2}}(2EF_1 - EE_2 - FE_1).$$

Formula (15.9) with $t = s$ can be simplified by means of the identity $u'U(s) + v'V(s) = 0$. Substituting for either V or U and using the fact that $u'(\partial T/\partial u') + v'(\partial T/\partial v') = 2T = 1$ when s is the parameter,

$$\kappa_g = -\frac{1}{H}\frac{U(s)}{v'} = \frac{1}{H}\frac{V(s)}{u'}. \tag{15.10}$$

EXERCISE 15.2. Prove that if (λ, μ) is the geodesic curvature vector, then

$$\kappa_g = \frac{-H\lambda}{Fu' + Gv'} = \frac{H\mu}{Eu' + Fv'}.$$

Geodesic curvature may be regarded as the intrinsic generalization of curvature of plane curves, as can be seen from the following result which will not be proved here.

Let P be a point of a given curve C on a surface and Q the point of C at a distance δs from P along C. If the geodesics which are tangent to C at P and Q meet at the point R, let $\delta\psi$ be the angle between the tangents to these geodesics at R. Then the geodesic curvature of C at P is $\lim\limits_{\delta s \to 0} \dfrac{\delta\psi}{\delta s}$.

For a plane curve, $\delta\psi$ is the angle between the tangents at P and Q and $\lim\limits_{\delta s \to 0}\dfrac{\delta\psi}{\delta s}$ is the curvature $\dfrac{d\psi}{ds}$ in the usual notation.

The above would be a satisfactory intrinsic definition of geodesic curvature except for the difficulty of proving that the tangent geodesics at P and Q do in fact meet at a point R near P. A more straightforward intrinsic generalization of curvature is as follows.

Let P be a point of a given curve C on a surface and Q the point of C at a distance δs from P along C. Let $\bar{C}$ be the geodesic arc PQ, of length $\delta\bar{s}$. Then if $\delta\theta$ is the angle between C and $\bar{C}$ at P and if $\delta\phi$ is the angle between $\bar{C}$ and C at Q, the geodesic curvature of C at P is

$\lim\limits_{\delta s \to 0} \dfrac{\delta\theta + \delta\phi}{\delta s}$ (see Fig. 4; note that for this figure κ_g is negative).

There is no difficulty about this construction because of the existence theorem for a geodesic joining two neighbouring points. To prove the result, let (u', v'), (u_0', v_0') be unit tangent vectors to C at Q and P respectively. Let $(\bar{u}', \bar{v}')$, $(\bar{u}_0', \bar{v}_0')$ be unit tangent vectors to $\bar{C}$ at Q and P respectively. Then

$$\sin \delta\theta = H(u_0, v_0)\{u_0' \bar{v}_0' - v_0' \bar{u}_0'\}, \quad \sin \delta\phi = H(u, v)\{\bar{u}' v' - u' \bar{v}'\}.$$

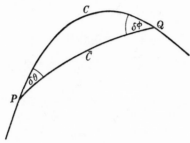

FIG. 4

We have

$$u' = u_0' + \delta s\, u_0'' + O(\delta s^2) \quad \text{as } \delta s \to 0,$$
$$v' = v_0' + \delta s\, v_0'' + O(\delta s^2) \quad \text{as } \delta s \to 0,$$
$$\bar{u}' = \bar{u}_0' + \delta\bar{s}\, f(u_0, v_0, \bar{u}_0, \bar{v}_0) + O(\delta\bar{s}^2) \quad \text{as } \delta\bar{s} \to 0,$$
$$\bar{v}' = \bar{v}_0' + \delta\bar{s}\, g(u_0, v_0, \bar{u}_0, \bar{v}_0) + O(\delta\bar{s}^2) \quad \text{as } \delta\bar{s} \to 0,$$

where in the last two equations we have used the geodesic equations (13.1).

Also we have

$$\delta\bar{s} = \delta s + O(\delta s^2) \quad \text{as } \delta s \to 0.$$

We write

$$H(u_0, v_0) = H_0, \qquad H(u, v) = H_0 + \delta H,$$

where

$$\delta H = O(\delta s) \quad \text{as } \delta s \to 0.$$

Then we have

$$\sin \delta\theta + \sin \delta\phi$$
$$= H_0\, \delta s [u_0'\{v_0'' - g(u_0, v_0, \bar{u}_0', \bar{v}_0')\} - v_0'\{u_0'' - f(u_0, v_0, \bar{u}_0', \bar{v}_0')\}] +$$
$$+ \delta H(\bar{u}_0' v_0' - \bar{v}_0' u_0') + O(\delta s^2).$$

Also, as $\delta s \to 0$ we have

$$\sin \delta\theta = \delta\theta + O(\delta s^2), \qquad \sin \delta\phi = \delta\phi + O(\delta s^2),$$
$$\bar{u}_0' = u_0' + O(\delta s), \qquad\qquad \bar{v}_0' = v_0' + O(\delta s).$$

Then

$$\delta\theta + \delta\phi = H_0\,\delta s[u_0'\{v_0''-g(u_0,v_0,u_0',v_0')\}-$$
$$-v_0'\{u_0''-f(u_0,v_0,u_0',v_0')\}]+O(\delta s^2) \quad \text{as } \delta s \to 0.$$

From (15.4), the geodesic curvature vector (λ,μ) of C at P is given by

$$\lambda = u_0''-f(u_0,v_0,u_0',v_0'), \qquad \mu = v_0''-g(u_0,v_0,u_0',v_0').$$

Hence $\dfrac{\delta\theta+\delta\phi}{\delta s} = H_0(u_0'\mu-v_0'\lambda)+O(\delta s).$

Thus, proceeding to the limit as $\delta s \to 0$ and dropping the suffix, we get

$$\lim_{\delta s \to 0}\frac{\delta\theta+\delta\phi}{\delta s} = H(u'\mu-v'\lambda) = \kappa_g \quad \text{from (15.5).}$$

Liouville's formula for κ_g. This is an expression for κ_g involving the angle θ which the curve under consideration makes with the parametric curves $v = $ constant. Regarding θ as a function of s along the curve, then Liouville's formula is

$$\kappa_g = \theta'+Pu'+Qv', \tag{15.11}$$

where

$$P = \frac{1}{2HE}(2EF_1-FE_1-EE_2), \qquad Q = \frac{1}{2HE}(EG_1-FE_2).$$

The direction coefficients of the curve $v = $ constant and the given curve are $(1/\sqrt{E},0)$ and (u',v') so that

$$\cos\theta = \sqrt{E}u'+\frac{F}{\sqrt{E}}v' = \frac{1}{\sqrt{E}}\frac{\partial T}{\partial u'}, \qquad \sin\theta = \frac{H}{\sqrt{E}}v'.$$

Differentiating $\cos\theta$,

$$-\sin\theta\frac{d\theta}{ds} = \frac{1}{\sqrt{E}}\frac{d}{ds}\left(\frac{\partial T}{\partial u'}\right)-\frac{1}{2E^{\frac{3}{2}}}(E_1u'+E_2v')\frac{\partial T}{\partial u'};$$

multiplying by $\sqrt{E}$ and substituting

$$\frac{d}{ds}\left(\frac{\partial T}{\partial u'}\right) = U+\frac{\partial T}{\partial u},$$

$$-Hv'\theta' = U+\tfrac{1}{2}(E_1u'^2+2F_1u'v'+G_1v'^2)-$$
$$-\frac{1}{2E}(E_1u'+E_2v')(Eu'+Fv')$$
$$= U+\frac{1}{2E}\{(2EF_1-FE_1-EE_2)u'v'+(EG_1-FE_2)v'^2\}.$$

Liouville's formula now appears on dividing by Hv' and substituting $U = -\kappa_g Hv'$ from (15.10).

EXAMPLE 15.2. Prove that if the orthogonal trajectories of the curves $v =$ constant are geodesics, then H^2/E is independent of u.

The orthogonal trajectories satisfy $\theta = \frac{1}{2}\pi$ and are geodesics if $\kappa_g = 0$. From Liouville's formula, $Pu' + Qv' = 0$. Also $\cos\theta = 0$, i.e. $Eu' + Fv' = 0$, and the trajectories will be geodesics if
$$EQ - FP = 0.$$
On substituting for P and Q the condition becomes
$$F^2E_1 - 2EFF_1 + E^2G_1 = 0,$$
i.e. $\partial(G - F^2/E)/\partial u = 0$ as required.

EXERCISE 15.3. Prove that if a curve C on a surface is projected orthogonally on to the tangent plane at a point P of C, it becomes a plane curve whose curvature at P is the geodesic curvature of C at P.

16. Gauss–Bonnet theorem

Consider a surface of class 3, with parameter system u, v, and let a closed curve C be the boundary of a simply connected region R of the surface. (By *simply connected* we mean that every closed curve lying in R can be contracted continuously into a point without leaving R.) Suppose that C consists of n arcs
$$A_0 A_1, \ A_1 A_2, ..., \ A_{n-1} A_n \quad (A_n = A_0),$$
where n is finite, and that each arc is of class 2. The vertices $A_0, A_1, ...$ are taken in order along C to agree with the positive sense of description of C; this is usually described as the sense which 'leaves the interior on the left', i.e. a positive rotation of $\frac{1}{2}\pi$ from the tangent gives the normal which points to the interior region R. At the vertex A_r $(r = 1, ..., n)$ let α_r be the angle between the tangents to the arcs $A_{r-1} A_r$ and $A_r A_{r+1}$, measured with the usual convention at A_r so that $-\pi < \alpha_r < \pi$; at A_n, α_n is the angle between the tangents to $A_{n-1} A_n$ and $A_n A_1$. Regarding C as a 'curvilinear polygon', $\alpha_1, ..., \alpha_n$ are the exterior angles at the vertices $A_1, ..., A_n$ (see Fig. 5 where $n = 6$).

The geodesic curvature exists at every point of C except possibly at the vertices, and the line integral $\int_C \kappa_g \, ds$ can therefore be calculated. The *excess* of C is defined as
$$\mathrm{ex}\, C = 2\pi - \sum_{r=1}^{n} \alpha_r - \int_C \kappa_g \, ds.$$
This is an invariant, independent of the particular parameter

system for the surface. The only possible effect of a change of parameter system is to reverse at every point the sense in which angles are measured; this would reverse the sense along C and therefore the description of the polygon, but κ_g at each point and α_r at each vertex would remain unchanged.

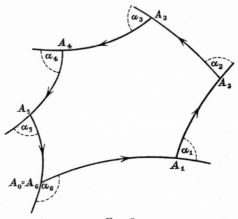

Fig. 5

For a curvilinear polygon C on the plane, κ_g is the ordinary curvature $d\psi/ds$ and $\int \kappa_g \, ds + \sum_{r=1}^{n} \alpha_r$ is the total angle through which the tangent turns in describing C. This angle is clearly 2π, so that the excess of C is zero. In particular, for a rectilinear polygon, $\kappa_g = 0$ at every point and $\sum \alpha_r$ is the sum of the exterior angles, i.e. 2π, giving ex $C = 0$. Since excess, as defined above, is intrinsic, it follows, that *on any surface isometric with the plane, the excess of a simple closed curve is zero.*

This result suggests that for a surface which is *not* isometric with the plane, the excess of a simple curve C enclosing a region R is in some sense a measure of the intrinsic difference between R and a region of the plane. The excess may therefore lead to an intrinsic definition of the *curvature* of a surface, based on the convention that a plane has zero curvature. This is in fact the case, and it will be shown that from the excess can be derived the important invariant known as the *Gaussian curvature* of a surface.

From Liouville's formula for κ_g,

$$\int_C \kappa_g \, ds = \int_C (d\theta + P \, du + Q \, dv),$$

where θ is the angle which C makes with the parametric curve $v = \text{constant}$ and P and Q are certain functions of u and v. Since the curves $v = \text{constant}$ form a family in the region R bounded by C, the tangent to C turns through 2π relative to these curves, i.e.

$$\int_C d\theta + \sum_{r=1}^n \alpha_r = 2\pi.$$

Hence
$$\operatorname{ex} C = -\int_C (P\,du + Q\,dv).$$

By Green's theorem, since R is simply connected and P and Q are differentiable functions of u and v in R,

$$\int_C (P\,du + Q\,dv) = \int_R \left(\frac{\partial Q}{\partial u} - \frac{\partial P}{\partial v}\right) du\,dv.$$

Hence, writing $dS = H\,du\,dv$ for the surface element,

$$\operatorname{ex} C = \int_R K\,dS, \tag{16.1}$$

where K is a function of u and v, independent of the curve C, given by

$$K = -\frac{1}{H}\left(\frac{\partial Q}{\partial u} - \frac{\partial P}{\partial v}\right). \tag{16.2}$$

Equation (16.1) shows that there is a certain function K of u and v which is determined by E, F, and G, and that the excess of any curve C which encloses a simply connected region R is equal to the surface integral of K over R. We shall now show that the function K is uniquely determined. Let $\bar{K}$ be a second function which also satisfies (16.1) and is independent of C. Then for every region R,

$$\int_R (\bar{K} - K)\,dS = 0. \tag{16.3}$$

Now suppose $\bar{K} \neq K$ at some point P, say $\bar{K} > K$. Then since $\bar{K} - K$ is continuous, there is a region R which contains P and in which $\bar{K} - K > 0$ at every point. For this R, $\int_R (\bar{K} - K)\,dS > 0$ which contradicts (16.3). A similar contradiction exists if $\bar{K} < K$ at P. Hence $\bar{K} = K$ at every point, i.e. K is uniquely determined as a function of u and v.

From this uniqueness property and from the form of (16.1) it follows that K is an invariant; at every point the value of K is independent of the parameter system. Also K is intrinsic, since it can be calculated when the metric is known. Thus K is an intrinsic geometrical invariant; it is called the *Gaussian curvature* of the surface.

For any region R, whether simply connected or not, $\int_{R} K\, dS$ is called the *total curvature* of R. Equation (16.1) now gives the

GAUSS–BONNET THEOREM. *For any curve C which encloses a simply connected region R, the excess of C is equal to the total curvature of R.*

For a *geodesic triangle ABC*, formed by geodesic arcs AB, BC, CA and enclosing a simply connected region R, the excess is

$$2\pi-(\pi-A)-(\pi-B)-(\pi-C) = A+B+C-\pi,$$

where A, B, C are the interior angles of the triangle. Thus the excess is the excess of $A+B+C$ over its Euclidean value π, a fact which accounts historically for our use of the word 'excess'. The total curvature of a geodesic triangle ABC is therefore equal to $A+B+C-\pi$.

More generally, for a geodesic polygon of any number of sides (geodesic arcs) the total curvature is equal to 2π minus the sum of the exterior angles, i.e. the excess of the sum of the interior angles over $(n-2)\pi$ where n is the number of sides.

EXERCISE 16.1. By first considering the region of the anchor ring of section 3 bounded by two meridians and the two parallels $u = 0$, $u = \pi$, prove that the total curvature of the whole surface is zero.

17. Gaussian curvature

An historical definition of Gaussian curvature K follows from the Gauss–Bonnet theorem for a geodesic triangle. If P is a given point and Δ the area of a geodesic triangle ABC which contains P, then at P,

$$K = \lim \frac{A+B+C-\pi}{\Delta}, \tag{17.1}$$

where the limit is taken as all vertices tend to P.

On a sphere of radius a, for example, the geodesics are great circles, and the area of a geodesic triangle ABC is $a^2(A+B+C-\pi)$. The Gaussian curvature at every point is therefore $1/a^2$. That K is constant over the sphere is to be expected from the fact that there is an isometric mapping of the sphere on itself in which any given point P corresponds to any other given point Q, so that $(K)_P = (K)_Q$ since K is an intrinsic invariant.

The formula $K = 1/a^2$ at a point of a sphere of radius a illustrates the fact that the dimensions of K are $(\text{length})^{-2}$. This follows more generally from the Gauss–Bonnet equation, in which the excess of a curve is clearly dimensionless.

The total curvature $\int_R K\,dS$ for any region R is dimensionless. On a sphere of radius a, for example, the total curvature for the whole sphere is $\text{area}/a^2 = 4\pi$. It will be seen in a later chapter that the total curvature of a compact surface depends only upon the topology of the surface.

The formula for K in terms of E, F, and G is given by (16.2), where P and Q are given by (15.11). Hence, at any point and in any parameter system,

$$K = \frac{1}{H}\frac{\partial}{\partial u}\left(\frac{FE_2 - EG_1}{2HE}\right) + \frac{1}{H}\frac{\partial}{\partial v}\left(\frac{2EF_1 - FE_1 - EE_2}{2HE}\right). \quad (17.2)$$

When the parametric curves are orthogonal, $F = 0$ and the formula for K can be written in the simpler and symmetric form

$$K = -\frac{1}{2H}\left\{\frac{\partial}{\partial u}\left(\frac{G_1}{H}\right) + \frac{\partial}{\partial v}\left(\frac{E_2}{H}\right)\right\} \quad (17.3)$$

where now $H = \sqrt{(EG)}$.

For example, on a sphere of radius a parameters can be chosen as in section 3 so that $E = a^2$, $F = 0$, $G = a^2\sin^2 u$. Then $H = a^2\sin u$ since $0 < u < \pi$, and the above formula gives

$$K = -\frac{1}{2a^2\sin u}\frac{\partial}{\partial u}(2\cos u) = \frac{1}{a^2}.$$

In Chapter III a very different kind of formula (non-intrinsic) for K will be given in terms of the second fundamental coefficients to be defined later. This formula is appropriate when the position vector $\mathbf{r}(u, v)$ for the surface is given and is generally simpler than the above for the purpose of calculation. It cannot, however, be applied when only the metric is given.

EXERCISE 17.1. Find the Gaussian curvature at the point (u, v) of the anchor ring of section 3 and verify that the total curvature of the whole surface is zero. (Cf. Exercise 16.1.)

EXERCISE 17.2. Prove that the Gaussian curvature of the surface given (in Monge form) by $z = f(x, y)$ is $(rt-s^2)(1+p^2+q^2)^{-2}$, where p, q, r, s, and t denote respectively

$$\partial z/\partial x, \quad \partial z/\partial y, \quad \partial^2 z/\partial x^2, \quad \partial^2 z/\partial x\partial y, \quad \text{and} \quad \partial^2 z/\partial y^2.$$

Geodesic polar form

With geodesic polar parameters the metric takes the form

$$du^2 + g^2\, dv^2,\tag{17.4}$$

where for convenience g is written for $\sqrt{G}$. In section 14 it was shown that $g(u, v)$ satisfies the condition $g = u + O(u^2)$ as $u \to 0$.

The Gaussian curvature at the point (u, v) is given by (17.3) with $E = 1$, $G = g^2$, and $H = g$. Hence

$$K = -g_{11}/g.\tag{17.5}$$

The centre (origin) of the geodesic polar parameters is excluded from the domain of u, v because it is a singularity, but since this is only artificial the Gaussian curvature exists there; suppose it has the value K_0 at the origin. Then as $u \to 0$, $g_{11} \sim -K_0 g \sim -K_0 u$; on integrating twice,

$$g(u, v) \sim u - K_0 \frac{u^3}{6} \quad \text{as } u \to 0.\tag{17.6}$$

Thus for small u, the parameter v does not enter $g(u, v)$ until terms of order smaller than u^3.

EXAMPLE 17.1. To calculate the circumference of a geodesic circle of small radius r and to see how it differs from the Euclidean formula $2\pi r$.

In geodesic polars the circle is the parallel $u = r$. Hence $ds = g\, dv$ and the circumference C is

$$C = \int_0^{2\pi} g(r, v)\, dv \sim \int_0^{2\pi} \left(r - \frac{K_0}{6} r^3\right) dv = 2\pi\left(r - \frac{K_0}{6} r^3\right).$$

Hence
$$C - 2\pi r \sim -\tfrac{1}{3}\pi K_0 r^3$$

to the first significant term, where K_0 is the Gaussian curvature at the centre of the circle.

This suggests another intrinsic formula for K. Let C be the circumference of the geodesic circle of centre P and radius r. Then

$$(K)_P = \lim_{r \to 0} \frac{2\pi r - C}{\tfrac{1}{3}\pi r^3}.$$

EXERCISE 17.3. Prove that, if A is the area of a geodesic disk of centre P and radius r, then

$$(K)_P = \lim_{r \to 0} \frac{\pi r^2 - A}{\tfrac{1}{12}\pi r^4}.$$

18. Surfaces of constant curvature

If K has the same value K_0 at every point of a surface, the surface is said to have constant curvature K_0.

MINDING'S THEOREM. *Two surfaces of the same constant curvature are locally isometric.*

Strictly, if P is any point of one of these surfaces and $\bar{P}$ is any point of the other, then $\bar{P}$ has a neighbourhood which is isometric with a neighbourhood of P, the points P and $\bar{P}$ being corresponding points. In what follows, 'surface' means a sufficiently small region.

We prove this theorem by showing that if S is a surface with constant curvature K_0, then

(1) if $K_0 = 0$, S is isometric with a plane;

(2) if $K_0 = 1/a^2$, S is isometric with a sphere of radius a; and

(3) if $K_0 = -1/a^2$, S is isometric with a certain surface of revolution, called a *pseudo-sphere*, determined by the value of a.

In each case a given point of S can be mapped into a prescribed point of the plane, sphere, or pseudo-sphere.

The theorem for two surfaces S and $\bar{S}$ with the same K then follows by mapping each surface isometrically on to the same plane, or sphere, or surface of revolution, so that given points P and $\bar{P}$ correspond to the same point.

Let P be a given point of the surface S of constant curvature K_0, and let C be a geodesic through P. Take as parametric curves the geodesics orthogonal to C together with their orthogonal trajectories. Let $v = c$ be the geodesic orthogonal to C at a point distance c from P measured along C, and let $u = c$ be the parallel orthogonal to the curves $v = $ constant and at a distance c from the parallel C measured along the geodesic. Then u, v is a parameter system in the neighbourhood of P, and the metric of the surface is of the form

$$du^2 + g^2\, dv^2$$

for some $g(u,v)$. Since $u = 0$ is the geodesic C, it follows from (10.8) that $\partial g^2/\partial u = 0$ when $u = 0$. Also, v is the arcual distance along C, i.e. $ds = dv$ when $u = 0$, so that $g = 1$ when $u = 0$. Hence,
$$(g)_{u=0} = 1, \qquad (g_1)_{u=0} = 0. \tag{18.1}$$

Using now the formula $K = -g_{11}/g$ proved in section 17, $g(u,v)$ satisfies the partial differential equation

$$g_{11} + K_0 g = 0 \tag{18.2}$$

with boundary conditions (18.1); these are sufficient to determine g when K_0 is given.

Case (1), $K_0 = 0$

When $g_{11} = 0$, g_1 is a function of v only and therefore $g_1 = 0$ since $(g_1)_{u=0} = 0$. From $g_1 = 0$ it follows that g is a function of v only and is therefore 1 since $(g)_{u=0} = 1$. With $g = 1$, the metric is

$$du^2 + dv^2,$$

i.e. the metric of a plane when u, v are taken as Cartesian coordinates. Hence, the surface S in the neighbourhood of P is isometric with a region of the plane.

This confirms that K is a satisfactory measure of curvature for a surface since its vanishing is both necessary and sufficient for the surface to be isometric with a plane.

Case (2), $K_0 = 1/a^2$

Equation (18.2) integrates to give

$$g(u, v) = A(v)\sin\frac{u}{a} + B(v)\cos\frac{u}{a}.$$

Hence $(g_1)_{u=0} = (1/a)A(v) = 0$, and $(g)_{u=0} = B(v) = 1$, so that $g = \cos(u/a)$ and the metric is

$$du^2 + \cos^2\frac{u}{a}\,dv^2.$$

This is the metric of a sphere of radius a. (The more usual metric is given by the transformation $u = a(\tfrac{1}{2}\pi - \bar{u})$, $v = a\bar{v}$.) The surface S in the neighbourhood of P is therefore isometric with a region of a sphere of radius a.

Case (3), $K_0 = -1/a^2$

By arguments similar to those for case (2), $g = \cosh(u/a)$ and the metric of S in the neighbourhood of P is

$$du^2 + \cosh^2\frac{u}{a}\,dv^2.$$

This form, in which E, F, G are functions of u only, shows that S is isometric with a certain surface of revolution (cf. Exercise 8.3).

Writing $u = a\bar{u}$, $v = a\bar{v}$, the metric becomes

$$a^2(d\bar{u}^2 + \cosh^2\bar{u}\,d\bar{v}^2).$$

This is the metric of the surface obtained by revolving the curve

$$x = a \cosh \bar{u}, \quad y = 0, \quad z = a \int_0^{\bar{u}} \sqrt{(1 - \sinh^2\theta)}\, d\theta \quad (|\bar{u}| < \log(1 + \sqrt{2}))$$

about the z-axis.

This completes the proof of the theorem on the isometries of surfaces of constant curvature.

The metrics and surfaces constructed above are special, chosen to prove the theorem as simply as possible. There are, however, other surfaces of revolution with constant curvature, since any function $g(u)$ which satisfies (18.2) (but not the boundary conditions (18.1)) gives a metric which can be transformed into the standard metric of a surface of revolution. For example, when $K_0 = -1/a^2$, g can be taken to be $ae^{u/a}$. Writing $u = a\bar{u}$, the metric becomes

$$a^2(d\bar{u}^2 + e^{2\bar{u}}\, dv^2),$$

which is therefore the metric of a surface of constant curvature $-1/a^2$.

An important example of a surface of constant zero curvature is the surface generated by the tangents to any space curve. If $\mathbf{r}(s)$ is the position vector of a point on the curve, in terms of the arc s as parameter, then a point on the surface is given by $\mathbf{r}(s) + v\mathbf{t}(s)$ where s and v are the parameters. The fundamental coefficients are

$$E = 1 + \kappa^2 v^2, \quad F = 1, \quad G = 1, \quad H = |\kappa v|$$

where κ, the curvature of the curve, is a function of s only. Substituting in (17.2) it can be verified that $K = 0$ at every point. (Note: every point on the given curve ($v = 0$) and on any inflexional tangent ($\kappa = 0$) is a singularity.) This example is important because it will be proved in Chapter III that every surface of zero curvature is either part of the tangent surface to some curve or part of a cone or cylinder.

19. Conformal mapping

Isometric mappings of a surface S on another surface S^* have already been mentioned in section 8. It is often convenient to consider mappings which are more general than isometries—for example, it is useful to map parts of the earth's surface on to a flat atlas. Certain of these maps are particular cases of conformal mappings discussed in this section.

A surface S is said to be conformally mapped onto a surface S^* if there is a differentiable homeomorphism of S on S^* such that the angle between any two curves at an arbitrary point P on S is equal to the angle between the corresponding curves on S^*.

An isometric mapping preserves both distances and angles, whereas a conformal mapping just preserves angles. Conditions will now be obtained for the conformal mapping of S on S^*.

Let S, S^* have respectively fundamental forms

$$ds^2 = E\,du^2 + 2F\,du\,dv + G\,dv^2,$$

$$ds^{*2} = E^*\,du^2 + 2F^*\,du\,dv + G^*\,dv^2,$$

the correspondence being such that corresponding points P, P^* have the same parametric values.

Let (l_1, m_1), (l_2, m_2) be direction coefficients of two directions through P, and let (l_1^*, m_1^*), (l_2^*, m_2^*) be corresponding direction coefficients at P^*. If θ, θ^* are the angles between these two pairs of directions, then

$$\cos\theta = El_1 l_2 + F(l_1 m_2 + m_1 l_2) + Gm_1 m_2, \qquad (19.1)$$

$$\cos\theta^* = E^* l_1^* l_2^* + F^*(l_1^* m_2^* + m_1^* l_2^*) + G^* m_1^* m_2^*. \qquad (19.2)$$

Since
$$l_1^* = \frac{du}{ds^*} = \frac{du}{ds}\frac{ds}{ds^*} = l_1\rho,$$

where $\rho = ds/ds^*$, equation (19.2) may be written,

$$\cos\theta^* = \rho^2[E^* l_1 l_2 + F^*(l_1 m_2 + m_1 l_2) + G^* m_1 m_2]. \qquad (19.3)$$

Since equations (19.1) and (19.3) must be identical for arbitrary direction coefficients,

$$E/E^* = F/F^* = G/G^* = \rho^2. \qquad (19.4)$$

Equation (19.4) is evidently a necessary and sufficient condition for the differentiable homeomorphism to be conformal.

A conformal mapping is thus seen as a generalization of a similarity mapping, but the scale factor ρ is in general a function of the parameters u and v. When ρ is a constant, the conformal mapping becomes a proper similarity mapping. In particular when $\rho = 1$, the mapping is isometric.

A fundamental result is the following:

Every point on a surface has a neighbourhood which can be mapped conformally on a region of the plane.

This has an important corollary:

Every point on a surface has a neighbourhood which can be mapped conformally on some neighbourhood of any other surface.

These results will now be established.

Let S be the given surface with metric

$$ds^2 = E\,du^2 + 2F\,dudv + G\,dv^2$$

in some coordinate domain. At any point P there are two (imaginary) directions such that $ds^2 = 0$. These are called the *isotropic directions* at P, and since $EG - F^2 \neq 0$ it follows that these directions are always distinct.

When curves along these directions are chosen as parametric curves the metric assumes the form $ds^2 = \lambda\,dudv$. The change of parameters

$$u = U + iV, \quad v = U - iV,$$

where U and V are real, leads to a metric of the form

$$ds^2 = \Lambda^2[dU^2 + dV^2]. \tag{19.5}$$

If this is compared with the metric of the plane

$$ds^2 = d\bar{u}^2 + d\bar{v}^2,$$

it is readily seen that the mapping $\bar{u} = U, \bar{v} = V$ gives a conformal mapping of a region of the given surface on a region of the plane. This proves the first result, and the corollary follows immediately.

EXERCISE 19.1. Obtain an alternative proof which avoids the use of complex parameters.

When the metric of a surface assumes the form (19.5) the parameters are said to form an *isothermic* system. It will now be shown that there are an infinite number of systems of isothermic parameters, each system corresponding to an analytic function of a single complex variable.

Suppose that (u, v), (U, V) are two different isothermic systems of the same surface so that

$$ds^2 = \lambda^2(du^2 + dv^2) = \Lambda^2(dU^2 + dV^2).$$

Since $\quad dU = U_1\,du + U_2\,dv, \quad\quad dV = V_1\,du + V_2\,dv,$

it follows that

$$\lambda^2(du^2 + dv^2)$$
$$= \Lambda^2\{(U_1^2 + V_1^2)\,du^2 + 2(U_1U_2 + V_1V_2)\,dudv + (U_2^2 + V_2^2)\,dv^2\},$$

from which
$$U_1^2 + V_1^2 = U_2^2 + V_2^2,$$
$$U_1 U_2 + V_1 V_2 = 0.$$

From these equations it follows that

either $U_1 = V_2, \quad U_2 = -V_1$
or $U_2 = V_1, \quad U_1 = -V_2$ $\left.\right\}$ (19.6)

These are, however, just the Cauchy–Riemann equations, and express the condition that $U+iV$ or $U-iV$ is an analytic function of the complex variable $u+iv$. It follows that if the curves $u =$ constant, $v =$ constant form an isothermic system, all other isothermic systems are given by

$$\mathrm{re}\, f(u+iv) = \text{constant},$$

$$\mathrm{im}\, f(u+iv) = \text{constant},$$

where f is any analytic function of the complex variable $u+iv$. Corresponding to each isothermic system, there is a natural conformal mapping of the surface on the plane.

Consider the general surface of revolution given by (3.2). The change of parameters

$$U = \int (f'^2 + g'^2)^{\frac{1}{2}} g^{-1}\, du, \qquad V = v, \qquad (19.7)$$

leads to the metric
$$ds^2 = g^2[dU^2 + dV^2],$$

from which it follows that the meridians and parallels form an isothermic parametric system.

The mapping $x = V, \qquad y = U$

maps the meridians into lines parallel to the y-axis, and the parallels into lines parallel to the x-axis.

In particular, for the sphere (3.1), equation (19.7) becomes

$$U = \log \tan \tfrac{1}{2} u, \qquad V = v.$$

This conformal mapping is a modification of Mercator's projection, and is sometimes used in geographical atlases. The representation is quite faithful near the equator, but the size of the polar regions is considerably exaggerated on the map.

Another common mapping used in geography is obtained by projecting the points of a sphere from the north pole onto the tangent plane at the south pole. This is the well-known stereographic projection, which is easily shown to be conformal.

20. Geodesic mapping

A surface S is mapped geodesically onto a surface S* if there is a differentiable homeomorphism of S onto S* such that geodesics on S go over into geodesics on S*.

A simple example of such a mapping is obtained by projecting the points of a sphere from the centre onto a tangent plane. The great circles (geodesics) on the sphere map into straight lines on the plane. This type of mapping of the earth's surface onto a flat atlas has the advantage that the path of shortest distance between two points P, Q on the earth's surface is given by the straight line P^*Q^* on the map.

If a mapping is both geodesic and conformal, it is necessarily an isometry or else a similarity mapping. To prove this, consider a system of geodesic coordinates on S so that the metric is

$$ds^2 = du^2 + G\,dv^2.$$

From the conformality condition, the metric of S^* will be of the form

$$ds^{*2} = \lambda(u,v)[du^2 + G\,dv^2]. \qquad (20.1)$$

Since the mapping is geodesic, it follows that the curves $v = $ constant are geodesics with respect to the metric (20.1). From (10.7) it follows that

$$\lambda_2 = 0, \quad \text{so that} \quad \lambda = \lambda(u).$$

Consider now any geodesic on S making an angle θ with the curves $v = $ constant. From Liouville's formula (15.11) it follows that

$$0 = d\theta + \frac{1}{2}\frac{G_1}{H}\,dv.$$

Since the mapping is both geodesic and conformal, the corresponding equation for S^* is

$$0 = d\theta + \frac{1}{2}\frac{G_1^*}{H^*}\,dv.$$

Hence
$$\frac{G_1}{\sqrt{G}} = \frac{(\lambda G)_1}{\lambda\sqrt{G}} = \frac{G_1}{\sqrt{G}} + \sqrt{G}\frac{\lambda_1}{\lambda},$$

from which $\lambda_1 = 0$ and λ is thus a constant. This shows that the mapping is a similarity mapping, and the assertion is justified.

It was shown in the previous section that any surface can be mapped conformally on any other surface. The requirements for a geodesic mapping are much more restrictive, as is shown by the following theorem due to Dini:

THEOREM. *Two surfaces which are mapped geodesically on each other by a non-conformal mapping must have line elements which can be written in the forms*

$$
\left.
\begin{aligned}
ds^2 &= (U-V)(du^2+dv^2) \\
ds^{*2} &= (V^{-1}-U^{-1})(U^{-1}\,du^2+V^{-1}\,dv^2)
\end{aligned}
\right\},
\qquad (20.2)
$$

where $U = U(u)$ *and* $V = V(v)$.

In order to prove this theorem, use will be made of the following result.

TISSOT'S THEOREM. *In any non-conformal mapping of a surface S on a surface S^*, given by a differentiable homeomorphism regular at each point, there exists at each point P of S a uniquely determined pair of real orthogonal directions such that the corresponding directions on S^* are also orthogonal.*

To prove this theorem, suppose parameters are chosen on S and S^* so that corresponding points have the same parameters. Let (l_1, m_1), (l_2, m_2) be two orthogonal directions at P on S, and let (l_1^*, m_1^*), (l_2^*, m_2^*) be the corresponding directions at P^* on S^*.

Since (l_1, m_1), (l_2, m_2) are orthogonal,

$$
El_1 l_2 + F(l_1 m_2 + l_2 m_1) + Gm_1 m_2 = 0. \qquad (20.3)
$$

Moreover, the corresponding directions will be orthogonal on S^* if $E^* l_1^* l_2^* + F^*(l_1^* m_2^* + l_2^* m_1^*) + G^* m_1^* m_2^* = 0$, which implies

$$
E^* l_1 l_2 + F^*(l_1 m_2 + l_2 m_1) + G^* m_1 m_2 = 0, \qquad (20.4)
$$

because $l_1^* = l_1 \, ds/ds^*$, $m_1^* = m_1 \, ds/ds^*$, $n_1^* = n_1 \, ds/ds^*$.

Eliminate l_1, m_1 from (20.3), (20.4) to obtain

$$
(EF^* - E^*F)l_2^2 + (EG^* - E^*G)l_2 m_2 + (FG^* - F^*G)m_2^2 = 0, \qquad (20.5)
$$

an equation which from (7.4) determines two orthogonal directions at each point. The discriminant of (20.5) is

$$
(EG^* - E^*G)^2 - 4(EF^* - E^*F)(FG^* - F^*G),
$$

which is identically equal to

$$
\frac{4(EG - F^2)}{E^2}(EF^* - FE^*)^2 + \left[EG^* - GE^* - \frac{2F}{E}(EF^* - FE^*) \right]^2
$$

since $E \neq 0$. Since the mapping is not conformal, the discriminant is strictly positive and the roots of (20.5) are real and distinct. It has therefore been proved that at each point P on S a pair of real orthogonal directions is uniquely determined such that the corresponding directions on S^* are also orthogonal. This completes the proof of Tissot's theorem.

In order to prove Dini's theorem we make use of the above result by choosing parameters so that the metrics of S and S^* assume the form

$$ds^2 = E\,du^2 + G\,dv^2,$$

$$ds^{*2} = E^*\,du^2 + G^*\,dv^2.$$

From the analysis following equation (10.8), it follows that the equation of a geodesic on S with parameter u is

$$EG\ddot{v} + \tfrac{1}{2}GG_1\dot{v}^3 + (\tfrac{1}{2}EG_2 - GE_2)\dot{v}^2 + (EG_1 - \tfrac{1}{2}GE_1)\dot{v} - \tfrac{1}{2}EE_2 = 0. \tag{20.6}$$

Similarly, the equation of a geodesic on S^* is

$$E^*G^*\ddot{v} + \tfrac{1}{2}G^*G_1^*\dot{v}^3 + (\tfrac{1}{2}E^*G_2^* - G^*E_2^*)\dot{v}^2 +$$
$$+ (E^*G_1^* - \tfrac{1}{2}G^*E_1^*)\dot{v} - \tfrac{1}{2}E^*E_2^* = 0. \tag{20.7}$$

Since the mapping is geodesic, these two equations must be identical and by comparing coefficients the following relations are obtained:

$$G_1/E = G_1^*/E^*, \tag{20.8}$$

$$E_2/E - \tfrac{1}{2}G_2/G = E_2^*/E^* - \tfrac{1}{2}G_2^*/G^*, \tag{20.9}$$

$$\tfrac{1}{2}E_1/E - G_1/G = \tfrac{1}{2}E_1^*/E^* - G_1^*/G^*, \tag{20.10}$$

$$E_2/G = E_2^*/G^*. \tag{20.11}$$

Since
$$\frac{\partial}{\partial v}(E^2/G) = 2E^2/G\,.\,(E_2/E - \tfrac{1}{2}G_2/G),$$

it follows from (20.9) that

$$\frac{\partial}{\partial v}\log(E^2/G) = \frac{\partial}{\partial v}\log(E^{*2}/G^*),$$

from which
$$E^2/G = E^{*2}/G^*\,.\,U^3, \tag{20.12}$$

for some function $U(u)$.

Similarly from (20.10) it follows that

$$G^2/E = G^{*2}/E^*\,.\,V^3 \tag{20.13}$$

for some function $V(v)$.

From (20.12) and (20.13) it follows that

$$E = E^*U^2V,$$

$$G = G^*UV^2.$$

When these results are substituted in (20.8), this becomes

$$\frac{G_1}{G} = \frac{U_1}{U-V}$$

from which it follows that, if $U > V$, $G = (U-V)\bar{V}^2$, where $\bar{V}$ is an arbitrary function of v. Similarly from (20.11) we obtain $E = (U-V)\bar{U}^2$, where $\bar{U}$ is an arbitrary function of u. The metric of S is thus

$$ds^2 = (U-V)[\bar{U}^2 du^2 + \bar{V}^2 dv^2],$$

and that of S^*,

$$(V^{-1} - U^{-1})[U^{-1}\bar{U}^2 du^2 + V^{-1}\bar{V}^2 dv^2].$$

A change of parameters $u^* = \int \bar{U} du$, $v^* = \int \bar{V} dv$ will reduce these metrics to the required forms (20.2). This completes the proof of Dini's Theorem.

An important theorem due to Beltrami states that the only surfaces in geodesic correspondence with the plane are those of constant curvature. The proof of this theorem is left as an exercise for the reader.

APPENDIX II. 1

The second existence theorem

Let S be a region of a surface of class r ($r \geqslant 3$), represented parametrically by equations $\mathbf{r} = \mathbf{r}(u, v)$.

Let Q be any point of S with parameters (u_0, v_0). Then there exists a neighbourhood N_Q of Q with a system of parameters U, V valid in N_Q, with the property that the geodesics through Q are given by linear equations of the form

$$U = p^1 s, \qquad V = p^2 s.$$

The transformation of parameters $(u, v) \rightarrow (U, V)$ is proper, and the functions giving this parametric transformation are of class $(r-2)$.

It is convenient to use the equations of the geodesics in the form (12.8), i.e.

$$\left.\begin{aligned}
\frac{d^2 u}{ds^2} + \Gamma^1_{11}\left(\frac{du}{ds}\right)^2 + 2\Gamma^1_{12}\frac{du}{ds}\frac{dv}{ds} + \Gamma^1_{22}\left(\frac{dv}{ds}\right)^2 = 0 \\
\frac{d^2 v}{ds^2} + \Gamma^2_{11}\left(\frac{du}{ds}\right)^2 + 2\Gamma^2_{12}\frac{du}{ds}\frac{dv}{ds} + \Gamma^2_{22}\left(\frac{dv}{ds}\right)^2 = 0
\end{aligned}\right\}, \tag{1}$$

where the Christoffel symbols of the second kind Γ^i_{jk} are given by equations (12.9). It follows from the form of (12.9) that these symbols are functions of class $(r-2)$.

Now, from the first existence theorem of section 13 it follows that equations (1) admit a unique solution

$$
\left. \begin{aligned}
u &= f^1(u_0, v_0, p^1, p^2, s) \\
v &= f^2(u_0, v_0, p^1, p^2, s)
\end{aligned} \right\}, \tag{2}
$$

which satisfy the initial conditions

$$
\left. \begin{aligned}
f^1(u_0, v_0, p^1, p^2, 0) &= u_0 \\
f^2(u_0, v_0, p^1, p^2, 0) &= v_0 \\
\left(\frac{\partial f^1}{\partial s} \right)_{s=0} &= p^1 \\
\left(\frac{\partial f^2}{\partial s} \right)_{s=0} &= p^2
\end{aligned} \right\}. \tag{3}
$$

If λ is any constant, it follows from the particular form of equations (1) that

$$
\left. \begin{aligned}
u &= f^1(u_0, v_0, p^1, p^2, \lambda s) \\
v &= f^2(u_0, v_0, p^1, p^2, \lambda s)
\end{aligned} \right\} \tag{4}
$$

are solutions of (1) which satisfy the initial conditions

$$
\left. \begin{aligned}
f^1(u_0, v_0, p^1, p^2, 0) &= u_0 \\
f^2(u_0, v_0, p^1, p^2, 0) &= v_0 \\
\left(\frac{\partial f^1}{\partial s} \right)_{s=0} &= \lambda p^1 \\
\left(\frac{\partial f^2}{\partial s} \right)_{s=0} &= \lambda p^2
\end{aligned} \right\}. \tag{5}
$$

Thus we have

$$
\left. \begin{aligned}
f^1(u_0, v_0, p^1, p^2, \lambda s) &= f^1(u_0, v_0, \lambda p^1, \lambda p^2, s) \\
f^2(u_0, v_0, p^1, p^2, \lambda s) &= f^2(u_0, v_0, \lambda p^1, \lambda p^2, s)
\end{aligned} \right\}, \tag{6}
$$

and the solutions of (1) may therefore be written

$$
\left. \begin{aligned}
u &= F^1(u_0, v_0, U, V) \\
v &= F^2(u_0, v_0, U, V)
\end{aligned} \right\}, \tag{7}
$$

where

$$
\left. \begin{aligned}
U &= p^1 s \\
V &= p^2 s
\end{aligned} \right\}. \tag{8}
$$

It also follows that $F^1(u_0, v_0, U, V)$, $F^2(u_0, v_0, U, V)$ are of class $(r-2)$ for values of (U, V) in some neighbourhood of the origin $(0, 0)$ (see, for example, L. Bieberbach (1926) pp. 39–41).

Since
$$\left(\frac{\partial u}{\partial U}\right)_0 p^1 + \left(\frac{\partial u}{\partial V}\right)_0 p^2 = p^1 \left.\right\}$$
$$\left(\frac{\partial v}{\partial U}\right)_0 p^1 + \left(\frac{\partial v}{\partial V}\right)_0 p^2 = p^2 \left.\right\}$$
(9)

for all values of p^1, p^2, we have

$$\left(\frac{\partial u}{\partial U}\right)_0 = 1, \quad \left(\frac{\partial u}{\partial V}\right)_0 = 0, \quad \left(\frac{\partial v}{\partial U}\right)_0 = 0, \quad \left(\frac{\partial v}{\partial V}\right)_0 = 1,$$

and hence the Jacobian $\partial(u,v)/\partial(U,V)$ does not vanish at the origin and hence for values of U, V near the origin. Therefore the equations
$$u = F^1(u_0, v_0, U, V)$$
$$v = F^2(u_0, v_0, U, V)$$

can be solved for U, V as functions of u, v, and the solution will be valid for points in some neighbourhood N_Q of Q.

From the form of equations (8) it follows that each point P in this neighbourhood can be joined to Q by one and only one geodesic arc which lies entirely in N_Q. This completes the proof of the theorem.

[It may be noted that a similar argument can be used to give a similar existence theorem applying to the system of n equations

$$\frac{d^2u^i}{ds^2} + \sum_{j,k=1}^{n} \Gamma_{jk}^i \frac{du^j}{ds} \frac{du^k}{ds} = 0, \quad i,j,k = 1,2,...,n,$$

where the coefficients Γ_{jk}^i are arbitrary functions of class $(r-2)$ in the n coordinates u^i.]

REFERENCES

BIEBERBACH, L., *Theorie der Differentialgleichungen*, Berlin (1926).

GOURSAT, E., *Cours d'analyse mathématique*, Gauthier–Villars (5th edition 1933).

WHITEHEAD, J. H. C., Convex regions in the geometry of paths, *Quart. J. Math.* (Oxford), **3** (1932) 33–42.

MISCELLANEOUS EXERCISES II

1. Show that the curves bisecting the angles between the parametric curves are given by
$$E\,du^2 - G\,dv^2 = 0.$$

2. Show that the parametric curves on the sphere given by

$x = a\sin u\cos v, \quad y = a\sin u\sin v, \quad z = a\cos u, \quad 0 < u < \tfrac{1}{2}\pi, \quad 0 < v < 2\pi,$

form an orthogonal system. Determine the two families of curves which meet the curves v = constant at angles of $\frac{1}{4}\pi$ and $\frac{3}{4}\pi$. Find the metric of the surfaces referred to these two families as parametric curves.

3. Show that on a right helicoid, the family of curves orthogonal to the curves $u \cos v$ = constant is the family $(u^2+a^2)\sin^2 v$ = constant.

4. The metric of a surface is $v^2\,du^2+u^2\,dv^2$. Find the equation of the family of curves orthogonal to the curves uv = constant, and find the metric referred to new parameters so that these two families are parametric.

$$\left[\frac{u}{v} = \text{constant}; \quad ds^2 = \frac{1}{2}\frac{\bar{v}^2}{\bar{u}^2}\,d\bar{u}^2+\frac{1}{2}\,d\bar{v}^2.\right]$$

5. Determine $\phi(v)$ so that the surface given by

$$x = u\cos v, \quad y = u\sin v, \quad z = \phi(v)$$

shall be locally isometric to a surface of revolution.

6. Find the differential equations for geodesics on the catenoid of revolution obtained by rotating the curve $x = c\cosh(z/c)$ about the z-axis.

7. *Liouville surfaces* have the property that their line element may be written in the form

$$ds^2 = [U(u)+V(v)](du^2+dv^2).$$

Show that the geodesics can be obtained by a quadrature.

8. Find the geodesics on the surface with metric

$$ds^2 = e^{2v}(du^2+U^2\,dv^2),$$

where U is a function of u alone.

9. If two families of geodesics on a surface intersect at a constant angle, prove that the surface has zero Gaussian curvature.

10. Prove that every space curve is a geodesic on its rectifying developable.

11. The curves u = constant, v = constant are orthogonal parametric curves.

    κ_a is the geodesic curvature along v = constant,

    κ_b is the geodesic curvature along u = constant.

Prove that
$$K = \frac{1}{\sqrt{(EG)}}\left\{\frac{\partial}{\partial v}(\kappa_a\sqrt{E})-\frac{\partial}{\partial u}(\kappa_b\sqrt{G})\right\},$$

and
$$K = \frac{d\kappa_a}{ds_b}-\frac{d\kappa_b}{ds_a}-\kappa_a^2-\kappa_b^2,$$

where s_a, s_b denote arc length respectively along v = constant, u = constant.

12. Prove that the geodesic curvature of a curve given by $\phi(u,v)$ = constant is

$$\kappa_g = \frac{1}{H}\frac{\partial}{\partial u}\frac{F\phi_2-G\phi_1}{\sqrt{(E\phi_2^2-2F\phi_1\phi_2+G\phi_1^2)}}+\frac{1}{H}\frac{\partial}{\partial v}\frac{F\phi_1-E\phi_2}{\sqrt{(E\phi_2^2-2F\phi_1\phi_2+G\phi_1^2)}}.$$

13. Show that on a surface of negative curvature two geodesics cannot meet in two points and enclose a simply connected area.

14. Find the curves which bisect the angles between the parametric curves on the surface given by

$$\frac{x}{a} = \tfrac{1}{2}(u+v), \qquad \frac{y}{b} = \tfrac{1}{2}(u-v), \qquad z = \tfrac{1}{2}uv,$$

and show that they form an isothermal system.

15. Show that the surface given by

$$\frac{x}{a} = \sqrt{\left(\frac{(a^2-u)(a^2-v)}{(a^2-b^2)(a^2-c^2)}\right)}, \quad \frac{y}{b} = \sqrt{\left(\frac{(b^2-u)(b^2-v)}{(b^2-a^2)(b^2-c^2)}\right)}, \quad \frac{z}{c} = \sqrt{\left(\frac{(c^2-u)(c^2-v)}{(c^2-a^2)(c^2-b^2)}\right)}$$

is an ellipsoid, and that the parametric curves form an isothermal system.

III

THE SECOND FUNDAMENTAL FORM: LOCAL NON-INTRINSIC PROPERTIES OF A SURFACE

1. The second fundamental form

THE previous chapter was concerned essentially with the intrinsic properties of a surface, while this chapter deals with properties of a surface relative to the Euclidean space in which it is embedded.

The normal curvature of a curve at a point P on a surface has already been defined in Chapter II, section 15. By multiplying equation (15.1) by $\mathbf{N}$ we obtain

$$\kappa_n = \mathbf{N}.\mathbf{r}''. \tag{1.1}$$

Since $\mathbf{r}'' = \mathbf{r}_1 u'' + \mathbf{r}_2 v'' + \mathbf{r}_{11} u'^2 + 2\mathbf{r}_{12} u'v' + \mathbf{r}_{22} v'^2$, equation (1.1) can be written

$$\kappa_n = (\mathbf{N}.\mathbf{r}_{11})u'^2 + 2(\mathbf{N}.\mathbf{r}_{12})u'v' + (\mathbf{N}.\mathbf{r}_{22})v'^2.$$

It follows that

$$\kappa_n = \frac{L\,du^2 + 2M\,du\,dv + N\,dv^2}{E\,du^2 + 2F\,du\,dv + G\,dv^2} \tag{1.2}$$

where L, M, N are defined by the relations

$$L = \mathbf{N}.\mathbf{r}_{11}, \quad M = \mathbf{N}.\mathbf{r}_{12}, \quad N = \mathbf{N}.\mathbf{r}_{22}. \tag{1.3}$$

Alternative expressions for L, M, N will now be obtained. By differentiating the equation $\mathbf{N}.\mathbf{r}_1 = 0$ we obtain $\mathbf{N}_1.\mathbf{r}_1 + \mathbf{N}.\mathbf{r}_{11} = 0$ and $\mathbf{N}_2.\mathbf{r}_1 + \mathbf{N}.\mathbf{r}_{12} = 0$. Similarly, the equation $\mathbf{N}.\mathbf{r}_2 = 0$ leads to $\mathbf{N}_2.\mathbf{r}_2 + \mathbf{N}.\mathbf{r}_{22} = 0$ and $\mathbf{N}_1.\mathbf{r}_2 + \mathbf{N}.\mathbf{r}_{21} = 0$. Substitution in equation (1.3) gives the relations

$$L = -\mathbf{N}_1.\mathbf{r}_1, \quad M = -\mathbf{N}_1.\mathbf{r}_2 = -\mathbf{N}_2.\mathbf{r}_1, \quad N = -\mathbf{N}_2.\mathbf{r}_2. \tag{1.4}$$

The quadratic form

$$L\,du^2 + 2M\,du\,dv + N\,dv^2 \tag{1.5}$$

is called the *second fundamental form*, and the functions of u and v denoted by L, M, and N are called the *second fundamental coefficients*. It follows from (1.2) that all curves having the same direction at P have the same normal curvature; hence normal curvature is a property of a surface and a direction at a point on the surface.

If ϕ denotes the angle between the principal normal **n** to a curve on the surface and the surface normal **N**, it follows from II (15.1) that

$$\kappa \cos\phi = \kappa_n, \tag{1.6}$$

a result known as Meusnier's theorem.

Since the denominator of the right-hand member of (1.2) is positive definite, it follows that the sign of κ_n depends only upon the sign of the form (1.5). If at a point P on the surface this form is definite (i.e. if $LN - M^2 > 0$), then κ_n maintains the same sign for all directions at P. In this case P is called an *elliptic point*. When $LN - M^2 = 0$, κ_n retains the same sign for all directions through P except one for which the curvature is zero; P is then called a *parabolic point*. When $LN - M^2 < 0$, κ_n is positive for directions lying within a certain angle, negative for directions lying outside this angle, and zero along the directions which form the angle; P is then called a *hyperbolic point*, and the critical directions are called *asymptotic directions*.

EXERCISE 1.1. Show that when the parameters are transformed, the discriminant $LN - M^2$ is multiplied by the square of the Jacobian determinant of the transformation, and deduce that the conditions for an elliptic, parabolic, or hyperbolic point are thus independent of the particular parametric representation chosen.

We now obtain a geometrical interpretation of the second fundamental form. In Fig 6, let $P(u,v)$ and $Q(u+h, v+k)$ be near points on a surface, and let d be the perpendicular distance from Q onto the tangent plane to the surface at P. If $\mathbf{r}_P$, $\mathbf{r}_Q$ are the position vectors of P and Q, then

$$\begin{aligned}
d &= (\mathbf{r}_Q - \mathbf{r}_P).\mathbf{N} \\
&= (h\mathbf{r}_1 + k\mathbf{r}_2).\mathbf{N} + \tfrac{1}{2}(h^2\mathbf{r}_{11} + 2hk\mathbf{r}_{12} + k^2\mathbf{r}_{22}).\mathbf{N} + O(h^3, k^3) \\
&= \tfrac{1}{2}(Lh^2 + 2Mhk + Nk^2) + O(h^3, k^3). \tag{1.7}
\end{aligned}$$

At an elliptic point d retains the same sign, and this implies that the surface near P lies entirely on one side of the tangent plane at P. At a hyperbolic point the surface crosses over the tangent plane. It follows that any point on an ellipsoidal surface is elliptic, any point on a circular cylinder is parabolic, and any point on the hyperbolic paraboloid II (1.5) is hyperbolic.

EXERCISE 1.2. Show that the anchor ring of Chapter II, section 3, contains all three types of points.

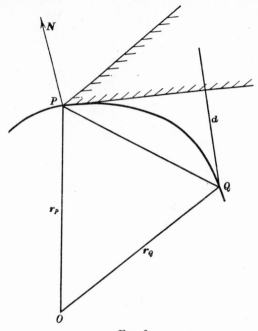

FIG. 6

2. Principal curvatures

The normal curvature at P in a direction specified by direction coefficients (l, m) is given by

$$\kappa = Ll^2 + 2Mlm + Nm^2, \tag{2.1}$$

where
$$El^2 + 2Flm + Gm^2 = 1. \tag{2.2}$$

As l, m vary, subject to (2.2), the normal curvature will vary. Its extreme values may be found by making use of Lagrange's multipliers. Write

$$\kappa = Ll^2 + 2Mlm + Nm^2 - \lambda(El^2 + 2Flm + Gm^2 - 1);$$

then when κ is stationary,

$$\frac{1}{2}\frac{\partial \kappa}{\partial l} = Ll + Mm - \lambda El - \lambda Fm = 0, \tag{2.3}$$

$$\frac{1}{2}\frac{\partial \kappa}{\partial m} = Ml + Nm - \lambda Fl - \lambda Gm = 0. \tag{2.4}$$

Multiply (2.3) by l, (2.4) by m, and add to get $\lambda = \kappa$. Eliminate l and m from (2.3) and (2.4) to get

$$\kappa^2(EG-F^2)-\kappa(EN+GL-2FM)+LN-M^2 = 0. \qquad (2.5)$$

The roots κ_a, κ_b of this equation are called the *principal curvatures*. Associated with these are the *mean curvature* μ defined by

$$\mu = \tfrac{1}{2}(\kappa_a+\kappa_b) = \frac{EN+GL-2FM}{2(EG-F^2)}, \qquad (2.6)$$

and the *Gaussian curvature* K defined by

$$K = \kappa_a\kappa_b = \frac{LN-M^2}{EG-F^2}. \qquad (2.7)$$

An important theorem due to Gauss, which will be proved later, allows us to identify the curvature defined by (2.7) with the curvature defined in Chapter II, section 16, from a consideration of the first fundamental form alone. This result justifies our use of the same terminology.

The *principal directions* corresponding to the principal curvatures are obtained by eliminating λ between (2.3) and (2.4), so that

$$(EM-FL)l^2+(EN-GL)lm+(FN-GM)m^2 = 0. \qquad (2.8)$$

The discriminant of this equation is

$$(EN-GL)^2-4(EM-FL)(FN-GM),$$

which is identically equal to

$$4\left(\frac{EG-F^2}{E^2}\right)(EM-FL)^2+\left\{EN-GL-\frac{2F}{E}(EM-FL)\right\}^2.$$

Since $EG-F^2 > 0$ it follows that the roots of equation (2.8) are real and distinct, provided that the coefficients E, F, G and L, M, N are not proportional. When the values of these coefficients are proportional, the principal directions are indeterminate and the normal curvature is the same in all directions. Such a point where

$$\frac{L}{E} = \frac{M}{F} = \frac{N}{G}$$

is called an *umbilic*.

At a point which is not an umbilic the two directions determined by equation (2.8) are orthogonal. This follows by direct application of the condition of orthogonality, II (7.4).

3. Lines of curvature

A curve on a surface whose tangent at each point is along a principal direction is called a *line of curvature*. Since the principal directions are given by (2.3) and (2.4), the equations of a line of curvature are

$$
\left.
\begin{aligned}
(L-\kappa E)\,du+(M-\kappa F)\,dv &= 0 \\
(M-\kappa F)\,du+(N-\kappa G)\,dv &= 0
\end{aligned}
\right\}, \tag{3.1}
$$

where κ is one of the principal curvatures. Replacing the fundamental coefficients by their expression in terms of derivatives of $\mathbf{r}$ and $\mathbf{N}$, we have

$$
(\kappa \mathbf{r}_1^2+\mathbf{N}_1.\mathbf{r}_1)\,du+(\kappa \mathbf{r}_1.\mathbf{r}_2+\mathbf{N}_2.\mathbf{r}_1)\,dv = 0,
$$

$$
(\kappa \mathbf{r}_1.\mathbf{r}_2+\mathbf{N}_1.\mathbf{r}_2)\,du+(\kappa \mathbf{r}_2^2+\mathbf{N}_2.\mathbf{r}_2)\,dv = 0,
$$

i.e.
$$
(\kappa\,d\mathbf{r}+d\mathbf{N}).\mathbf{r}_1 = 0,
$$

$$
(\kappa\,d\mathbf{r}+d\mathbf{N}).\mathbf{r}_2 = 0.
$$

Since the vector $(\kappa\,d\mathbf{r}+d\mathbf{N})$ is tangential to the surface, we have

$$
\kappa\,d\mathbf{r}+d\mathbf{N} = 0. \tag{3.2}
$$

Conversely, if (3.2) holds along a curve for any function κ, then equations (3.1) follow and the curve is thus a line of curvature. Equation (3.2) characterizes the lines of curvature and is known as *Rodrigues' formula*.

If the lines of curvature are taken as parametric curves, two of the fundamental coefficients—viz. F and M—become zero. We have $F = 0$ because the principal directions are orthogonal, and $M = 0$ because equation (2.8) must here reduce to $lm = 0$, i.e. $du\,dv = 0$.

Let κ_a, κ_b be the principal curvatures at a point P. With the lines of curvature as parametric curves, the normal curvature in a direction ψ with direction coefficients (l, m) is

$$
\kappa = Ll^2+Nm^2. \tag{3.3}
$$

It follows that
$$
\kappa_a = L/E, \qquad \kappa_b = N/G. \tag{3.4}
$$

Since also $\cos\psi = l\sqrt{E}$, $\sin\psi = m\sqrt{G}$, we find

$$
\kappa = \kappa_a \cos^2\psi+\kappa_b \sin^2\psi, \tag{3.5}
$$

a result due to Euler.

The Dupin Indicatrix

Suppose at a point O on a given surface a set of rectangular Cartesian coordinates are chosen so that Ox, Oy are along the principal directions at O, and Oz is along the normal at O. It follows from equation (1.7) that the equation of the surface near O is $z = ax^2+by^2$, provided that terms of higher order are neglected. The plane $z = 2h$ intersects the surface near O in the conic $z = 2h$, $2h = ax^2+by^2$. The normal curvature at O in the direction Ox is $\lim_{x\to0} 2h/x^2 = a = \kappa_a$. Similarly, $b = \kappa_b$. It follows that if we define R_a, R_b to be the reciprocals of κ_a, κ_b, the curve of section is the conic

$$x^2/R_a+y^2/R_b = 2h, \quad z = 2h. \tag{3.6}$$

This conic is known as *Dupin's indicatrix*, and it gives an immediate geometrical interpretation of the variation of normal curvature with direction.

If κ_a, κ_b have the same sign, the conic is an ellipse with semi-axes of length $(2hR_a)^{\frac{1}{2}}$, $(2hR_b)^{\frac{1}{2}}$, and is real or imaginary according to the sign of h. If κ_a, κ_b have different signs, the conic is one of two conjugate hyperbolas according to the sign of h. In this case the directions of the asymptotes at O are called the *asymptotic directions* at O. The normal curvature changes sign as the direction passes through an asymptotic position. Evidently the angles between the asymptotic directions are bisected by the principal directions.

EXERCISE 3.1. Prove that at any point P on a surface there is a paraboloid such that the normal curvature of the surface at P in any direction is the same as that of the paraboloid.

Two directions at P are said to be *conjugate* if the corresponding diameters of the Dupin indicatrix are conjugate.

In terms of general curvilinear coordinates, the equations of the indicatrix are $z = 2h$, $2h = Lx^2+2Mxy+Ny^2$. It follows that the directions (l_1, m_1), (l_2, m_2) will be conjugate if

$$Ll_1l_2+M(l_1m_2+l_2m_1)+Nm_1m_2 = 0. \tag{3.7}$$

In particular, the directions of the parametric curves will be conjugate if $M = 0$. Thus the lines of curvature are in conjugate directions at every point, as is otherwise obvious.

If $d\mathbf{r}/ds$ is a tangent vector in the direction (l_1, m_1) and $\delta\mathbf{N}/\delta s$ is the rate of change of surface normal $\mathbf{N}$ with arc length in the direction (l_2, m_2), then equation (3.7) may be written

$$\frac{d\mathbf{r}}{ds} \cdot \frac{\delta \mathbf{N}}{\delta s} = 0.$$

An *asymptotic line* is a curve whose direction at every point is asymptotic. The equation of such lines is

$$\frac{d\mathbf{r}}{ds} \cdot \frac{d\mathbf{N}}{ds} = 0, \quad \text{i.e. } L\,du^2 + 2M\,dudv + N\,dv^2 = 0,$$

from which it follows that asymptotic lines are self-conjugate.

4. Developables

A developable is a surface enveloped by a one-parameter family of planes. Such a family is given by the equation

$$\mathbf{r} \cdot \mathbf{a} = p, \tag{4.1}$$

where $\mathbf{a}$ and p are functions of a real parameter u. It will be convenient to refer to that plane determined by the value u of the parameter as the plane u.

The planes u, v $(u < v)$ will intersect in a line provided they are not parallel. If $f(u) = \mathbf{r} \cdot \mathbf{a}(u) - p(u)$, the equations of this line are $f(u) = 0$, $f(v) = 0$. From Rolle's theorem it follows that there is a value u_1, $u < u_1 < v$ such that $\dot{f}(u_1) = 0$. As $v \to u$, $u_1 \to u$ and the equations of the limiting position of the line become

$$\left. \begin{array}{l} \mathbf{r} \cdot \mathbf{a} = p \\ \mathbf{r} \cdot \dot{\mathbf{a}} = \dot{p} \end{array} \right\}. \tag{4.2}$$

This line is called the *characteristic line* corresponding to the plane u. It can be regarded as the line of intersection of the plane u with an infinitesimally near plane.

In a similar way, the three planes, u, v, w $(u < v < w)$, will generally intersect in one point, and the limiting position of this point as $v \to u$ and $w \to u$ independently is called the *characteristic point* corresponding to u. By using Rolle's theorem, the equations which determine this point are

$$\left. \begin{array}{l} \mathbf{r} \cdot \mathbf{a} = p \\ \mathbf{r} \cdot \dot{\mathbf{a}} = \dot{p} \\ \mathbf{r} \cdot \ddot{\mathbf{a}} = \ddot{p} \end{array} \right\}. \tag{4.3}$$

If $\mathbf{a}$, $\dot{\mathbf{a}}$, $\ddot{\mathbf{a}}$ are linearly dependent, these equations either have no solution or else the solution is indeterminate. For example, when the developable is a cylinder with generating lines all parallel to **a**. When the family of planes forms a pencil, the developable degenerates to the axis of the pencil, and hence the characteristic points are indeterminate. When the family of planes envelops a cone, all planes have the same characteristic point which is the vertex of the cone.

In general, the characteristic points corresponding to planes of the family determine a curve on the developable called the *edge of regression*, with equations given by (4.3) where **r** is regarded as a function of u. The tangent **t** to the edge of regression satisfies the equation $\dot{s}\mathbf{t}.\mathbf{a}+\mathbf{r}.\dot{\mathbf{a}} = \dot{p}$, which with the second equation of (4.3) reduces to

$$\mathbf{t}.\mathbf{a} = 0. \tag{4.4}$$

We differentiate the second equation of (4.3) to get

$$\dot{s}\mathbf{t}.\dot{\mathbf{a}}+\mathbf{r}.\ddot{\mathbf{a}} = \ddot{p},$$

which with the third equation of (4.3) reduces to

$$\mathbf{t}.\dot{\mathbf{a}} = 0. \tag{4.5}$$

It follows from (4.4) and (4.5) that **t** is parallel to the characteristic line (4.2). We differentiate equation (4.4) and use (4.5) to get

$$\mathbf{n}.\mathbf{a} = 0. \tag{4.6}$$

From (4.4) and (4.6) it follows that **a** is parallel to the binormal of the edge of regression, and that the osculating plane at any point on this curve is identical with the corresponding plane of the family (4.1).

It will now be shown that, in general, a developable consists of two sheets which are tangent to the edge of regression along a sharp edge, a property which justifies the terminology used.

Let O be the point $s = 0$ on the edge of regression C, and let Ox, Oy, Oz be a set of rectangular Cartesian axes chosen respectively along **t**, **n**, and **b** at O. Then any point on the developable has position vector given by $\mathbf{R} = \mathbf{r}+v\mathbf{t}$. On expanding $\mathbf{R}$ in powers of s we get

$$\mathbf{R}(s) = \mathbf{r}(s)+v\mathbf{t}(s) = s\mathbf{t}+\tfrac{1}{2}s^2\kappa\mathbf{n}+\tfrac{1}{6}s^3(\kappa'\mathbf{n}+\kappa\tau\mathbf{b}-\kappa^2\mathbf{t})+O(s^4)+$$
$$+v\{\mathbf{t}+s\kappa\mathbf{n}+\tfrac{1}{2}s^2(\kappa'\mathbf{n}+\kappa\tau\mathbf{b}-\kappa^2\mathbf{t})+O(s^3)\}. \tag{4.7}$$

The normal plane $x = 0$ meets the surface where

$$v = -s-\tfrac{1}{3}s^3\kappa^2+O(s^4);$$

and substitution of this value in (4.7) gives

$$y = -\tfrac{1}{2}\kappa s^2 + O(s^3),$$
$$z = -\tfrac{1}{3}\kappa\tau s^3 + O(s^4).$$

We eliminate s to obtain

$$z^2 = -\frac{8}{9}\frac{\tau^2}{\kappa}y^3, \tag{4.8}$$

from which it follows that the developable cuts the normal plane to the edge of regression in a cusp whose tangent is along the principal normal. The two sheets of the developable are thus tangent to the edge of regression along a sharp edge.

5. Developables associated with space curves

The one-parameter family of osculating planes at points on a skew curve form the *osculating developable* of the curve. We now prove that the edge of regression is the curve itself.

The family of osculating planes has equation

$$(\mathbf{R} - \mathbf{r}(s)) . \mathbf{b} = 0. \tag{5.1}$$

Differentiation gives $-\mathbf{t} . \mathbf{b} + (\mathbf{R} - \mathbf{r}) . (-\tau \mathbf{n}) = 0$ which reduces to

$$(\mathbf{R} - \mathbf{r}) . \mathbf{n} = 0. \tag{5.2}$$

We differentiate (5.2) and use equation (5.1) to get

$$(\mathbf{R} - \mathbf{r}) . \mathbf{t} = 0. \tag{5.3}$$

Equations (5.1), (5.2), (5.3) imply $\mathbf{R} = \mathbf{r}$, giving the required result.

The family of normal planes to a skew curve form the *polar developable* of the given curve. We now prove that the edge of regression of the polar developable is the locus of centres of spherical curvature of the given curve.

Differentiation of the equation of the family of normal planes

$$(\mathbf{R} - \mathbf{r}) . \mathbf{t} = 0 \tag{5.4}$$

leads to

$$(\mathbf{R} - \mathbf{r}) . \mathbf{n} = \rho. \tag{5.5}$$

Differentiation of (5.5) and use of (5.4) leads to

$$(\mathbf{R} - \mathbf{r}) . \mathbf{b} = \sigma\rho'. \tag{5.6}$$

It follows from (5.4), (5.5), (5.6) that

$$\mathbf{R} = \mathbf{r} + \rho\mathbf{n} + \sigma\rho'\mathbf{b}, \tag{5.7}$$

so that $\mathbf{R}$ is the position vector of the centre of spherical curvature.

The rectifying planes to a skew curve determine the *rectifying developable* of the given curve. The principal normal to the curve coincides at points along the curve with the normal to the developable surface, and the curve is thus a geodesic in the rectifying developable.

We can now see that the argument in section 4 shows that *any developable*, which is not a cylinder or a cone, may be regarded as the osculating developable of its edge of regression.

EXERCISE 5.1. Prove that the edge of regression of the rectifying developable has equation

$$\mathbf{R} = \mathbf{r} + \kappa \frac{(\tau\mathbf{t} + \kappa\mathbf{b})}{\kappa'\tau - \kappa\tau'}.$$

We conclude this section by proving the following:

THEOREM 5.1. *A necessary and sufficient condition for a surface to be a developable is that its Gaussian curvature shall be zero.*

If the developable is a cylinder or a cone, the Gaussian curvature is evidently zero. If these cases are excluded, the developable may be regarded as the osculating developable of its edge of regression, and its equation may be written

$$\mathbf{R} = \mathbf{r}(s) + v\mathbf{t}(s). \tag{5.8}$$

The second fundamental coefficients M and N corresponding to (5.8) are readily calculated to be zero, so that

$$K = (LN - M^2)/(EG - F^2)$$

is zero. This proves that the Gaussian curvature of *any* developable is necessarily zero.

To prove the sufficiency, we consider

$$LN - M^2 = (\mathbf{r}_1 . \mathbf{N}_1)(\mathbf{r}_2 . \mathbf{N}_2) - (\mathbf{r}_1 . \mathbf{N}_2)(\mathbf{r}_2 . \mathbf{N}_1),$$

which from a vector identity is equal to

$$(\mathbf{r}_1 \times \mathbf{r}_2) . (\mathbf{N}_1 \times \mathbf{N}_2) \quad \text{i.e.} \quad H[\mathbf{N}, \mathbf{N}_1, \mathbf{N}_2].$$

By hypothesis, $K = 0$ so that $[\mathbf{N}, \mathbf{N}_1, \mathbf{N}_2] = 0$, and hence $\mathbf{N}, \mathbf{N}_1, \mathbf{N}_2$ are coplanar. From $\mathbf{N} . \mathbf{N}_1 = 0$, $\mathbf{N} . \mathbf{N}_2 = 0$ it follows that $\mathbf{N}_1 = \mathbf{0}$ or $\mathbf{N}_2 = \mathbf{0}$ or $\mathbf{N}_1 = k\mathbf{N}_2$.

Suppose, for example, that $\mathbf{N}_2 = \mathbf{0}$. The tangent plane at a point on the surface has equation $(\mathbf{R} - \mathbf{r}) . \mathbf{N} = 0$. Since

$$[(\mathbf{R} - \mathbf{r}) . \mathbf{N}]_2 = -(\mathbf{r} . \mathbf{N})_2 = -\mathbf{r} . \mathbf{N}_2 = 0,$$

the equation of the tangent plane does not involve v, and the surface is the envelope of a one-parameter family of planes, i.e. a developable. If the condition $\mathbf{N}_1 = k\mathbf{N}_2$ holds, we apply the proper parameter transformation

$$u = u'+v', \qquad v = u'-kv'$$

to obtain $\mathbf{N}_{1'} = \mathbf{N}_1+\mathbf{N}_2$, $\mathbf{N}_{2'} = \mathbf{N}_1-k\mathbf{N}_2 = \mathbf{0}$. Again the surface normal $\mathbf{N}$ depends on only one parameter, and the surface is consequently a developable. This completes the proof of the theorem.

6. Developables associated with curves on surfaces

The following theorem due to Monge characterizes lines of curvature on a surface.

A necessary and sufficient condition that a curve on a surface be a line of curvature is that the surface normals along the curve form a developable.

To prove the theorem, we consider the surface formed by the normals along the curve $\mathbf{r} = \mathbf{r}(s)$. Any point on this surface will have position vector

$$\mathbf{R} = \mathbf{r}(s)+v\mathbf{N}(s). \tag{6.1}$$

It is easily verified that for this surface $M = [\mathbf{t}, \mathbf{N}, \mathbf{N}']$, and $N = 0$. Hence its Gaussian curvature is zero if and only if $[\mathbf{t}, \mathbf{N}, \mathbf{N}'] = 0$. In view of the previous theorem, the surface normals along the curve form a developable if and only if $[\mathbf{t}, \mathbf{N}, \mathbf{N}'] = 0$. It remains to prove that this condition is satisfied if and only if the curve is a line of curvature.

Since $\mathbf{t} \times \mathbf{N}'$ is normal to the given surface, the equation $[\mathbf{t}, \mathbf{N}, \mathbf{N}'] = 0$ implies that $\mathbf{t} \times \mathbf{N}' = 0$, i.e. $\mathbf{N}' = -\kappa\mathbf{t}$ for some function κ. Conversely, if $\mathbf{N}' = -\kappa\mathbf{t}$ then $[\mathbf{t}, \mathbf{N}, \mathbf{N}'] = 0$. But the equation $\mathbf{N}' = -\kappa\mathbf{t}$ is Rodrigues' formula (cf. (3.2)) and holds if and only if the curve is a line of curvature. This completes the proof of the theorem.

We now obtain an alternative interpretation of the conjugate directions defined in section 3. The tangent planes at points on a curve C lying on a surface form a developable, and we now prove that the characteristic line of the developable at any point P on C is in a direction conjugate to that of the tangent to C at P.

On differentiating the equation of the family of tangent planes $(\mathbf{R}-\mathbf{r}).\mathbf{N} = 0$, we get

$$(\mathbf{R}-\mathbf{r}).\frac{d\mathbf{N}}{ds} = 0. \tag{6.2}$$

If (l, m) are direction coefficients of the characteristic line at a point P,
$$(\mathbf{R}-\mathbf{r}) = l\mathbf{r}_1 + m\mathbf{r}_2. \tag{6.3}$$

We substitute in (6.2) to get
$$(\mathbf{N}_1 u' + \mathbf{N}_2 v') \cdot (l\mathbf{r}_1 + m\mathbf{r}_2) = 0,$$

which gives
$$Llu' + M(lv' + mu') + Nmv' = 0. \tag{6.4}$$

But this is precisely the condition that the direction (l, m) is conjugate to the direction (u', v') of the tangent at P. This proves the theorem.

7. Minimal surfaces

Surfaces whose mean curvature is zero at all points are called *minimal surfaces*, and these have many interesting properties. The adjective minimal is justified by the following variational property:

If there is a surface of minimum area passing through a closed space curve, it is necessarily a minimal surface, i.e. a surface of zero mean curvature.

To prove this let Σ be a surface bounded by a closed curve C, and let Σ' be another surface derived from Σ by a small displacement ϵ in the direction of the normal. We assume that ϵ_1 and ϵ_2 are both small; more precisely
$$\epsilon_1 = O(\epsilon), \qquad \epsilon_2 = O(\epsilon) \quad \text{as } \epsilon \to 0.$$

Then, if $\mathbf{R}$ denotes the position vector of the displaced surface,
$$\mathbf{R} = \mathbf{r} + \epsilon\mathbf{N},$$
$$\mathbf{R}_1 = \mathbf{r}_1 + \epsilon_1\mathbf{N} + \epsilon\mathbf{N}_1,$$
$$\mathbf{R}_2 = \mathbf{r}_2 + \epsilon_2\mathbf{N} + \epsilon\mathbf{N}_2.$$

Let E^*, F^*, G^* denote the first fundamental coefficients of Σ'. Then
$$E^* = \mathbf{R}_1^2 = E - 2\epsilon L + O(\epsilon^2),$$
$$F^* = \mathbf{R}_1 \cdot \mathbf{R}_2 = F - 2\epsilon M + O(\epsilon^2),$$
$$G^* = \mathbf{R}_2^2 = G - 2\epsilon N + O(\epsilon^2), \quad \text{as } \epsilon \to 0.$$

Then
$$H^{*2} = E^* G^* - F^{*2} = EG - F^2 - 2\epsilon(EN + GL - 2FM) + O(\epsilon^2),$$

and so
$$H^* = H\left\{1 - 2\epsilon\,\frac{EN + GL - 2FM}{EG - F^2}\right\}^{\frac{1}{2}} + O(\epsilon^2),$$

i.e.
$$H^* = H - 2\epsilon\mu H + O(\epsilon^2), \quad \text{as } \epsilon \to 0.$$

Let

$$A = \int_{\Sigma} H \, du \, dv, \qquad A^* = \int_{\Sigma'} H^* \, du \, dv = \int_{\Sigma} (H - 2\epsilon\mu H) \, du \, dv + O(\epsilon^2),$$

i.e.
$$A^* = A - \int_{\Sigma} 2\epsilon\mu H \, du \, dv + O(\epsilon^2).$$

If A is stationary, evidently $\mu = 0$, i.e. the surface is necessarily of zero mean curvature, which completes the proof.

EXERCISE 7.1. Show that the asymptotic lines on a minimal surface form an isothermal net. Show that the lines of curvature also form an isothermal net on a minimal surface.

8. Ruled surfaces

A *ruled surface* is generated by the motion of a straight line with one degree of freedom, the various positions of the line being called *generators*. The developable surfaces discussed in section 4 belong to the family of ruled surfaces, but these surfaces are very special and have properties not characteristic of ruled surfaces in general. An example of a ruled surface which is not a developable is a hyperboloid of revolution shown in Fig. 7.

Let C be any curve on a ruled surface having the property that it meets each generator precisely once. Such a curve will be called a *base curve*. It is evident that such a curve is by no means uniquely determined. Then the surface is determined by any base curve C and the direction of the generators at each point of C. Let $\mathbf{r}(u)$ be the position vector of a current point P on C and let $\mathbf{g}(u)$ be a unit vector along the generator at P.

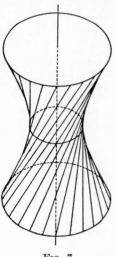

FIG. 7

Then the position vector of a general point on the ruled surface is given by
$$\mathbf{R} = \mathbf{r} + v\mathbf{g}, \tag{8.1}$$

where v is a parameter which measures directed distance along the generator from C.

Using a dot to denote differentiation with respect to the parameter u of a generator, the metric of the surface is

$$ds^2 = (\dot{\mathbf{r}}^2 + 2v\dot{\mathbf{g}} \cdot \dot{\mathbf{r}} + v^2\dot{\mathbf{g}}^2) \, du^2 + 2\mathbf{g} \cdot \dot{\mathbf{r}} \, du \, dv + dv^2. \tag{8.2}$$

The unit normal $\mathbf{N}$ is given by

$$H\mathbf{N} = (\dot{\mathbf{r}}+v\dot{\mathbf{g}})\times\mathbf{g}, \tag{8.3}$$

from which it follows that the tangent plane to the surface varies at points on the same generator unless $[\dot{\mathbf{r}},\dot{\mathbf{g}},\mathbf{g}] = 0$.

The second fundamental coefficients of the surface are found to be

$$HL = [\ddot{\mathbf{r}},\dot{\mathbf{r}},\mathbf{g}]+[\ddot{\mathbf{g}},\dot{\mathbf{r}},\mathbf{g}]v+[\ddot{\mathbf{r}},\dot{\mathbf{g}},\mathbf{g}]v+[\ddot{\mathbf{g}},\dot{\mathbf{g}},\mathbf{g}]v^2,$$

$$HM = [\dot{\mathbf{r}},\mathbf{g},\dot{\mathbf{g}}], \tag{8.4}$$

$$HN = 0.$$

The asymptotic lines are given by $du(L\,du+2M\,dv) = 0$, from which it follows that the generators are asymptotic lines. The other family of asymptotic lines is given by an equation of the form

$$\frac{dv}{du} = A+Bv+Cv^2,$$

where A, B, C are functions of u alone. This is a Riccati type differential equation, and it is known from the theory of such equations that the most general solution is of the form

$$v = (cP+Q)/(cR+S), \tag{8.5}$$

where P, Q, R, S are functions of u and c is an arbitrary constant.

Let four asymptotic lines of this family be specified by values c_1, c_2, c_3, c_4, and let these lines be met by the generator $u = u_0$ in four points whose v parameter has values v_1, v_2, v_3, v_4. From the form of (8.5) it follows that the cross-ratio $(v_1 v_2 v_3 v_4)$ is equal to the cross-ratio $(c_1 c_2 c_3 c_4)$ and is thus independent of u_0. Thus *the cross-ratio of the four points in which four given asymptotic lines are met by any generator is the same for all generators.*

From (8.4) it follows that the Gaussian curvature is $-[\dot{\mathbf{r}},\mathbf{g},\dot{\mathbf{g}}]^2/H^4$. It is convenient to define a function $p(u)$, called *the parameter of distribution*, by writing

$$p(u) = [\dot{\mathbf{r}},\mathbf{g},\dot{\mathbf{g}}]/\dot{\mathbf{g}}^2. \tag{8.6}$$

Evidently this function is constant along each generator. By replacing $\mathbf{r}$ by $\mathbf{r}+w\mathbf{g}$, $p(u)$ is clearly seen to be independent of the particular base curve chosen. It is also independent of the choice of parameter u. In terms of p the Gaussian curvature is

$$K = -p^2\dot{\mathbf{g}}^4/H^4, \tag{8.7}$$

so K is always negative except along those generators where $p = 0$. Since $K = 0$ for a developable surface, it follows that a developable surface is a ruled surface for which the parameter of distribution is identically zero.

We now show that on each generator of a general ruled surface there is a special point, called the *central point* of the generator, which is determined as follows. Let P, Q be two points on some base curve C, and let the common perpendicular to the generating lines through P, Q meet these generators in P_1, Q_1 respectively. As Q tends to P the point P_1 will tend to some point called the central point of the generator. A formula will now be obtained which determines the position of the central point on each generator.

The limiting direction of the vector $\overline{P_1 Q_1}$ must lie in the surface and hence be perpendicular to $\mathbf{N}$; also it must be perpendicular to the generator through P, and hence parallel to the vector $\mathbf{g} \times \mathbf{N}$. This direction must be perpendicular to the generators through P and Q, and proceeding to the limit as $Q \to P$ we have

$$\dot{\mathbf{g}} . (\mathbf{g} \times \mathbf{N}) = 0,$$

which may be written

$$(\dot{\mathbf{g}} \times \mathbf{g}) . \mathbf{N} = 0. \tag{8.8}$$

Substitute for $\mathbf{N}$ in (8.8) from (8.3) to get

$$(\dot{\mathbf{g}} \times \mathbf{g}) . ((\dot{\mathbf{r}} + v\dot{\mathbf{g}}) \times \mathbf{g}) = 0.$$

Appealing to the vector identity

$$(\mathbf{a} \times \mathbf{b}) . (\mathbf{c} \times \mathbf{d}) = (\mathbf{a} . \mathbf{c})(\mathbf{b} . \mathbf{d}) - (\mathbf{a} . \mathbf{d})(\mathbf{b} . \mathbf{c}),$$

with relations $\mathbf{g}^2 = 1$, $\mathbf{g} . \dot{\mathbf{g}} = 0$, we get

$$\dot{\mathbf{g}} . \dot{\mathbf{r}} + v\dot{\mathbf{g}}^2 = 0, \tag{8.9}$$

from which v is uniquely determined provided that $\dot{\mathbf{g}}^2 \neq 0$.

The central points on all the generators form a locus called the *line of striction*, which is a well-determined curve naturally associated with the ruled surface. If we choose this as base curve, it follows from (8.9) that $\dot{\mathbf{g}} . \dot{\mathbf{r}} = 0$. Since in addition $\dot{\mathbf{g}} . \mathbf{g} = 0$, it follows that the vector $\dot{\mathbf{r}} \times \mathbf{g}$ must be parallel to $\dot{\mathbf{g}}$. Write $\dot{\mathbf{r}} \times \mathbf{g} = \alpha \dot{\mathbf{g}}$ for some function α. Then scalar multiplication by $\dot{\mathbf{g}}$ gives

$$[\dot{\mathbf{r}}, \mathbf{g}, \dot{\mathbf{g}}] = \alpha \dot{\mathbf{g}}^2 = p\dot{\mathbf{g}}^2,$$

so we have $\alpha = p$. Thus

$$\dot{\mathbf{r}} \times \mathbf{g} = p\dot{\mathbf{g}}. \tag{8.10}$$

Equation (8.3) can now be written

$$H\mathbf{N} = p\dot{\mathbf{g}} + v\dot{\mathbf{g}} \times \mathbf{g}. \tag{8.11}$$

From the form of the metric (8.2) with $\dot{\mathbf{g}}.\dot{\mathbf{r}} = 0$ it follows that

$$H^2 = v^2\dot{\mathbf{g}}^2 + \dot{\mathbf{r}}^2 - (\dot{\mathbf{g}}.\dot{\mathbf{r}})^2$$
$$= v^2\dot{\mathbf{g}}^2 + \dot{\mathbf{r}}^2\mathbf{g}^2 - (\dot{\mathbf{g}}.\dot{\mathbf{r}})^2.$$

Using the vector identity

$$(\mathbf{a} \times \mathbf{b})^2 = \mathbf{a}^2\mathbf{b}^2 - (\mathbf{a}.\mathbf{b})^2,$$

this becomes $\qquad H^2 = v^2\dot{\mathbf{g}}^2 + (\dot{\mathbf{r}} \times \mathbf{g})^2,$

and using (8.10) this gives

$$H^2 = (p^2 + v^2)\dot{\mathbf{g}}^2. \tag{8.12}$$

Hence (8.11) can be written

$$\mathbf{N} = \frac{p}{(p^2 + v^2)^{\frac{1}{2}}}\mathbf{a} + \frac{v}{(p^2 + v^2)^{\frac{1}{2}}}\mathbf{a} \times \mathbf{g}, \tag{8.13}$$

where $\mathbf{a}$ is the unit vector along $\dot{\mathbf{g}}$.

Let ϕ denote the angle between the directions of $\mathbf{N}$ at points on a generator distant v and 0 from the central point. Then, if $p \neq 0$,

$$\sin\phi = v/(p^2 + v^2)^{\frac{1}{2}},$$
$$\cos\phi = p/(p^2 + v^2)^{\frac{1}{2}},$$
$$\tan\phi = v/p. \tag{8.14}$$

Thus *the tangent of the angle through which the normal* $\mathbf{N}$ *rotates as the point* P *moves along a generator varies directly with the distance moved from the central point.* As v increases from $-\infty$ to ∞ the angle ϕ increases from $-\frac{1}{2}\pi$ to $\frac{1}{2}\pi$ if $p > 0$, and decreases from $\frac{1}{2}\pi$ to $-\frac{1}{2}\pi$ if $p < 0$. When the central point is reached the normal has rotated through an angle $\frac{1}{2}\pi$, and this fact justifies the use of the word 'central'.

Equations (8.7), (8.12) now give the simple formula

$$K = -p^2/(p^2 + v^2)^2 \tag{8.15}$$

for the Gaussian curvature at the point distant v from the central point on a generator of parameter p.

The geometrical significance of the parameter of distribution p may be obtained by calculating the length of the common perpendicular between successive generators. From the previous analysis it follows that the common perpendicular is parallel to the unit vector $\mathbf{g} \times \mathbf{a}$. Thus the shortest distance d is the projection of the arc element PQ on this common perpendicular. Writing s for the length of the arc PQ and choosing as parameter u the arc length of the base curve, we have by neglecting s^2

$$d = s\dot{\mathbf{r}}.(\dot{\mathbf{g}} \times \mathbf{a}) = s[\dot{\mathbf{r}}, \mathbf{g}, \dot{\mathbf{g}}]/|\dot{\mathbf{g}}| = sp|\dot{\mathbf{g}}|. \tag{8.16}$$

It follows that d is of the same order as the arc element s if $p \neq 0$.

This explains what might at first sight seem remarkable, that *although the line of striction appears as the locus of the feet of perpendiculars to consecutive generators it is not perpendicular to the generators themselves.*

This is in marked contrast to the properties of a developable surface. In this case the central point of a generator may be interpreted as the point of contact of the generator with the edge of regression, and the edge of regression itself becomes the line of striction. The surface normal $\mathbf{N}$ remains constant along all points of a generator, and the interpretation of central point in terms of the rotation of $\mathbf{N}$ no longer applies. Moreover, the formula for the perpendicular distance between consecutive tangents (generators) to a curve (edge of regression) found in I, Example 4.7, gives

$$d = \tfrac{1}{12}\kappa\tau s^3, \tag{8.17}$$

showing that when the ruled surface is a developable d involves s only to the third order.

EXERCISE 8.1. Show that the ruled surface generated by the binormals of a skew curve has the curve itself as line of striction.

EXERCISE 8.2. Show that the parameter of distribution of the ruled surface generated by the principal normals of a skew curve is equal to $\tau(\tau^2 + \kappa^2)^{-1}$, where κ and τ are the curvature and torsion of the curve.

9. The fundamental equations of surface theory

In this section relations are obtained between the fundamental coefficients E, F, G and L, M, N. These arise as conditions of integrability of certain partial differential equations, and they express the fact that the surfaces considered are embedded in three-dimensional Euclidean space.

It will be convenient to introduce at this stage notations used in tensor calculus, though the actual calculus of tensors will not be used in this chapter. Greek suffixes α, β, γ,... will take the values 1 or 2. The summation convention will be used, i.e. in any term a repeated suffix implies summation over the range of values of the suffix.

Since at each point on the surface the vectors $\mathbf{N}$, $\mathbf{r}_\alpha$ form a basis for the three-dimensional vector space, there exist uniquely defined coefficients $\Gamma^\delta_{\alpha\beta}$ and $\Omega_{\alpha\beta}$ such that

$$\mathbf{r}_{\alpha\beta} = \Gamma^\delta_{\alpha\beta}\,\mathbf{r}_\delta + \Omega_{\alpha\beta}\,\mathbf{N}. \qquad (9.1)$$

Evidently the symbols $\Gamma^\delta_{\alpha\beta}$ and $\Omega_{\alpha\beta}$ are symmetric in α and β. We now calculate explicit formulae for these coefficients. We multiply (9.1) scalarly by $\mathbf{N}$ to get

$$\Omega_{\alpha\beta} = \mathbf{N}.\mathbf{r}_{\alpha\beta}, \qquad (9.2)$$

from which it follows that $\Omega_{11} = L, \Omega_{12} = M, \Omega_{22} = N$. Hence the second fundamental form may be written as $\Omega_{\alpha\beta}\,du^\alpha\,du^\beta$, with $u^1 = u,\, u^2 = v$. We multiply (9.1) scalarly by $\mathbf{r}_\gamma$ to get

$$\mathbf{r}_\gamma.\mathbf{r}_{\alpha\beta} = \Gamma^\delta_{\alpha\beta}\,\mathbf{r}_\delta.\mathbf{r}_\gamma. \qquad (9.3)$$

We define new symbols $g_{\delta\gamma} = \mathbf{r}_\delta.\mathbf{r}_\gamma$. Evidently $g_{\delta\gamma}$ is symmetric in δ and γ, and $g_{11} = E,\, g_{12} = F,\, g_{22} = G$; thus the first fundamental form may be written $ds^2 = g_{\alpha\beta}\,du^\alpha\,du^\beta$. Since the matrix $(g_{\gamma\delta})$ is non-singular, we may define its reciprocal matrix $(g^{\alpha\beta})$ by the equations

$$g^{\alpha\beta}g_{\beta\gamma} = \delta^\alpha{}_\gamma,$$

where the symbol on the right is unity if $\alpha = \gamma$, and is zero otherwise.

Equation (9.3) may now be written

$$\mathbf{r}_\gamma.\mathbf{r}_{\alpha\beta} = g_{\gamma\delta}\,\Gamma^\delta_{\alpha\beta},$$

or more simply as $\qquad \mathbf{r}_\gamma.\mathbf{r}_{\alpha\beta} = \Gamma_{\gamma\alpha\beta}, \qquad (9.4)$

where the new symbols $\Gamma_{\gamma\alpha\beta}$ are defined by

$$\Gamma_{\gamma\alpha\beta} = g_{\gamma\delta}\,\Gamma^\delta_{\alpha\beta}. \qquad (9.5)$$

Multiplying (9.5) by the matrix $g^{\gamma\epsilon}$ gives the equation

$$\Gamma^\epsilon_{\alpha\beta} = g^{\gamma\epsilon}\Gamma_{\gamma\alpha\beta}, \qquad (9.6)$$

which expresses the old symbols in terms of the new. It follows that it will be sufficient to calculate the symbols $\Gamma_{\gamma\alpha\beta}$, and this is readily achieved as follows. Differentiate the relation $\mathbf{r}_\alpha.\mathbf{r}_\gamma = g_{\alpha\gamma}$ with respect to u^β to get

$$\mathbf{r}_\gamma.\mathbf{r}_{\alpha\beta} + \mathbf{r}_\alpha.\mathbf{r}_{\gamma\beta} = g_{\alpha\gamma.\beta}, \qquad (9.7)$$

which may be written

$$\Gamma_{\gamma\alpha\beta} + \Gamma_{\alpha\gamma\beta} = g_{\alpha\gamma.\beta}, \qquad (9.8)$$

where $\qquad\qquad g_{\alpha\gamma.\beta} = \partial g_{\alpha\gamma}/\partial u^\beta.$

Consider now the two further equations obtained from (9.8) by cyclically interchanging α, β, γ, given by

$$\Gamma_{\alpha\beta\gamma} + \Gamma_{\beta\alpha\gamma} = g_{\beta\alpha.\gamma}, \tag{9.9}$$

$$\Gamma_{\beta\gamma\alpha} + \Gamma_{\gamma\beta\alpha} = g_{\gamma\beta.\alpha}. \tag{9.10}$$

Adding equations (9.9), (9.10), subtracting equation (9.8), and using the symmetry of $\Gamma_{\alpha\beta\gamma}$ and $g_{\alpha\beta}$, we get

$$\Gamma_{\beta\alpha\gamma} = \tfrac{1}{2}\{g_{\alpha\beta.\gamma} + g_{\beta\gamma.\alpha} - g_{\alpha\gamma.\beta}\}. \tag{9.11}$$

Equation (9.6) now gives an explicit formula for the required coefficients. Equations (9.1) with these values for the coefficients are called *the equations of Gauss*.

Another set of equations, due to Weingarten, expresses the first derivatives of the surface normal in terms of the fundamental coefficients.

Since $\mathbf{N}^2 = 1$, $\mathbf{N}.\mathbf{N}_\alpha = 0$, and hence the vector $\mathbf{N}_\alpha$ lies in the tangent plane. Thus there must exist a matrix $(b^\beta{}_\alpha)$ such that

$$\mathbf{N}_\alpha = b^\beta{}_\alpha \mathbf{r}_\beta. \tag{9.12}$$

Multiply (9.12) scalarly by $\mathbf{r}_\gamma$ to obtain

$$\mathbf{r}_\gamma.\mathbf{N}_\alpha = b^\beta{}_\alpha g_{\beta\gamma}. \tag{9.13}$$

Since $\mathbf{r}_\gamma.\mathbf{N} = 0$, differentiating with respect to u^α gives

$$\mathbf{r}_{\gamma\alpha}.\mathbf{N} + \mathbf{r}_\gamma.\mathbf{N}_\alpha = 0,$$

so that $\qquad \mathbf{r}_\gamma.\mathbf{N}_\alpha = -\mathbf{r}_{\gamma\alpha}.\mathbf{N} = -\Omega_{\alpha\gamma}.$

Thus (9.13) becomes $\qquad -\Omega_{\alpha\gamma} = b^\beta{}_\alpha g_{\beta\gamma}. \tag{9.14}$

Multiplication of (9.14) by the reciprocal matrix $g^{\gamma\epsilon}$ leads to the result

$$b^\epsilon{}_\alpha = -g^{\gamma\epsilon}\Omega_{\alpha\gamma}.$$

Substitute in (9.12) to obtain *the equations of Weingarten*

$$\mathbf{N}_\alpha = -g^{\beta\gamma}\Omega_{\gamma\alpha}\mathbf{r}_\beta. \tag{9.15}$$

We now obtain conditions of integrability of the partial differential equations (9.1) which arise from the identities

$$\mathbf{r}_{\alpha\beta.\gamma} - \mathbf{r}_{\alpha\gamma.\beta} = 0. \tag{9.16}$$

Differentiate (9.1) with respect to u^γ to get

$$\mathbf{r}_{\alpha\beta.\gamma} = \Gamma^\delta_{\alpha\beta.\gamma}\mathbf{r}_\delta + \Gamma^\delta_{\alpha\beta}\mathbf{r}_{\delta\gamma} + \Omega_{\alpha\beta.\gamma}\mathbf{N} + \Omega_{\alpha\beta}\mathbf{N}_\gamma. \tag{9.17}$$

Substituting for $\mathbf{r}_{\delta\gamma}$ from (9.1) and $\mathbf{N}_\gamma$ from (9.15), the resulting equation is

$$\mathbf{r}_{\alpha\beta.\gamma} = \mathbf{r}_\mu\{\Gamma^\mu_{\alpha\beta.\gamma} + \Gamma^\delta_{\alpha\beta}\Gamma^\mu_{\delta\gamma} - g^{\lambda\mu}\Omega_{\alpha\beta}\Omega_{\gamma\lambda}\} + \mathbf{N}\{\Omega_{\alpha\beta.\gamma} + \Gamma^\delta_{\alpha\beta}\Omega_{\delta\gamma}\}.$$

Substitution in (9.16), and use of the fact that the vectors $\mathbf{r}_\mu$, $\mathbf{N}$ form a basis, gives

$$\Gamma^\mu_{\alpha\beta.\gamma} - \Gamma^\mu_{\alpha\gamma.\beta} + \Gamma^\delta_{\alpha\beta}\,\Gamma^\mu_{\delta\gamma} - \Gamma^\delta_{\alpha\gamma}\,\Gamma^\mu_{\delta\beta} = g^{\lambda\mu}(\Omega_{\alpha\beta}\Omega_{\gamma\lambda} - \Omega_{\alpha\gamma}\Omega_{\beta\lambda}),$$
$$(9.18)$$

together with

$$\Omega_{\alpha\beta.\gamma} - \Omega_{\alpha\gamma.\beta} = \Gamma^\delta_{\alpha\gamma}\Omega_{\delta\beta} - \Gamma^\delta_{\alpha\beta}\Omega_{\gamma\delta}. \qquad (9.19)$$

It is convenient to introduce the symbol $R^\mu{}_{\alpha\gamma\beta}$ for the left-hand member of (9.18), and a further symbol defined by

$$R_{\mu\alpha\beta\gamma} = g_{\mu\epsilon}\,R^\epsilon{}_{\alpha\beta\gamma}.$$

Multiplication of (9.18) by $g_{\epsilon\mu}$ then gives

$$R_{\alpha\epsilon\beta\gamma} = (\Omega_{\alpha\beta}\Omega_{\gamma\epsilon} - \Omega_{\alpha\gamma}\Omega_{\beta\epsilon}) \qquad (9.20)$$

which is known as *the equation of Gauss*.

Equations (9.19) are known as *the Mainardi–Codazzi equations*. From (9.20) it follows that $R_{1212} = LN - M^2$, and hence the Gaussian curvature $K = (LN - M^2)/(EG - F^2)$ is expressible in terms of E, F, G, and their derivatives. It follows that the Gaussian curvature is an intrinsic invariant. Gauss regarded this result as 'a most excellent theorem', but for us the result is not surprising since we defined Gaussian curvature intrinsically in Chapter II. It merely remains for us to verify that the Gaussian curvature defined in Chapter II is precisely that given in this section. Since K is an invariant, it is sufficient to prove the result in a special coordinate system, e.g. with geodesic polar coordinates.

Direct calculation of the left-hand member of (9.20) from the metric
$$ds^2 = du^2 + \lambda^2\,dv^2,$$

gives
$$R_{1212} = -\lambda\lambda_{11},$$

and so
$$R_{1212}/\lambda^2 = -\lambda_{11}/\lambda,$$

agreeing with formula (17.5) of Chapter II. This establishes the compatibility of the two definitions of Gaussian curvature.

The question arises whether the condition of integrability of the Weingarten equations (9.15) introduces further relations apart from (9.19) and (9.20). We now prove that this is not the case.

Differentiate (9.15) with respect to u^β, and use (9.1) to get

$$\mathbf{N}_{\alpha\beta} = -\mathbf{r}_\delta\{g^{\gamma\delta}{}_{.\beta}\Omega_{\alpha\gamma} + g^{\gamma\delta}\Omega_{\alpha\gamma.\beta} + g^{\gamma\epsilon}\Omega_{\alpha\gamma}\,\Gamma^\delta_{\beta\epsilon}\} - g^{\gamma\epsilon}\Omega_{\alpha\gamma}\Omega_{\beta\epsilon}\,\mathbf{N}.$$
$$(9.21)$$

The condition $N_{\alpha\beta}-N_{\beta\alpha}=0$ leads to the condition

$$g^{\epsilon\gamma}\Omega_{\alpha\gamma}\Omega_{\beta\epsilon}-g^{\epsilon\gamma}\Omega_{\beta\epsilon}\Omega_{\alpha\gamma}=0$$

which is satisfied identically, together with the condition that the expression

$$g^{\gamma\delta}{}_{.\beta}\Omega_{\alpha\gamma}+g^{\gamma\delta}\Omega_{\alpha\gamma.\beta}+g^{\gamma\epsilon}\Omega_{\alpha\gamma}\Gamma^{\delta}_{\beta\epsilon} \tag{9.22}$$

is symmetric in α and β.

We now appeal to the following result which is proved a little later.

LEMMA. *The functions $g^{\gamma\delta}$ satisfy the identity*

$$g^{\gamma\delta}{}_{.\beta}+\Gamma^{\gamma}_{\epsilon\beta}g^{\epsilon\delta}+\Gamma^{\delta}_{\epsilon\beta}g^{\gamma\epsilon}=0.$$

Substituting in (9.22) for $g^{\gamma\delta}{}_{.\beta}$ from this identity we get the condition that

$$-\Omega_{\alpha\gamma}g^{\epsilon\delta}\Gamma^{\gamma}_{\epsilon\beta}-\Omega_{\alpha\gamma}g^{\gamma\epsilon}\Gamma^{\delta}_{\epsilon\beta}+g^{\gamma\delta}\Omega_{\alpha\gamma.\beta}+g^{\gamma\epsilon}\Omega_{\alpha\gamma}\Gamma^{\delta}_{\beta\epsilon}$$

must be symmetric in α and β. The second and fourth terms cancel, and the condition reduces to the symmetry in α and β of the expression

$$g^{\gamma\delta}\Omega_{\alpha\gamma.\beta}-\Omega_{\alpha\gamma}g^{\epsilon\delta}\Gamma^{\gamma}_{\epsilon\beta}.$$

Multiplication by $g_{\delta\mu}$ gives the expression $\Omega_{\alpha\mu.\beta}-\Omega_{\alpha\gamma}\Gamma^{\gamma}_{\mu\beta}$, and the condition for symmetry of this is precisely the Mainardi–Codazzi equations (9.19).

To prove the lemma used above, differentiate with respect to u^{γ} the relation $g^{\alpha\beta}g_{\beta\delta}=\delta^{\alpha}_{\delta}$ to obtain

$$g^{\alpha\beta}{}_{.\gamma}g_{\beta\delta}+g^{\alpha\beta}g_{\beta\delta.\gamma}=0. \tag{9.23}$$

From the definition of $\Gamma_{\beta\gamma\delta}$ given in (9.11) it follows as an identity that

$$g_{\gamma\delta.\beta}=\Gamma_{\gamma\beta\delta}+\Gamma_{\delta\beta\gamma}, \tag{9.24}$$

Substitute in (9.23) and multiply by the reciprocal matrix to get

$$g^{\alpha\epsilon}{}_{.\gamma}=-g^{\alpha\beta}g^{\delta\epsilon}\{\Gamma_{\beta\gamma\delta}+\Gamma_{\delta\gamma\beta}\}=-g^{\delta\epsilon}\Gamma^{\alpha}_{\gamma\delta}-g^{\alpha\beta}\Gamma^{\epsilon}_{\gamma\beta},$$

which proves the lemma.

EXAMPLE 9.1. Show that in terms of E, F, G, L, M, N, the Weingarten equations are

$$H^2N_1=(FM-GL)\mathbf{r}_1+(FL-EM)\mathbf{r}_2,$$

$$H^2N_2=(FN-GM)\mathbf{r}_1+(FM-EN)\mathbf{r}_2,$$

and deduce that $H^2N_1\times N_2=(LN-M^2)\mathbf{N}$.

The first two equations may be deduced from (9.15) by merely changing the notation, but the following proof is more direct.

Since $N^2 = 1$, the vectors N_1, N_2 lie in the tangent plane and hence

$$N_1 = ar_1 + br_2,$$
$$N_2 = cr_1 + dr_2,$$

for some coefficients a, b, c, d.

Multiply both these equations in turn by r_1, r_2 to give

$$-L = aE + bF, \qquad -M = cE + dF,$$
$$-M = aF + bG, \qquad -N = cF + dG.$$

The coefficients a, b, c, d obtained from these four equations lead to the required equations.

To obtain the last equation, take the vector product of the first two equations to get

$$H^4 N_1 \times N_2$$
$$= \{(FM - GL)(FM - EN) - (FL - EM)(FN - GM)\} r_1 \times r_2,$$

which simplifies to the required equation.

EXERCISE 9.1. Show that when the lines of curvature are chosen as parametric curves, the Codazzi relations expressed in terms of E, G, L, N, and their derivatives are

$$L_2 = \tfrac{1}{2} E_2 \left(\frac{L}{E} + \frac{N}{G} \right),$$

$$N_1 = \tfrac{1}{2} G_1 \left(\frac{L}{E} + \frac{N}{G} \right).$$

Show also that the equation of Gauss may be written

$$\frac{LN}{\sqrt{(EG)}} + \frac{\partial}{\partial u} \left(\frac{1}{\sqrt{E}} \frac{\partial \sqrt{G}}{\partial u} \right) + \frac{\partial}{\partial v} \left(\frac{1}{\sqrt{G}} \frac{\partial \sqrt{E}}{\partial v} \right) = 0.$$

10. Parallel surfaces

A surface $\bar{S}$ whose points are at a constant distance along the normal from another surface S is said to be *parallel to* S. For example, two concentric spheres are parallel surfaces. If r is the position vector of a point P on S, the corresponding point $\bar{P}$ on $\bar{S}$ has position vector $\bar{r}$ where

$$\bar{r} = r - aN \tag{10.1}$$

and a is a constant, positive or negative.

Choose lines of curvature of S as parametric curves. Then we have

$$\left.\begin{array}{l} \bar{\mathbf{r}}_1 = \mathbf{r}_1 - a\mathbf{N}_1 \\ \bar{\mathbf{r}}_2 = \mathbf{r}_2 - a\mathbf{N}_2 \end{array}\right\}. \tag{10.2}$$

Substitute for $\mathbf{N}_1$ and $\mathbf{N}_2$ from the Weingarten equations in Example 9.1 with $F = M = 0$ to get

$$\left.\begin{array}{l} \bar{\mathbf{r}}_1 = \mathbf{r}_1(1+aL/E) = \mathbf{r}_1(1+a\kappa_a) \\ \bar{\mathbf{r}}_2 = \mathbf{r}_2(1+aN/G) = \mathbf{r}_2(1+a\kappa_b) \end{array}\right\}, \tag{10.3}$$

using (3.4).

Denote by $\bar{E}$, $\bar{F}$, $\bar{G}$, $\bar{L}$, $\bar{M}$, $\bar{N}$, the fundamental coefficients of $\bar{S}$. Then

$$\bar{E} = E(1+a\kappa_a)^2, \qquad \bar{F} = 0, \qquad \bar{G} = G(1+a\kappa_b)^2. \tag{10.4}$$

From (10.3),

$$\bar{H}\mathbf{N} = \bar{\mathbf{r}}_1 \times \bar{\mathbf{r}}_2 = H\{1+a(\kappa_a+\kappa_b)+a^2\kappa_a\kappa_b\}\mathbf{N}$$

$$= H(1+2\mu a+Ka^2)\mathbf{N}, \tag{10.5}$$

where μ and K are respectively the mean curvature and Gaussian curvature of S.

The expression in the bracket is $(1+a\kappa_a)(1+a\kappa_b)$ and this will vanish if and only if $a = -\kappa_a^{-1}$ or $-\kappa_b^{-1}$. We can exclude this possibility by assuming that a does not take these particular values, and we can denote by $e\,(=\pm1)$ the sign of the expression. Equation (10.5) then gives

$$\bar{\mathbf{N}} = e\mathbf{N}, \tag{10.6}$$

showing that although S and $\bar{S}$ have the same normal lines, they may be directed in opposite directions.

The coefficients of the second fundamental form of $\bar{S}$ are

$$\bar{L} = e(1+a\kappa_a)L, \qquad \bar{M} = 0, \qquad \bar{N} = e(1+a\kappa_b)N. \tag{10.7}$$

Since $\bar{F} = \bar{M} = 0$, it follows that the curves on $\bar{S}$ corresponding to lines of curvature of S are lines of curvature on $\bar{S}$.

We now calculate the Gaussian curvature $\bar{K}$ and mean curvature $\bar{\mu}$ of $\bar{S}$ in terms of the corresponding values of K and μ. We have

$$\bar{K} = \frac{\bar{L}\bar{N}}{\bar{H}^2} = \frac{LN\{1+a(\kappa_a+\kappa_b)+a^2\kappa_a\kappa_b\}}{H^2\{1+2\mu a+Ka^2\}^2} = \frac{K}{(1+2\mu a+Ka^2)}. \tag{10.8}$$

Also $\quad 2\bar\mu = \bar L/\bar E + \bar N/\bar G = e\left\{\dfrac{L(1+a\kappa_a)}{E(1+a\kappa_a)^2} + \dfrac{N(1+a\kappa_b)}{G(1+a\kappa_b)^2}\right\}$

$$= e\left\{\frac{\kappa_a}{1+a\kappa_a} + \frac{\kappa_b}{1+a\kappa_b}\right\}$$

$$= e\left\{\frac{2\mu+2Ka}{1+2\mu a+Ka^2}\right\},$$

so we have $\qquad\qquad \bar\mu = e\left(\dfrac{\mu+Ka}{1+2\mu a+Ka^2}\right).$ \hfill (10.9)

When S has constant positive Gaussian curvature equal to A^{-2} and we take $a = \pm A$, we obtain for the mean curvature of $\bar S$

$$\bar\mu = \pm e/2A.$$

When S has constant mean curvature equal to $(2A)^{-1}$ and we take $a = -A$, we obtain for the Gaussian curvature of $\bar S$

$$\bar K = A^{-2}.$$

We thus have the following theorem due to Bonnet.

THEOREM 10.1. *In general, with every surface of constant positive Gaussian curvature A^{-2} there are associated two surfaces of constant mean curvature $\pm(2A)^{-1}$ which are parallel to the former and distant $\pm A$ from it.*

With every surface of constant mean curvature $(2A)^{-1}$ there is a parallel surface of constant Gaussian curvature at a distance A from it.

Our proof is not yet complete since we have excluded from our analysis the possibility $a = -\kappa_a^{-1}$ or $a = -\kappa_b^{-1}$. This case occurs certainly when S is a sphere of radius A, for then one parallel surface of constant mean curvature is the concentric sphere of radius $2A$ while the other surface degenerates into a single point, the centre. It is always possible to prevent $\bar S$ from having singularities by choosing a to have the same sign as that of the principal curvatures. If this be done, then there will *always* be *one* parallel surface of constant mean curvature while the second surface will necessarily have singularities at the umbilics of the given surface. We can thus remove the words 'in general' from the hypothesis of the first part of the theorem by asserting that *with every surface* of constant positive Gaussian curvature there exists at least one (*non-singular*) *parallel surface with constant mean curvature.*

11. Fundamental existence theorem for surfaces

In this section we shall prove the following fundamental existence theorem first proved by Bonnet in 1867.

When the coefficients of the two quadratic differential forms

$$E\,du^2+2F\,dudv+G\,dv^2, \qquad L\,du^2+2M\,dudv+N\,dv^2$$

are such that the first form is positive definite and the six coefficients satisfy the Codazzi equations and the equation of Gauss, then there exists a surface, uniquely determined to within a Euclidean displacement, for which these forms are respectively the first and second fundamental forms.

We shall first prove the theorem in the particular case when $F = 0$, $M = 0$ and deduce the more general theorem from this.

It is convenient to deal first with the uniqueness part of theorem. We do this by showing that if a surface exists having the prescribed fundamental forms, then at each point the three mutually orthogonal unit vectors consisting of the two unit tangent vectors to the parametric curves and the unit vector normal to the surface must satisfy a certain set of linear differential equations. A uniqueness theorem from the theory of differential equations shows that there is at most one solution for which these vectors assume prescribed values for initial values of the parameters. If two surfaces existed with the prescribed fundamental forms, then by a Euclidean motion we could arrange that the triads of vectors coincide at some initial point. The previous uniqueness theorem then implies that the triads of vectors coincide at all points, and the two surfaces therefore differ at most by a Euclidean motion.

Suppose that a surface exists having

$$E\,du^2+G\,dv^2, \qquad L\,du^2+N\,dv^2$$

as fundamental differential forms, and let its equation referred to a system of rectangular Cartesian coordinates be $\mathbf{r} = \mathbf{r}(u,v)$. Denote by $\mathbf{l}$, $\boldsymbol{\lambda}$ the unit tangent vectors along the parametric curves $v = $ constant, $u = $ constant. Then

$$\mathbf{l} = \frac{1}{\sqrt{E}}\mathbf{r_1}, \qquad \boldsymbol{\lambda} = \frac{1}{\sqrt{G}}\mathbf{r_2}. \tag{11.1}$$

The equations of Gauss are

$$\left.\begin{aligned}
\mathbf{r}_{11} &= \frac{E_1}{2E}\mathbf{r}_1 - \frac{E_2}{2G}\mathbf{r}_2 + L\mathbf{N} \\[2mm]
\mathbf{r}_{12} &= \frac{E_2}{2E}\mathbf{r}_1 + \frac{G_1}{2G}\mathbf{r}_2 \\[2mm]
\mathbf{r}_{22} &= -\frac{G_1}{2E}\mathbf{r}_1 + \frac{G_2}{2G}\mathbf{r}_2 + N\mathbf{N}
\end{aligned}\right\}, \qquad (11.2)$$

and the equations of Weingarten are

$$\left.\begin{aligned}
\mathbf{N}_1 &= -\frac{L}{E}\mathbf{r}_1 \\[2mm]
\mathbf{N}_2 &= -\frac{N}{G}\mathbf{r}_2
\end{aligned}\right\}. \qquad (11.3)$$

Differentiate equations (11.1) and use (11.2), (11.3) to obtain

$$\left.\begin{aligned}
\frac{\partial \mathbf{l}}{\partial u} &= -\frac{1}{\sqrt{G}}\frac{\partial \sqrt{E}}{\partial v}\boldsymbol{\lambda} + \frac{L}{\sqrt{E}}\mathbf{N}, & \frac{\partial \mathbf{l}}{\partial v} &= \frac{1}{\sqrt{E}}\frac{\partial \sqrt{G}}{\partial u}\boldsymbol{\lambda} \\[2mm]
\frac{\partial \boldsymbol{\lambda}}{\partial u} &= \frac{1}{\sqrt{G}}\frac{\partial \sqrt{E}}{\partial v}\mathbf{l}, & \frac{\partial \boldsymbol{\lambda}}{\partial v} &= -\frac{1}{\sqrt{E}}\frac{\partial \sqrt{G}}{\partial u}\mathbf{l} + \frac{N}{\sqrt{G}}\mathbf{N} \\[2mm]
\frac{\partial \mathbf{N}}{\partial u} &= -\frac{L}{\sqrt{E}}\mathbf{l}, & \frac{\partial \mathbf{N}}{\partial v} &= -\frac{N}{\sqrt{G}}\boldsymbol{\lambda}
\end{aligned}\right\} \ (11.4)$$

These differential equations are necessary conditions to be satisfied by $\mathbf{l}, \boldsymbol{\lambda}, \mathbf{N}$ in order that there shall exist a surface with the prescribed fundamental forms.

Write $\mathbf{l} = (l_x, l_y, l_z)$, $\boldsymbol{\lambda} = (\lambda_x, \lambda_y, \lambda_z)$, $\mathbf{N} = (N_x, N_y, N_z)$.

Then we have three sets of solutions (l_x, λ_x, N_x), (l_y, λ_y, N_y), (l_z, λ_z, N_z) for the dependent variables (l^*, λ^*, N^*) which satisfy the equations

$$\left.\begin{aligned}
\frac{\partial l^*}{\partial u} &= -\frac{1}{\sqrt{G}}\frac{\partial \sqrt{E}}{\partial v}\lambda^* + \frac{L}{\sqrt{E}}N^*, & \frac{\partial l^*}{\partial v} &= \frac{1}{\sqrt{E}}\frac{\partial \sqrt{G}}{\partial u}\lambda^* \\[2mm]
\frac{\partial \lambda^*}{\partial u} &= \frac{1}{\sqrt{G}}\frac{\partial \sqrt{E}}{\partial v}l^*, & \frac{\partial \lambda^*}{\partial v} &= -\frac{1}{\sqrt{E}}\frac{\partial \sqrt{G}}{\partial u}l^* + \frac{N}{\sqrt{G}}N^* \\[2mm]
\frac{\partial N^*}{\partial u} &= -\frac{L}{\sqrt{E}}l^*, & \frac{\partial N^*}{\partial v} &= -\frac{N}{\sqrt{G}}\lambda^*
\end{aligned}\right\}.$$

$$(11.4')$$

It is well known† that equations of the form (11.4′) admit at most one solution (l^*, λ^*, N^*) which assumes prescribed values when $u = u_0$, $v = v_0$.

† See, for example, A. R. Forsyth (1903), § 173.

Now, if there were two surfaces S, S' with the prescribed fundamental forms, then by a Euclidean motion we could arrange that the triads $(\mathbf{l}, \boldsymbol{\lambda}, \mathbf{N})$, $(\mathbf{l}', \boldsymbol{\lambda}', \mathbf{N}')$ coincide when $u = u_0$, $v = v_0$. The above uniqueness theorem shows that $(\mathbf{l}, \boldsymbol{\lambda}, \mathbf{N})$, $(\mathbf{l}', \boldsymbol{\lambda}', \mathbf{N}')$ must then coincide for all values of u and v, and hence the surfaces S, S' differ by at most a Euclidean motion. This completes the uniqueness part of the theorem.

We now prove the existence of a surface having the prescribed fundamental forms. We are given functions E, G, L, N, of which E, G are positive, with the property that they satisfy the equation of Gauss and the Codazzi equations. As seen from Exercise 9.1, these equations are

$$\frac{LN}{\surd(EG)} + \frac{\partial}{\partial u}\left(\frac{1}{\sqrt{E}}\frac{\partial \sqrt{G}}{\partial u}\right) + \frac{\partial}{\partial v}\left(\frac{1}{\sqrt{G}}\frac{\partial \sqrt{E}}{\partial v}\right) = 0, \qquad (11.5)$$

$$\left. \begin{array}{l} \dfrac{\partial}{\partial v}\left(\dfrac{L}{\sqrt{E}}\right) - \dfrac{N}{G}\dfrac{\partial \sqrt{E}}{\partial v} = 0 \\[3mm] \dfrac{\partial}{\partial u}\left(\dfrac{N}{\sqrt{G}}\right) - \dfrac{L}{E}\dfrac{\partial \sqrt{G}}{\partial u} = 0 \end{array} \right\}. \qquad (11.6)$$

Consider now the set of equations (11.4′). It is readily verified that these equations are completely integrable because of conditions (11.5) and (11.6). For example, from (11.4′) we have

$$\frac{\partial}{\partial v}\left(\frac{\partial l^*}{\partial u}\right) = -\frac{1}{\sqrt{G}}\frac{\partial \sqrt{E}}{\partial v}\frac{\partial \lambda^*}{\partial v} - \lambda^*\frac{\partial}{\partial v}\left(\frac{1}{\sqrt{G}}\frac{\partial \sqrt{E}}{\partial v}\right) + \frac{L}{\sqrt{E}}\frac{\partial N^*}{\partial v} + N^*\frac{\partial}{\partial v}\left(\frac{L}{\sqrt{E}}\right)$$

$$= -\frac{1}{\sqrt{G}}\frac{\partial \sqrt{E}}{\partial v}\left\{-\frac{1}{\sqrt{E}}\frac{\partial \sqrt{G}}{\partial u}l^* + \frac{N}{\sqrt{G}}N^*\right\} -$$

$$-\lambda^*\frac{\partial}{\partial v}\left(\frac{1}{\sqrt{G}}\frac{\partial \sqrt{E}}{\partial v}\right) + \frac{L}{\sqrt{E}}\left\{-\frac{N}{\sqrt{G}}\lambda^*\right\} + N^*\frac{\partial}{\partial v}\left(\frac{L}{\sqrt{E}}\right).$$

Also

$$\frac{\partial}{\partial u}\left(\frac{\partial l^*}{\partial v}\right) = \frac{1}{\sqrt{E}}\frac{\partial \sqrt{G}}{\partial u}\frac{\partial \lambda^*}{\partial u} + \lambda^*\frac{\partial}{\partial u}\left(\frac{1}{\sqrt{E}}\frac{\partial \sqrt{G}}{\partial u}\right)$$

$$= \frac{1}{\sqrt{E}}\frac{\partial \sqrt{G}}{\partial u}\frac{1}{\sqrt{G}}\frac{\partial \sqrt{E}}{\partial v}l^* + \lambda^*\frac{\partial}{\partial u}\left(\frac{1}{\sqrt{E}}\frac{\partial \sqrt{G}}{\partial u}\right).$$

Hence

$$\frac{\partial}{\partial v}\left(\frac{\partial l^*}{\partial u}\right) - \frac{\partial}{\partial u}\left(\frac{\partial l^*}{\partial v}\right) = N^*\left\{\frac{\partial}{\partial v}\left(\frac{L}{\sqrt{E}}\right) - \frac{N}{G}\frac{\partial \sqrt{E}}{\partial v}\right\} -$$

$$-\lambda^*\left\{\frac{LN}{\surd(EG)} + \frac{\partial}{\partial u}\left(\frac{1}{\sqrt{E}}\frac{\partial \sqrt{G}}{\partial u}\right) + \frac{\partial}{\partial v}\left(\frac{1}{\sqrt{G}}\frac{\partial \sqrt{E}}{\partial v}\right)\right\} = 0$$

from (11.5) and (11.6).

Similarly we may prove

$$\frac{\partial}{\partial v}\left(\frac{\partial \lambda^*}{\partial u}\right) - \frac{\partial}{\partial u}\left(\frac{\partial \lambda^*}{\partial v}\right) = 0,$$

and

$$\frac{\partial}{\partial v}\left(\frac{\partial N^*}{\partial u}\right) - \frac{\partial}{\partial u}\left(\frac{\partial N^*}{\partial v}\right) = 0.$$

It follows that† the complete system (11.4') admits three particular sets of solutions (l_x, λ_x, N_x), (l_y, λ_y, N_y), (l_z, λ_z, N_z), which for initial values $u = u_0$, $v = v_0$ assume the values $(1, 0, 0)$, $(0, 1, 0)$, and $(0, 0, 1)$. The vectors (l_x, λ_x, N_x), (l_y, λ_y, N_y), (l_z, λ_z, N_z) are thus three mutually orthogonal unit vectors when $u = u_0$, $v = v_0$. We shall now prove that this is also the case for all values of u and v.

From equations (11.4') we have

$$l_x dl_x + \lambda_x d\lambda_x + N_x dN_x$$

$$= l_x\left[\left\{-\frac{1}{\sqrt{G}}\frac{\partial\sqrt{E}}{\partial v}\lambda_x + \frac{L}{\sqrt{E}}N_x\right\}du + \frac{1}{\sqrt{E}}\frac{\partial\sqrt{G}}{\partial u}\lambda_x dv\right] +$$

$$+ \lambda_x\left[\frac{1}{\sqrt{G}}\frac{\partial\sqrt{E}}{\partial v}l_x du + \left\{\frac{NN_x}{\sqrt{G}} - \frac{1}{\sqrt{E}}\frac{\partial\sqrt{G}}{\partial u}l_x\right\}dv\right] +$$

$$+ N_x\left[-\frac{L}{\sqrt{E}}l_x du - \frac{N}{\sqrt{G}}\lambda_x dv\right] = 0.$$

Hence $(l_x^2 + \lambda_x^2 + N_x^2)$ has a constant value for all values of u and v, and from the conditions when $u = u_0$, $v = v_0$ it follows that this constant value is 1.

Similarly, $l_y dl_y + \lambda_y d\lambda_y + N_y dN_y = 0$ and $l_z dl_z + \lambda_z d\lambda_z + N_z dN_z = 0$
so that

$$\left. \begin{aligned} l_x^2 + \lambda_x^2 + N_x^2 &= 1 \\ l_y^2 + \lambda_y^2 + N_y^2 &= 1 \\ l_z^2 + \lambda_z^2 + N_z^2 &= 1 \end{aligned} \right\}. \tag{11.7}$$

In a similar way we may prove that

$$l_x dl_y + l_y dl_x + \lambda_x d\lambda_y + \lambda_y d\lambda_x + N_x dN_y + N_y dN_x = 0,$$

so that $(l_x l_y + \lambda_x \lambda_y + N_x N_y)$ takes the constant value 0 for all values of u and v. Hence

$$\left. \begin{aligned} l_x l_y + \lambda_x \lambda_y + N_x N_y &= 0 \\ l_y l_z + \lambda_y \lambda_z + N_y N_z &= 0 \\ l_z l_x + \lambda_z \lambda_x + N_z N_x &= 0 \end{aligned} \right\}. \tag{11.8}$$

† A. R. Forsyth, loc. cit.

The six conditions (11.7), (11.8) are together equivalent to the condition that

$$\begin{pmatrix} l_x & \lambda_x & N_x \\ l_y & \lambda_y & N_y \\ l_z & \lambda_z & N_z \end{pmatrix}$$

is an orthogonal matrix. It follows that the three vectors $\mathbf{l}$, $\boldsymbol{\lambda}$, $\mathbf{N}$ with components (l_x, l_y, l_z), $(\lambda_x, \lambda_y, \lambda_z)$, (N_x, N_y, N_z) are mutually orthogonal unit vectors for all values of u and v.

From equations (11.4) we have

$$\frac{\partial}{\partial u}(\sqrt{G}\boldsymbol{\lambda}) - \frac{\partial}{\partial v}(\sqrt{E}\mathbf{l}) = \boldsymbol{\lambda}\frac{\partial\sqrt{G}}{\partial u} - \mathbf{l}\frac{\partial\sqrt{E}}{\partial v} + \sqrt{G}\frac{\partial\boldsymbol{\lambda}}{\partial u} - \sqrt{E}\frac{\partial\mathbf{l}}{\partial v}$$

$$= \boldsymbol{\lambda}\frac{\partial\sqrt{G}}{\partial u} - \mathbf{l}\frac{\partial\sqrt{E}}{\partial v} + \mathbf{l}\frac{\partial\sqrt{E}}{\partial v} - \boldsymbol{\lambda}\frac{\partial\sqrt{G}}{\partial u} = \mathbf{0}.$$

It follows that the set of Pfaffian forms represented by

$$\sqrt{E}\,\mathbf{l}\,du + \sqrt{G}\,\boldsymbol{\lambda}\,dv$$

are exact differentials.

Hence the equation

$$\mathbf{r} = \int (\sqrt{E}\,\mathbf{l}\,du + \sqrt{G}\,\boldsymbol{\lambda}\,dv) \tag{11.9}$$

defines a vector $\mathbf{r} = \mathbf{r}(u, v)$.

We assert that the surface defined by $\mathbf{r} = \mathbf{r}(u, v)$ *satisfies the requirements.*

Since

$$\mathbf{r}_1 = \sqrt{E}\,\mathbf{l}, \qquad \mathbf{r}_2 = \sqrt{G}\,\boldsymbol{\lambda},$$

and $\mathbf{l}$, $\boldsymbol{\lambda}$ are orthogonal unit vectors, we have

$$\mathbf{r}_1^2 = E, \qquad \mathbf{r}_1.\mathbf{r}_2 = 0, \qquad \mathbf{r}_2^2 = G$$

as required.

Moreover,

$$\mathbf{r}_{11} = \mathbf{l}\frac{\partial\sqrt{E}}{\partial u} + \sqrt{E}\frac{\partial\mathbf{l}}{\partial u} = \mathbf{l}\frac{\partial\sqrt{E}}{\partial u} - \frac{\sqrt{E}}{\sqrt{G}}\frac{\partial\sqrt{E}}{\partial v}\boldsymbol{\lambda} + L\mathbf{N},$$

from (11.4).

Similarly,

$$\mathbf{r}_{12} = \mathbf{l}\frac{\partial\sqrt{E}}{\partial v} + \sqrt{E}\frac{\partial\mathbf{l}}{\partial v} = \mathbf{l}\frac{\partial\sqrt{E}}{\partial v} + \boldsymbol{\lambda}\frac{\partial\sqrt{G}}{\partial u},$$

and

$$\mathbf{r}_{22} = \boldsymbol{\lambda}\frac{\partial\sqrt{G}}{\partial v} + \sqrt{G}\frac{\partial\boldsymbol{\lambda}}{\partial v} = \boldsymbol{\lambda}\frac{\partial\sqrt{G}}{\partial v} - \frac{\sqrt{G}}{\sqrt{E}}\frac{\partial\sqrt{G}}{\partial u}\mathbf{l} + N\mathbf{N}.$$

Hence

$$\mathbf{r}_{11}.\mathbf{N} = L, \qquad \mathbf{r}_{12}.\mathbf{N} = 0, \qquad \mathbf{r}_{22}.\mathbf{N} = N.$$

Thus the surface whose position vector is defined by (11.9) has

$$E\,du^2 + G\,dv^2,$$

$$L\,du^2 + N\,dv^2,$$

as fundamental forms, as required. This completes the proof of the theorem for the particular case when $F = 0$, $M = 0$.

We now proceed to prove the general theorem when we are given six functions E^*, F^*, G^*, L^*, M^*, N^*, of variables u^*, v^* which satisfy the Codazzi relations and the equation of Gauss. Consider the differential equation

$$(E^*M^* - F^*L^*)\,du^{*2} + (E^*N^* - G^*L^*)\,du^*dv^* +$$
$$+ (F^*N^* - G^*M^*)\,dv^{*2} = 0. \quad (11.10)$$

This equation yields solutions

$$\phi(u^*, v^*) = \text{constant},$$

$$\psi(u^*, v^*) = \text{constant},$$

and, on making the proper parameter transformation

$$\phi(u^*, v^*) = u,$$

$$\psi(u^*, v^*) = v,$$

we can certainly find a domain of the new parameters u, v where the given fundamental forms assume the special form with $F = 0$, $M = 0$. The previous argument shows the existence of a surface possessing these forms as fundamental differential forms, and a change back to the original parameters gives the existence of a surface with the prescribed fundamental forms. This completes the proof of the fundamental existence theorem except for consideration of the very special case when the coefficients of (11.10) vanish simultaneously for all u^*, v^*. It will be seen later (p. 129) that in this case the surface is necessarily locally plane or spherical.

REFERENCE

FORSYTH, A. R., *Treatise on Differential Equations*, Macmillan 1903 (3rd edition).

MISCELLANEOUS EXERCISES III

1. Show that if L, M, N vanish everywhere on a surface, then the surface is part of a plane.

2. Show that the meridians and parallels of a surface of revolution are its lines of curvature.

3. Show that the principal radii of curvature of the surface

$$y \cos(z/a) = x \sin(z/a)$$

are equal to $\pm(x^2+y^2+a^2)/a$. Find the lines of curvature.

4. Find the asymptotic lines on the surface

$$x = a(1+\cos u)\cot v, \qquad y = a(1+\cos u), \qquad z = a\cos u \operatorname{cosec} v.$$

5. If $\kappa_1, \kappa_2, ..., \kappa_m$ denote the normal curvatures of m sections of a surface which make equal angles $2\pi/m$ with one another, prove that, if $m > 2$,

$$\kappa_1 + \kappa_2 + ... + \kappa_m = m\mu.$$

6. The surface of revolution given by

$$x = u\cos v, \qquad y = u\sin v, \qquad z = a\log\{u+\sqrt{(u^2-a^2)}\}$$

is generated by rotating a catenary about its axis. Prove that it is a minimal surface. Show also that it is the only minimal surface of revolution.

7. Prove that a surface for which E, F, G, L, M, N are non-zero constants, is necessarily a circular cylinder.

8. Prove that, except for the plane, the right helicoid is the only ruled surface which is minimal.

9. Show that the surface $e^z \cos x = \cos y$ is minimal.

10. Show that all straight lines on a surface are asymptotic lines, and that along a curved asymptotic line the osculating plane coincides with the tangent plane to the surface.

11. A developable surface D is enveloped by the tangent planes to a surface S at points on a curve C lying on S. Prove that at any point P on C the generating line of D is in a direction conjugate to the direction of C at P.

12. Show that a hyperboloid of revolution of one sheet is a ruled surface whose line of striction is the minimal circle and whose parameter of distribution is constant.

13. Prove that the Gaussian curvature is the same at two points of a generator which are equidistant from the central point.

14. If a curve on a ruled surface satisfies any two of the three conditions (i) of being a geodesic, (ii) of being the line of striction, (iii) of intersecting the generators at a constant angle, show that the remaining condition is also satisfied.

15. Find the umbilics of the ellipsoid

$$\frac{x^2}{a^2}+\frac{y^2}{b^2}+\frac{z^2}{c^2} = 1$$

and prove that the tangent planes at these points are parallel to the circular sections of the ellipsoid.

16. Two surfaces intersect each other along a curve at a constant angle. If the curve is a line of curvature on one surface show that it is a line of curvature on the other. Is the converse theorem true?

17. The third fundamental form of a surface, denoted by III, is defined by

$$\text{III} = d\mathbf{N}.d\mathbf{N}.$$

If the first and second fundamental forms are denoted by I and II respectively, prove that the three fundamental forms are related by the identity

$$K\text{I} - 2\mu\text{II} + \text{III} = 0.$$

18. At a point on a surface where the Gaussian curvature is negative and equal to K, show that the torsion of the asymptotic lines is $\pm\sqrt{(-K)}$.

IV

DIFFERENTIAL GEOMETRY OF SURFACES IN THE LARGE

1. Introduction

THE previous chapters were concerned with the properties of a region of a surface defined by suitably restricting the parameters u and v. These are essentially local properties, the word *local* indicating that in order to obtain the property at a point P it is necessary to have information about the surface only in the neighbourhood of P. In the present chapter we shall be concerned with properties involving the surface as a whole; for example whether like a spherical cap it has a boundary or whether it is compact like a sphere. Differential geometry of surfaces in the large is the study of relations between the local and global properties of surfaces.

This chapter is included in a book of this nature in order to give the student at an early stage some idea of the types of problems which arise in global theory. To treat these problems completely in full detail would require a knowledge of topology which the student probably does not possess. No attempt at completeness has been made, some of the theorems are stated without proof, while in section 8 a proof of one theorem is merely outlined. It may be argued that to deal with the subject at this stage is premature and that one should wait until the student is sufficiently equipped topologically to tackle these problems with due regard to rigour. The author does not share this viewpoint because he suggests that an early intuitive treatment which enables one to grasp the essential features of a subject is a valuable introduction to a more critical and detailed study.

This chapter is independent of the rest of the book, and it can be omitted without prejudice to understanding the content of Part 2. Students wishing to read this chapter should be familiar with the notions of compactness and connectedness in metric spaces, Hausdorff space, homeomorphism, covering space, and Euler characteristic. A suitable introductory account of the relevant ideas is contained, for example, in the books by E. M. Patterson (1956) and S. Wylie and P. Hilton (1960). In particular, frequent use will be made of the theorem that a real-valued continuous function defined on a compact space attains its extreme values.

In sections 3, 4, and 5 attention will be restricted to those surfaces which are defined as embedded in three-dimensional Euclidean space, and in this case many results can be obtained by applying the methods discussed in Chapters II and III. In particular, the second fundamental form is then available, together with the equations of Gauss and Codazzi. Later in the chapter a surface is defined *in abstracto*, without any reference to a containing space.

2. Compact surfaces whose points are umbilics

For the first few theorems of this chapter we shall use the definition of surface given in Chapter II, and assume that each point has a neighbourhood (homeomorphic to an open 2-cell) which can be described by parametric equations $\mathbf{r} = \mathbf{r}(u, v)$.

As our first theorem of differential geometry in the large we shall prove

THEOREM 2.1. *The only compact surfaces of class $\geqslant 2$ for which every point is an umbilic are spheres.*

This is an example of a global theorem since part of the hypothesis—viz. the compactness of the surface considered as a set of points in E_3—evidently involves the surface as a whole. A useful technique in proving global theorems in differential geometry is first to establish the result locally in some neighbourhood of an arbitrary point, and then to try to extend the result so that it applies globally. We employ precisely this technique in proving Theorem 2.1. By means of the local differential geometry developed in Chapters II and III we shall prove that in the neighbourhood of any point the surface is either spherical or plane. We then use the property of compactness to reject one alternative, and show that the surface must in fact be a sphere.

Let S be a compact surface of class $\geqslant 2$ for which every point is an umbilic. Let P be any point on S, and let V be a coordinate neighbourhood of S containing P, in which part of S is represented parametrically by $\mathbf{r} = \mathbf{r}(u, v)$.

Since every point of V is an umbilic, it follows that every curve lying in V must be a line of curvature. Hence, from Rodrigues' formula, at all points of V,

$$d\mathbf{N} + \kappa \, d\mathbf{r} = 0, \tag{2.1}$$

where κ is the normal curvature of S in the direction $d\mathbf{r}$. From (2.1) we get
$$\mathbf{N}_1 = -\kappa\mathbf{r}_1, \qquad \mathbf{N}_2 = -\kappa\mathbf{r}_2,$$
which with the identity $\quad \mathbf{N}_{12} = \mathbf{N}_{21},$

gives $\qquad\qquad \kappa_2\mathbf{r}_1 - \kappa_1\mathbf{r}_2 = \mathbf{0}.$

Since $\mathbf{r}_1$, $\mathbf{r}_2$ are linearly independent we obtain $\kappa_1 = \kappa_2 = 0$, so that κ is constant. Integration of (2.1) gives, for $\kappa \neq 0$,
$$\mathbf{r} = \mathbf{a} - \kappa^{-1}\mathbf{N}, \tag{2.2}$$
showing that V lies on the surface of a sphere of centre $\mathbf{a}$ and radius κ^{-1}. When $\kappa = 0$, (2.1) gives
$$\mathbf{N} = \mathbf{b}, \tag{2.3}$$
showing that V lies on a plane. This completes the local part of the theorem—so far all we have proved is that in the neighbourhood of any point the surface is spherical or plane.

Associate with each point P on the surface a neighbourhood V, having the above property. The set of all neighbourhoods V_P covers S, and from the compactness we deduce that S is covered by a finite sub-cover formed by V_i, $i = 1, 2, ..., N$. Consider two overlapping neighbourhoods V_i, V_j. From the previous local argument it follows that κ is constant in V_i and also in V_j. By considering the value of κ at points in $V_i \cap V_j$ it follows that κ takes the same value over the whole of the surface. Moreover, this value cannot be zero, otherwise the surface would contain a straight line and would not be compact. Hence the surface must be a sphere, and the theorem is proved.

3. Hilbert's lemma

We shall make use of the following lemma due to Hilbert.

In a closed region R of a surface of constant positive Gaussian curvature without umbilics, the principal curvatures take their extreme values at the boundary.

This lemma is purely local in character and we shall use results of Chapter III to prove it. We prefer to restate the lemma in a slightly different form suggested by W. F. Newns.

If at a point P_0 of any surface, the principal curvatures κ_a, κ_b are such that either (i) $\kappa_a > \kappa_b$, κ_a *has a maximum at P_0 and κ_b has a minimum at P_0, or* (ii) $\kappa_a < \kappa_b$, κ_a *has minimum at P_0 and κ_b has a maximum at P_0, then the Gaussian curvature K cannot be positive at P_0.*

We shall prove the lemma by the method of contradiction. Suppose that the lemma is false. Then there is a point P_0 at which the principal curvatures have distinct extreme values, one maximum and the other minimum. Taking the lines of curvature as parametric curves, the principal curvatures are

$$\kappa_a = L/E, \qquad \kappa_b = N/G \tag{3.1}$$

(cf. III, (3.4)). The Codazzi equations are (cf. Chapter III, Exercise 9.1)

$$L_2 = \tfrac{1}{2}E_2\left(\frac{L}{E}+\frac{N}{G}\right) \\ N_1 = \tfrac{1}{2}G_1\left(\frac{L}{E}+\frac{N}{G}\right) \tag{3.2}$$

from which we find

$$\frac{\partial\kappa_a}{\partial v} = \frac{1}{2}\frac{E_2}{E}(\kappa_b-\kappa_a) \\ \frac{\partial\kappa_b}{\partial u} = \frac{1}{2}\frac{G_1}{G}(\kappa_a-\kappa_b) \tag{3.3}$$

Since the principal curvatures have extrema, the left-hand members vanish at P_0. It follows that at P_0, $E_2 = G_1 = 0$, and hence that, at P_0,

$$\frac{\partial^2\kappa_a}{\partial v^2} = \frac{1}{2}\frac{E_{22}}{E}(\kappa_b-\kappa_a) \\ \frac{\partial^2\kappa_b}{\partial u^2} = \frac{1}{2}\frac{G_{11}}{G}(\kappa_a-\kappa_b) \tag{3.4}$$

There are now two possibilities: either

(i) κ_a has a maximum. In this case

$$\kappa_a-\kappa_b > 0, \qquad \partial^2\kappa_a/\partial v^2 \leqslant 0, \qquad \partial^2\kappa_b/\partial u^2 \geqslant 0; \tag{3.5}$$

or

(ii) κ_a has a minimum. Then

$$\kappa_b-\kappa_a > 0, \qquad \partial^2\kappa_a/\partial v^2 \geqslant 0, \qquad \partial^2\kappa_b/\partial u^2 \leqslant 0. \tag{3.6}$$

In either case we see that $E_{22} \geqslant 0$ and $G_{11} \geqslant 0$. (Notice that the signs of κ_a, κ_b are irrelevant.) But this contradicts the fact (see II (17.3)) that the Gaussian curvature K satisfies

$$K = -\frac{1}{2EG}(E_{22}+G_{11}),$$

since the right-hand member is negative or zero while K is assumed strictly positive. This contradiction completes the proof of the lemma.

4. Compact surfaces of constant Gaussian or mean curvature

We note that a compact surface must possess a 'highest point', and at this point the curvature is necessarily non-negative. From this, it follows that a compact surface cannot have constant negative curvature. Moreover, a compact surface cannot have constant zero curvature, for otherwise it would contain straight lines which would contradict the compactness. We shall now prove a theorem about compact surfaces of constant Gaussian curvature, and in view of the previous remarks we need consider only the case of constant positive curvature.

THEOREM 4.1. *The only compact surfaces with constant Gaussian curvature are spheres.*

Let S be a compact surface with constant positive Gaussian curvature K. Since S is compact, there is a point P_0 at which the maximum value of the principal curvature is attained. Since the product of the principal curvatures (i.e. the Gaussian curvature) is constant, if follows that the principal curvatures have respectively a maximum and a minimum value at P_0, with the maximum not less than the minimum. From Hilbert's lemma it follows that the two principal curvatures must be equal, i.e. at no point does either principal curvature exceed $\sqrt{K}$. Hence every point of S is an umbilic, and the theorem now follows from Theorem 2.1.

In a similar manner we prove

THEOREM 4.2. *The only compact surfaces whose Gaussian curvature is positive and mean curvature constant are spheres.*

Let S be a compact surface of positive Gaussian curvature and constant mean curvature, and denote by κ_a, κ_b respectively the larger and smaller principal curvatures. Since κ_a is continuous and S compact, there is a point P_0 at which κ_a attains its maximum value. Since the mean curvature is constant, it follows that κ_b attains its minimum value at P_0. Now we have the relation $\kappa_a \geqslant \kappa_b$ everywhere. If $\kappa_a > \kappa_b$ at P_0, then the lemma in section 3 would apply and we could conclude that $K \leqslant 0$, contrary to hypothesis. Hence we must have $\kappa_a = \kappa_b = \mu$ at P_0, and hence everywhere on S. This completes the proof of the theorem.

5. Complete surfaces

The surfaces considered in the previous theorems were restricted to be compact. This is a strong restriction which would exclude, for example, developable surfaces and many common surfaces like paraboloids. It will be seen that a restriction more suitable for the purpose of differential geometry is to require the surface to be complete in the sense of the theory of metric spaces. We first show that the surface can be regarded as a metric space in this sense. It will be remembered that a set of points S carries the structure of a metric space when there is a real-valued function $\rho: S \times S \to R_1$ with the properties

 (i) $\rho(A, B) = 0$ if and only if $A = B$,

 (ii) $\rho(A, B) = \rho(B, A)$,

 (iii) $\rho(A, C) \leqslant \rho(A, B) + \rho(B, C)$,

for all points A, B, C of S.

We assume that our surface S is connected so that any two points can be joined by arc-wise connected paths. If γ is any path joining A to B, then this path can be divided into a finite number of segments so that each segment lies entirely in one coordinate neighbourhood, and adjacent coordinate neighbourhoods overlap. The length of the segment whose equation relative to a coordinate neighbourhood is $u = u(t)$, $v = v(t)$ is given by

$$\int \sqrt{(E\dot{u}^2 + 2F\dot{u}\dot{v} + G\dot{v}^2)}\, dt$$

taken between appropriate limits. Then, the length of γ is defined to be the sum of the lengths of its segments. We now define the distance function ρ by

$\rho(A, B)$ *is the greatest lower bound of the lengths of all arc-wise connected C^1 paths joining A to B.* It is evident that the distance function so defined satisfies conditions (ii) and (iii) of the metric space axioms, while condition (i) is satisfied because the first fundamental form of the surface is positive definite.

A sequence of points $\{x_n\}$ on the surface is said to form a *Cauchy sequence* when, given a positive real number ϵ, an integer n_0 can be found such that $\rho(x_p, x_q) < \epsilon$ provided p and q both exceed n_0. Evidently, if $\{x_n\}$ converges to a limit x, then the sequence $\{x_n\}$ is a Cauchy sequence. The question naturally arises whether every Cauchy sequence of points converges in the surface. If the surface is such that *every Cauchy sequence converges*, then the metric space is said to be 'complete'.

The following simple example shows that not all surfaces are complete. Consider the surface formed by the two-dimensional

Cartesian plane of pairs of real numbers (x, y) when the origin $(0, 0)$ is removed. The distance function ρ is the Euclidean distance function defined by

$$\rho(A, B) = \sqrt{\{(x_A - x_B)^2 + (y_A - y_B)^2\}},$$

when (x_A, y_A), (x_B, y_B) are the rectangular coordinates of points A and B. The sequence of points $\{(1/n, 0)\}$ is easily seen to be a Cauchy sequence which does not converge in the surface, so this surface is not complete.

It may be noted that the two points $(a, 0)$, $(-a, 0)$ $(a > 0)$ cannot be joined by a geodesic (straight line) lying entirely on the surface. In the next section we shall prove that on a *complete* surface any two points can be joined by a geodesic which lies entirely on the surface.

6. Characterization of complete surfaces

In this section we consider three properties each of which can be used to characterize complete surfaces. That these three properties are equivalent was first proved by Hopf and Rinow (1931), but the proof given here follows closely a simplified proof given by de Rham (1952). The properties are:

(a) Every Cauchy sequence of points of S is convergent.

(b) Every geodesic can be prolonged indefinitely in either direction, or else it forms a closed curve.

(c) Every bounded set of points of S is relatively compact.

It is evident that condition (c) implies (a). We now prove that (a) implies (b). Let γ be a geodesic which cannot be extended indefinitely. If γ is a closed curve, then (b) is satisfied. If γ is not a closed curve and if $P(x)$ is some point on γ, then there is some number l such that γ can be prolonged for distances (measured along γ) less than l but cannot be prolonged for distances greater than l. Consider now the sequence of points $\{x_n\}$ lying on γ at distances from P along γ given by $l(1 - 1/n)$. Evidently $\{x_n\}$ is a Cauchy sequence, which by hypothesis (a) converges to some point Q on γ whose distance from P is precisely l. If $\{x'_n\}$ is another Cauchy sequence such that $\rho(x_1, x'_n) \to l$ then $\{x'_n\}$ tends to some limit Q'. Now, the sequence $x_1, x'_1, x_2, x'_2, x_3, x'_3, \ldots$ is also a Cauchy sequence tending to both Q and Q'. Hence $Q = Q'$, and there exists a unique end point Q distant l from P along γ. Consider now a coordinate neighbourhood of S which contains Q. At Q there is uniquely determined a direction **t** which is the direction of the geodesic $-\gamma$ which starts at Q. In this coordinate neighbourhood there is a unique geodesic at Q which has the direction $(-\mathbf{t})$, and this gives a continuation of γ beyond Q, contrary to hypothesis. It follows that γ must satisfy condition (*b*), so we have proved that (*a*) implies (*b*). Since (*c*) implies (*a*), we conclude that (*c*) implies (*b*). In order to prove that the three conditions are equivalent, it remains only to prove that (*b*) implies (*c*).

Suppose now that S has the property (*b*). Consider a point a of S, and geodesic arcs which start at a. We define that *initial vector* of a geodesic arc starting at a to be the tangent vector to this arc at a which has the same sense as the geodesic and *whose length is equal to the length of the geodesic arc*. Since S has the property (*b*), it follows that every tangent vector to S at a, whatever its length, is the initial vector of some geodesic arc starting at a which is uniquely determined. This arc may eventually cut itself or, if it forms part of a closed geodesic, may even cover part of itself.

Let S_r be the set of points x of S whose distance from a does not exceed r, i.e. $\rho(x, a) \leqslant r$, and let E_r be the set of points x of S_r which can be joined to a by a geodesic arc whose length is actually equal to $\rho(x, a)$.

We first prove that the set of points E_r is compact. Let $\{x_h\}$, $h = 1, 2, ...$, be a sequence of points of E_r, and let $\mathbf{T}_h$ be the initial vector of a geodesic arc of length $\rho(a, x_h)$ joining a to x_h. Then the sequence of vectors $\{\mathbf{T}_h\}$ regarded as a sequence of points in two-dimensional Euclidean space, admits at least one vector of accumulation $\mathbf{T}$. Moreover, this vector $\mathbf{T}$ is the initial vector of a geodesic arc whose extremity belongs to E_r and is a point of accumulation of the sequence $\{x_h\}$. This proves that E_r is compact.

We next show that $$E_r = S_r. \tag{6.1}$$

Certainly (6.1) is true when $r = 0$. Also, if it is true for $r = R > 0$, then it is certainly true for $r < R$. We shall now prove that, conversely, if (6.1) is true for $r < R$, then it is still true for $r = R$. Now, every point of S_R is the limit of a sequence of points whose distance from a is less than R. By hypothesis these points belong to E_R, and since E_R is closed, it follows that their limit belongs to E_r. Thus (6.1) is valid for $r = R$. In order to establish (6.1) completely, it is merely necessary to show that if it holds for $r = R$, then it still holds for $r = R+s, s > 0$. This follows because it would then be possible to extend the range of validity of (6.1) to an arbitrary extent by an appropriate number of extensions of the range by an amount s.

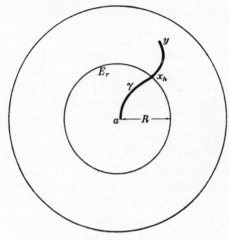

We next show that to any point y such that $\rho(a, y) > R$ there is a point x such that

$$\rho(a, x) = R, \tag{6.2}$$

and

$$\rho(a, y) = R + \rho(y, x). \tag{6.3}$$

Since $\rho(a, y)$ has been defined as the lowest bound of the lengths of arcs from a to y, it follows that we can join a to y by a curve γ whose length is less than $\rho(a, y) + h^{-1}$ for any integer h. Let x_h be the last point of this curve belonging to $E_R = S_R$ (see Fig. 8). Then we have from axiom 5 (iii),

$$\rho(a, y) \leqslant \rho(a, x_h) + \rho(x_h, y),$$

i.e.

$$\rho(a, y) \leqslant R + \rho(x_h, y),$$

since $\rho(a, x_h) = R$. Alternatively, we can write this inequality as

$$\rho(x_h, y) \geqslant \rho(a, y) - R. \tag{6.4}$$

Since the arc length of γ from a to y is the sum of the arc lengths from a to x_h and from x_h to y, we have

$$\rho(x_h, y) \leqslant \operatorname{arc}(x_h, y),$$

i.e.

$$\rho(x_h, y) \leqslant \operatorname{arc}(a, y) - \operatorname{arc}(a, x_h)$$
$$\leqslant \rho(a, y) + h^{-1} - \operatorname{arc}(a, x_h)$$
$$\leqslant \rho(a, y) + h^{-1} - R.$$

Now let $h \to \infty$; $\{x_h\}$ will have at least one point of accumulation x with the property

$$\rho(x, y) \leqslant \rho(a, y) - R. \tag{6.5}$$

Comparison of (6.4), (6.5) shows that at this point,

$$\rho(a, y) = R + \rho(y, x),$$

and we have therefore proved the existence of a point x satisfying (6.2) and (6.3).

We have already seen in Chapter II that provided the two points x, y are not too far apart, then the point y is the extremity of one and only one geodesic arc of origin x and of length $\rho(x, y)$. More precisely,† there exists a continuous function $s(x) > 0$ such that if $\rho(x, y) < s(x)$, the point y is the extremity of the unique geodesic arc of length $\rho(x, y)$ joining x to y. Moreover, the continuous function $s(x)$ attains a positive minimum value on the compact set E_R, and we take s to be this minimum.

Then, if (6.1) is true for $r = R$ and if $R < \rho(a, y) \leqslant R + s$, there exists an $x \in E_R$ such that $\rho(a, x) = R$ and $\rho(x, y) = \rho(a, y) - R \leqslant s$. Consequently there exists a geodesic arc L' of length $\rho(a, x)$ joining a to x, and a geodesic arc L'' of length $\rho(x, y)$ joining x to y. The composite arc formed by L' and L'' joins a to y and has as its length $\rho(a, y)$. It follows then that this composite arc is a geodesic arc, and y is thus joined to a by a geodesic arc whose length is equal to the distance of y from a. Hence y belongs to E_{R+s} and the range of validity of (6.1) is thus extended from E_R to E_{R+s}. We have incidentally proved that hypothesis (c) implies that any two points of S can be joined by a geodesic arc whose length is equal to their distance.

Suppose now that we are given a bounded set of points M on S. Evidently we can find some R such that M is contained in S_R, and since $S_R (= E_R)$ is compact, it follows that M is relatively compact. We have thus shown that (b) implies (c), and the equivalence of the three conditions (a), (b), and (c) is thus established.

We have also proved

THEOREM 6.1. *On a complete surface any two points can be joined by a geodesic arc whose length is equal to their distance.*

Since a compact surface evidently possesses property (c), it follows that all compact surfaces are complete. A simple example

† The reader is referred to a paper by O. Veblen and J. H. C. Whitehead (1931) for a detailed justification.

of a non-compact complete surface is the plane.

7. Hilbert's theorem

We shall devote the whole of this section to the proof of the following result due to Hilbert.

THEOREM 7.1. *A complete analytic surface, free from singularities, with constant negative Gaussian curvature, cannot exist in three-dimensional Euclidean space.*

We have already seen that a compact surface with these properties cannot exist, but when the condition of compactness is relaxed to completeness the proof is much more difficult. The following proof follows closely to that given in a paper by L. Bieberbach (1926).

In our proof we shall make use of the notion of universal covering space of a given space. The reader is probably already familiar with this notion, but if not he will find a lucid account of the relevant properties of universal covering spaces in Pontrjagin (1946). The essential idea is the following. Let p be a point on the surface S, and let Q be the set of all paths of S which begin at p. We divide the set Q into classes, putting into each class the totality of paths that are homotopically equivalent. We denote the set of these classes by S', so that a point of S' is an equivalence class of paths on S. There is a natural mapping ϕ of the set S' on the space S, for if A is a point on S', then all the equivalent paths in S belonging to A must end in the same point a, and we write $a = \phi(A)$. It is shown that the set of points S' can be considered as forming a surface called the *universal covering surface* which has the following important properties.

(1) The natural mapping of S' on S is a continuous open mapping. Moreover, ϕ is a locally homeomorphic mapping, i.e. for every point A of S' there exists a neighbourhood U^* such that the mapping ϕ is homeomorphic on the neighbourhood U^* (Pontrjagin, theorem 58).

(2) The universal covering surface S' of a surface S is always simply connected (Pontrjagin, theorem 59). Property (1) implies that S and S' are locally homeomorphic so that all the local properties of the space S are automatically true for the space S'. Moreover, the differential-geometric structure on S induces a differential-geometric structure on S'.

We now return to the proof of Hilbert's theorem. We assume that a surface S exists having the required properties and we obtain a proof by contradiction. Consider an arbitrary geodesic line on the surface S and take an arbitrary point O on this geodesic as origin. If s denotes the arc length of this geodesic measured from O, the completeness of S ensures that the geodesic can be continued in both directions from $-\infty$ to $+\infty$. It is possible that the geodesic will ultimately cross itself so that the same point on S will have two different s-values. However, if we consider instead of S its universal covering surface S', then different values of s will correspond to different points on S'. This follows because on a surface of negative Gaussian curvature two geodesic arcs cannot enclose a simply connected region (cf. Exercises II, 13).

At each point of parameter s on the given geodesic, consider the orthogonal geodesic line and let its arc length t be chosen as parameter. In this notation the equation of the given geodesic is $t = 0$. Now, two of these geodesic arcs at s_1, s_2 cannot meet on the surface S to form with the geodesic arc $s_1 s_2$ a simply connected region. For if this were the case, the sum of the angles of the geodesic triangle so formed would not be less than 2π, contrary to the results of Chapter II. We denote a point in the covering space S' by the pair of coordinates (s, t), and we observe that different pairs (s, t) correspond to different points on S'. We now show that *every point of S can be represented on the covering surface S' in this manner.*

It follows from Chapter II, section 18, that the line element of the surface assumes the form

$$ds^2 + G(s)\, dt^2.$$

Suppose now that a point P of the surface S remained uncovered by our construction (see Fig. 9). Join P to O ($s = 0$, $t = 0$) by some rectifiable curve γ. Then there must be some point Q on γ with the property that all points between O and Q can be covered, while points on γ arbitrarily near Q on the side of Q remote from O cannot be covered. If Q_1 lies on γ between O and Q it follows from the form of the metric that the length of the curve OQ_1 is greater than or equal to s_{Q_1} where s_{Q_1} is the s-coordinate of the corresponding point on S'. The set of values $\{s_{Q_1}\}$ is therefore bounded, and we define s_Q to be the least upper bound of this set. Let R be the point on the geodesic $t = 0$ distant s_Q from O, and consider the orthogonal geodesics along some interval on the geodesic $t = 0$ which contains R. These geodesics will cover a strip of the surface which certainly contains the point Q, and points beyond Q on the curve γ. This gives a contradiction, and we conclude that *every point of the surface S can be covered in this way.* There is thus a local homeomorphism between points of S and the $(s–t)$ plane, but this correspondence may not be (1–1) in the large.

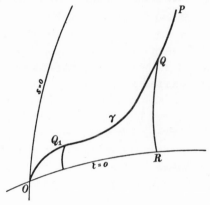

Fig. 9

However, the covering space S' is homeomorphic with the $(s–t)$ plane.

We now consider the asymptotic lines on the surface S. These are given by the differential equation

$$L\,ds^2 + 2M\,ds\,dt + N\,dt^2 = 0.$$

Since $K < 0$ we conclude that $LN - M^2 < 0$ and hence that at each point of S the asymptotic directions are real and different. Hence at each point of S' these determine two distinct directions, and similarly at each point of the $(s–t)$ plane. Since the $(s–t)$ plane is simply connected it follows that the differential equation gives rise to two vector fields which can be continued over the whole plane. The Lipschitz condition for the uniqueness of the solution of the differential equation is satisfied because we have assumed that S is of class ω. Thus throughout the whole $(s–t)$ plane there are two systems of asymptotic lines with the property that a curve from each system passes through an arbitrary point. Moreover, since S is free from singularities, the differential equation has no singularities. It follows from the theorem of Bendixon†

· † Cf. Bieberbach (1923).

that each asymptotic line can be prolonged to an arbitrary extent in both directions, and if τ denotes the arc length, then

$$\lim_{\tau \to -\infty} (s^2+t^2) = \infty, \qquad \lim_{\tau \to +\infty} (s^2+t^2) = \infty.$$

We next prove that *each asymptotic line of one system cuts each asymptotic line of the other system in exactly one point.* First we prove

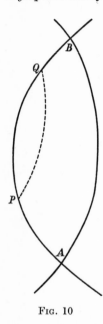

FIG. 10

that two such lines cut in *at most* one point. If this were not so, then there would be a region of the s–t plane bounded by two asymptotic lines of different systems. Consider first the case when the asymptotic lines meet at A and B such that the continuation of the lines does not contain any interior point of the region bounded by the two lines (Fig. 10). Let P be a point on one of the lines lying between A and B, and consider the asymptotic line of the second system which passes through P. Because this second line through P cannot intersect the line AB belonging to the same system, it follows that it must intersect the line AB of the opposite system in a further point Q. Moreover, as P moves from A towards the end B, so Q will move from B towards the end A. There must be one point where P and Q coincide, and at that point the asymptotic directions will coincide. This, however, contradicts the hypothesis $K < 0$.

Consider now the second case, where by continuation of the asymptotic lines at least one line penetrates the region bounded by the two asymptotic lines (Fig. 11). Then, this asymptotic line will meet the line of the opposite system at a second point C. Then the continuation BC together with the asymptotic line BC form a system of the type discussed under Fig. 10, and again we obtain a contradiction. We see then that in either case we obtain a contradiction, and so we have proved that each asymptotic line of one system cannot meet each asymptotic line of the other system in *more than one point*.

In order to prove that such lines must meet in *at least one point*, it is convenient to refer to the asymptotic lines as parametric lines. Suppose that N is a neighbourhood of S in which the *lines of curvature* are chosen as parametric lines. If κ_a, κ_b denote the principal curvatures at a point P on N and if $K = -1/a^2$ is the constant negative Gaussian curvature, we can write

$$\kappa_a = a^{-1}\cot\rho, \qquad \kappa_b = -a^{-1}\tan\rho,$$

where $0 < \rho < \frac{1}{2}\pi$.

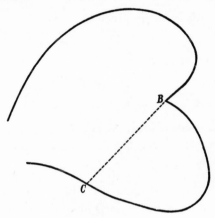

Fig. 11

An argument similar to that of section 3 leads to the equations

$$\left.\begin{aligned}
\frac{\partial\kappa_a}{\partial v} &= \frac{1}{2}\frac{E_2}{E}(\kappa_b - \kappa_a) \\
\frac{\partial\kappa_b}{\partial u} &= \frac{1}{2}\frac{G_1}{G}(\kappa_a - \kappa_b)
\end{aligned}\right\}. \tag{7.1}$$

Substitute for κ_a, κ_b to get

$$\left.\begin{aligned}
\frac{E_2}{E} &= 2\rho_2\cot\rho \\
\frac{G_1}{G} &= -2\rho_1\tan\rho
\end{aligned}\right\}. \tag{7.2}$$

These equations integrate to give

$$\left.\begin{aligned}
E &= U(u)\sin^2\rho \\
G &= V(v)\cos^2\rho
\end{aligned}\right\}, \tag{7.3}$$

where $U(u)$, $V(v)$ are certain functions of u and v respectively.

By means of a suitable reparametrization of the type previously considered in Chapter II, these functions may be taken as unity and the first fundamental form then becomes

$$\sin^2\rho\,du^2 + \cos^2\rho\,dv^2.$$

In terms of the new parameters,

$$L = \kappa_a E = a^{-1} \sin \rho \cos \rho,$$
$$N = \kappa_b G = -a^{-1} \sin \rho \cos \rho,$$
$$M = 0,$$

and the asymptotic lines are given by

$$du^2 - dv^2 = 0.$$

Choose new parameters σ, τ where

$$\sigma = \tfrac{1}{2}(v+u),$$
$$\tau = \tfrac{1}{2}(v-u).$$

Then, the parametric curves $\sigma = $ constant, $\tau = $ constant are asymptotic lines. Moreover, the metric assumes the form

$$d\sigma^2 + 2\cos 2\rho \, d\sigma d\tau + d\tau^2, \tag{7.4}$$

and σ, τ measure the arc lengths of the asymptotic lines.

Through the origin O of the $(s–t)$ plane there pass two asymptotic lines. Through each point on these two lines we draw the asymptotic lines of the opposite systems. Then we shall prove that each point of the $(s–t)$ plane lies on one asymptotic line of each system. Suppose that there is a point P on the plane which cannot be reached in this way. Join P to O by a continuous curve γ. Then there will be a point Q on γ with the property that every point on γ between O and Q can be reached in this way, but points on γ arbitrarily near to Q on the side remote from O (and possibly including the point Q itself) cannot be reached.

Consider now a neighbourhood of Q which is covered by the asymptotic lines and has the property that each pair of lines from different systems cut in a single point in this neighbourhood. Consider a point Q_0 lying in this neighbourhood, and let the asymptotic lines through Q_0 cut the coordinate curves $\tau = 0$, $\sigma = 0$ in two points $Q_0^{(1)}$, $Q_0^{(2)}$ respectively. Let Q_i denote a typical point which lies on γ between Q_0 and Q. Let the asymptotic lines through Q_i meet the coordinate curves at $Q_i^{(1)}$, $Q_i^{(2)}$, and let these lines meet the lines through Q_0 in $\bar{Q}_i^{(1)}$ and $\bar{Q}_i^{(2)}$ (see Fig. 12). Then

$$Q_0 \bar{Q}_i^{(1)} = Q_0^{(1)} Q_i^{(1)}, \qquad Q_0 \bar{Q}_i^{(2)} = Q_0^{(2)} Q_i^{(2)},$$

provided Q_i lies in a neighbourhood of Q_0 where the line element can be written in the form (7.4). Any asymptotic line which cuts $Q_0 \bar{Q}_1$ between Q_0 and $\bar{Q}_1$ and which is sufficiently near to Q_0 will

cut equal lengths from all asymptotic lines which meet $Q_0 Q_0^{(1)}$.
If this were not true for all the asymptotic lines meeting $Q_0 \bar{Q}_1$,
we could choose a point R on the asymptotic line joining Q_0 to $\bar{Q}_1$
such that all points between Q_0 and R possess this property, but
there are points arbitrarily near R (and possibly R itself) which
do not possess this property. Now the asymptotic line through R

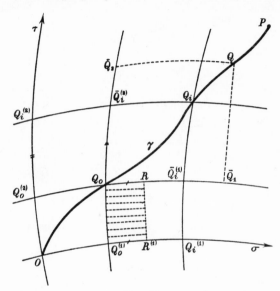

Fig. 12

will cut the coordinate line $\tau = 0$ in the point $R^{(1)}$ such that the
lengths $Q_0 R$, $Q_0^{(1)} R^{(1)}$ are equal, and, moreover, all the asymptotic
lines between $Q_0^{(1)}$ and Q_0 will have equal lengths intercepted by the
asymptotic line through R. We measure off from all these asymptotic lines the length $Q_0 R$, in the direction of increasing σ. Then
we assert that the end points of these segments form an asymptotic
line. This is certainly the case when we consider neighbourhoods of
points on the line $R R^{(1)}$ and make use of the net of asymptotic lines
in this neighbourhood. It is equally true for all asymptotic lines
which meet $Q_0 Q_1$ in a neighbourhood of R. In particular, it is
true for the asymptotic lines through $\bar{Q}_1$ and for those in a certain
neighbourhood of $\bar{Q}_1$, which is contrary to hypothesis. Thus, the
two asymptotic lines through O will cut an arbitrary asymptotic
line in the plane, and since the point O has been chosen arbitrarily,
it follows that each asymptotic line of one system meets every

asymptotic line of the other system in precisely one point. We can thus take (σ, τ) as coordinates for points in the whole plane, and so the metric for the whole plane assumes the form (7.4).

Let ω be the angle between the parametric curves. Then from formula II (5.3) we see that $\omega = 2\rho$ and hence $0 < \omega < \pi$.

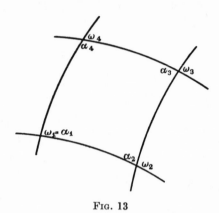

FIG. 13

Now, using equation II (17.3) for the Gaussian curvature $K = -a^{-2}$, we find

$$\frac{\partial^2 \omega}{\partial \sigma \partial \tau} = -K \sin \omega. \tag{7.5}$$

Consider now the quadrilateral formed by the asymptotic lines $\sigma = \pm \alpha$, $\tau = \pm \alpha$ (see Fig. 13).

We have for the total curvature

$$\iint K \, dS = \iint K \sin \omega \, d\sigma d\tau$$
$$= \omega_1 - \omega_2 + \omega_3 - \omega_4$$
$$= \alpha_1 + \alpha_2 + \alpha_3 + \alpha_4 - 2\pi.$$

It follows that *the absolute magnitude of the total curvature of an arbitrarily large region cannot exceed 2π.*

Consider now the first form of the metric

$$ds^2 + G(s) \, dt^2.$$

We have

$$K = -\frac{1}{2\sqrt{G}} \frac{\partial}{\partial s}\left(\frac{G_s}{\sqrt{G}}\right),$$

and

$$\sqrt{G} = \cosh(s/a).$$

The total curvature over a region bounded by parametric lines $s = \pm l$, $t = \pm l$ is

$$\iint K \, dS = \iint K \sqrt{G} \, dsdt = -\frac{1}{2} \iint \frac{\partial}{\partial s}\left(\frac{G_s}{\sqrt{G}}\right) dsdt = -\frac{4l}{a}\sinh\frac{l}{a}.$$

But in magnitude this tends to infinity as $l \to \infty$, *which contradicts the previous assertion that the absolute magnitude of the total curvature cannot exceed* 2π. This contradiction completes the proof of Hilbert's theorem.

Reviewing the proof given in this section, it will be seen that the real difficulty consists in proving that the asymptotic lines form a coordinate system valid for the whole plane. The 'proof' given by W. Blaschke appears to assume this as intuitively obvious, and is certainly incomplete.

8. Conjugate·points on geodesics

In Chapter II it was proved that if there exists a curve of shortest distance between two points on a surface, then that curve is necessarily a geodesic. In this section we discuss whether a given geodesic joining two points is necessarily the shortest distance between them. The following theorem shows that this is the case when the given geodesic can be embedded in a field of geodesics. By a field of geodesics is meant a one-parameter set of geodesics defined over a region R of a surface such that through each point of R passes one and only one curve of the set.

THEOREM 8.1. *If P and Q are two points of a geodesic which can be embedded in a field of geodesics, then the arc PQ of the geodesic is shorter than any other arc which joins P to Q and lies entirely in that region of the surface covered by the field.*

To prove the theorem choose parameters so that the geodesics of the family are the curves $v = $ constant, with $v = v_0$ as the given geodesic, and let the curves $u = $ constant be geodesic parallels orthogonal to them, so chosen that the metric reduces to the form $ds^2 = du^2 + \lambda^2 dv^2$. If the coordinates of P and Q are (u_1, v_0), (u_2, v_0) with $u_2 > u_1$, the length of the geodesic arc PQ is $(u_2 - u_1)$.

Let C be an arbitrary curve passing through P and Q, given by the equation $v = \phi(u)$ where $\phi(u_1) = v_0$, $\phi(u_2) = v_0$. Then the arc length of C is

$$l = \int_{u_1}^{u_2} \{1 + \lambda^2 (d\phi/du)^2\}^{\frac{1}{2}} \, du.$$

Evidently l exceeds u_2-u_1, unless $d\phi/du = 0$ when C is the given geodesic. This completes the proof of Theorem 8.1.

However, it is most unlikely that the region R of the geodesic field extends over the entire surface S, so the previous argument is in general inapplicable to complete surfaces. For example, the surface of a sphere cannot be covered by a geodesic field because any two great circles intersect in two points of the sphere. Moreover, if A, B are any two non-antipodal points, that geodesic arc which is the longer part of the great circle joining A, B is evidently not the shortest distance from A to B. We now prove

THEOREM 8.2. *When the surface S has negative curvature everywhere, the length of a geodesic which joins any two points A, B is always less than the lengths of neighbouring curves through A and B.*

This result is proved as follows. Let one system of parametric curves be the geodesics normal to the given geodesic AB, and the other system be the orthogonal trajectories. Let u denote the length of the geodesic normal PQ from P to AB, and let v denote the length AQ. The line element of the surface becomes

$$ds^2 = du^2 + \lambda^2 dv^2,$$

where $\qquad \lambda(0,v) = 1, \qquad \lambda_1(0,v) = 0.$

In terms of these parameters the Gaussian curvature is given by

$$K = -\lambda_{11}/\lambda, \quad \text{so that} \quad \lambda_{11} = -\lambda K.$$

The function λ may thus be expanded as a power series in u in the form

$$\lambda = 1 - K\frac{u^2}{2} - K_1\frac{u^3}{6} + O(u^4),$$

where K and K_1 are evaluated with $u = 0$.

A neighbouring curve APB which differs very little from AB will have an equation of the form $u = \phi(v)$, where u will be small. The length of this curve will be

$$l = \int_A^B \{\phi'^2 + \lambda^2\}^{\frac{1}{2}} dv = \int_A^B \{1 + \phi'^2 - K\phi^2 - \tfrac{1}{3}K_1\phi^3\}^{\frac{1}{2}} dv,$$

where terms of the fourth order are neglected. We now assume that ϕ' never becomes infinite and is thus of the same order of smallness as u. With this assumption the difference between l and the geodesic arc length s may be written

$$l-s = \tfrac{1}{2} \int_A^B \{\phi'^2 - K\phi^2 - \tfrac{1}{3}K_1\phi^3\}\, dv.$$

Now the sign of the variation of the arc length will be given by the second-order terms, provided that these do not vanish identically. If only these terms are retained the equation becomes

$$l-s = \tfrac{1}{2} \int_A^B (\phi'^2 - K\phi^2)\, dv. \tag{8.1}$$

Now, if K is always negative, the integrand is always positive and so $l > s$. This proves the required result.

The remainder of this section will consider the analogous problem when K is not always negative. Since the metric is of the form $ds^2 = du^2 + \lambda^2 dv^2$, it follows that the arc length of the orthogonal trajectory taken between the geodesics v and $v+\delta v$ is given by $\lambda \delta v$. Alternatively, $\lambda \delta v$ is the length of the segment of the normal from a typical point of the geodesic v cut off by the geodesic $v+\delta v$. If v and $v+\delta v$ are regarded as constants, then the arc length $\lambda \delta v$ will vary with the arc length u of the geodesic v. If $p = \lambda \delta v$, from $\lambda_{11} = -K\lambda$ it follows that $p_{11} = -Kp$, i.e.

$$d^2p/du^2 + Kp = 0,$$

a differential equation which was first obtained by Jacobi in 1836. Consider the solution of this differential equation which vanishes at the point A, and suppose that this solution vanishes again at the point A_1 on the geodesic, while maintaining a constant sign in the interval AA_1. Then all the geodesics which leave A in a direction infinitesimally near to the direction of AB will intersect AB again in the point A_1 or in points infinitesimally near A_1. Now if B lies *between* A and A_1, it follows that the geodesic segment AB is shorter than any neighbouring curves joining A and B. The point A_1 is called a *conjugate* point of A along the geodesic AB.

This leads to the result due to Jacobi:

THEOREM 8.3. *In order that the geodesic distance AB should be the shortest distance, it is necessary and sufficient that B lies between A and its conjugate point A_1.*

The sufficiency has been proved above. We now outline a proof of the necessity using a lemma due to Erdmann.

Consider the problem of finding a curve $y = y(x)$, which passes through two points (x_1, y_1), (x_2, y_2), has a discontinuity of slope on the line $x = x_3$, and is such that the integral

$$J = \int_{x_1}^{x_2} f(x, y, y')\, dx$$

assumes an extreme value (see Fig. 14).

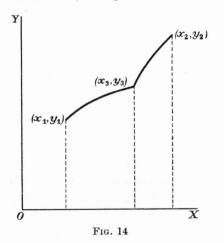

Fɪɢ. 14

Let $\qquad y'_+ = \lim_{\delta \to 0} y'(x_3 + \delta); \quad y'_- = \lim_{\delta \to 0} y'(x_3 - \delta)$

where δ is positive. Then Erdmann's lemma states that *for an extreme value, in addition to the equation of Euler, it is necessary that*

$$f_{+y'} = f_{-y'},$$

where $f_{+y'} = f_{y'}(x_3, y_3, y'_+)$, $f_{-y'} = f_{y'}(x_3, y_3, y'_-)$.

To prove the lemma, we note that the variation of the integral over the curves $y(x)$ and $y + \epsilon\eta(x)$, where $\eta(x_1) = 0$, $\eta(x_2) = 0$, is given by

$$J(\epsilon) = \int_{x_1}^{x_3} f(x, y + \epsilon\eta, y' + \epsilon\eta')\, dx + \int_{x_3}^{x_2} f(x, y + \epsilon\eta, y' + \epsilon\eta')\, dx,$$

it being assumed that the 'corner' still moves along the line $x = x_3$. In the usual manner, it follows that a necessary condition is $J'(0) = 0$. This reduces to

$$\int_{x_1}^{x_3} \left(f_y - \frac{d}{dx} f_{y'}\right)\eta\, dx + \int_{x_3}^{x_2} \left(f_y - \frac{d}{dx} f_{y'}\right)\eta\, dx + \eta_3(f_{-y'} - f_{+y'}) = 0.$$

From this it follows that, in addition to Euler's equation $f_y - df_{y'}/dx = 0$, it is necessary to have $f_{+y'} = f_{-y'}$, and the lemma is proved.

We now return to the proof of Theorem 8.3. From equation (8.1), it follows that the geodesic distance s is a minimum provided that

$$\delta^2(s) = \tfrac{1}{2} \int\limits_A^B (u'^2 - Ku^2) \, dv$$

is non-negative. Now, if $\delta^2(s) \geqslant 0$ for all u, it follows that the curve $u = 0$ must make the integral

$$\int\limits_A^B (u'^2 - Ku^2) \, dv$$

a minimum. It is easily verified that, except for notation, the Euler equation corresponding to this is Jacobi's differential equation.

Assume now that the geodesic distance AB still gives the shortest distance with B lying beyond A_1, i.e. $\delta^2(s) \geqslant 0$, and we hope to arrive at a contradiction. By hypothesis there is a solution of Jacobi's differential equation (and therefore of Euler's equation) which vanishes at A, and has its next zero at A_1. If $u = \phi(v)$ is such a solution, then, of course, so is $u = \epsilon\phi(v)$ for an arbitrary constant ϵ.

Now define a new function $\tilde{u}$ which coincides with $u = \phi(v)$ from A to A_1, and is identically zero from A_1 to B. The next step in the argument is to prove that such a function $\tilde{u}$ is a 'corner' solution of the problem of giving $\delta^2(s)$ an extreme value.

Since

$$\int\limits_A^{A_1} u u'' \, dv = [uu']_A^{A_1} - \int\limits_A^{A_1} u'^2 \, dv = - \int\limits_A^{A_1} u'^2 \, dv,$$

where $u = \phi(v)$, it follows that

$$\int\limits_A^B (\tilde{u}'^2 - K\tilde{u}^2) \, dv = \int\limits_A^{A_1} (u'^2 - Ku^2) \, dv = - \int\limits_A^{A_1} u(u'' + Ku) \, dv = 0,$$

since $u'' + Ku = 0$.

Since $\tilde{u}$ satisfies the condition $\delta^2(s) = 0$, and can be chosen as near to the curve $u = 0$ as we please since ϵ is arbitrary, it follows that $u = 0$ gives $\delta^2(s)$ its minimal value. Moreover, u must be a 'corner' solution of the problem of finding a minimum of $\delta^2(s)$. From Erdmann's lemma, $u'_+ = u'_- = 0$. But this is impossible because there is no non-trivial solution of the equation $u'' + Ku = 0$

which vanishes simultaneously with its derivative. This gives the required contradiction, and the theorem is completely proved.

Jacobi's theorem will now be used to prove the following interesting theorem due to Bonnet.

THEOREM 8.4. *If along a geodesic the Gaussian curvature exceeds a positive constant $1/a^2$, then the curve cannot be the shortest distance between its extremities along an arc length exceeding πa.*

The main lemma used in the proof of this result is a standard theorem from the theory of differential equations due to Sturm. This theorem is stated below without proof, but a very simple and elegant proof can be found in Darboux (1896).

STURM'S THEOREM. *Consider the two distinct differential equations*

$$\frac{d^2 V}{dx^2} = HV, \qquad \frac{d^2 V}{dx^2} = H'V,$$

where for all values of x in the range considered, $H'(x) \geqslant H(x)$. Then, if $\phi(x)$ is a solution of the first equation having two consecutive zeros at x_0 and x_1, a solution of the second equation which has a zero at x_0 cannot have another zero in the closed interval $[x_0, x_1]$.

As a corollary we have:

If for all values of x in the range considered, $H'(x) \leqslant H(x)$, and if $\phi(x)$ is a solution of the first equation having two consecutive zeros at x_0 and x_1, then any solution of the second equation which has a zero at x_0 must have at least one other zero in the inverval $[x_0, x_1]$.

Consider Jacobi's differential equation $(d^2 p / dv^2) + Kp = 0$, which is of the type considered by Sturm. Let p be a solution of this equation, and let v_0, v_1 be two consecutive zeros corresponding to the points A and A_1. It follows from Jacobi's theorem that the arc AB will be the shortest distance between A and B if and only if B lies between A and A_1.

Suppose now that the Gaussian curvature along the line AA_1 always exceeds the positive constant $1/a^2$, so that $K \geqslant 1/a^2$. The solution of the equation

$$\frac{d^2 p}{dv^2} = -\frac{p}{a^2}$$

which vanishes for $v = v_0$ is

$$C \sin \frac{v - v_0}{a},$$

and its next zero after v_0 is just $v_0 + \pi a$. It follows that if the arc length AB exceeds a, then B will not lie between A and A_1, and the theorem is proved.

An analogous result is the following:

THEOREM 8.5. *If at all points of a geodesic the Gaussian curvature is less than $1/b^2$, the curve is necessarily of shorter length than neighbouring curves along an arc length at least equal to πb.*

The proof follows easily from the hypothesis $K \leqslant 1/b^2$, and the fact that the interval between consecutive roots of the equation $d^2 p/dv^2 = -p/b^2$ is πb. As this cannot be smaller than the interval between consecutive roots of the previous equation, it follows in this case that if the arc length AB is less than πb, then B will certainly lie between A and A_1, thus giving the required result.

Suppose now the surface S is compact, and has the property that $K \geqslant 1/a^2$ everywhere. From section 6 it follows that if A and B are any two points on S, there is a geodesic joining A to B which is of shorter length than the neighbouring curves. It follows from Theorem 8.4 that the maximum distance between A and B cannot exceed πa. This proves the following:

THEOREM 8.6. *If on a compact surface S the curvature everywhere exceeds $1/a^2$, the maximum distance between any two points cannot exceed πa.*

EXERCISE 8.1. Prove that the Gaussian curvature at any point on the ellipsoid

$$\frac{x^2}{a^2} + \frac{y^2}{b^2} + \frac{z^2}{c^2} = 1$$

is given by

$$\frac{p^4}{a^2 b^2 c^2},$$

where p is the distance of the centre from the tangent plane.

Show that if $a \geqslant b \geqslant c$, every geodesic arc of length greater than $\pi ab/c$ cannot be the shortest distance between its extremities; but every geodesic arc of length less than $\pi bc/a$ is necessarily shorter than the neighbouring curves joining its extremities.

9. Intrinsically defined surfaces

So far a surface has been regarded as embedded in three-dimensional Euclidean space, whose metric induces the metric of the surface. Indeed, some of the arguments of the previous sections make direct use of the surrounding space, for example when the surface normal $\mathbf{N}$ is used.

We recall that no curve with non-zero torsion can be embedded in a plane, but curves with non-zero torsion do exist in three-dimensional space. In section 7 we proved that there are no complete surfaces with constant negative curvature in three-dimensional space, but we did not prove that there are none in higher dimensional space. In order to investigate whether such configurations are possible in *some* dimension, there are two courses open, viz.

(i) consider surfaces embedded in n-dimensional Euclidean space for a fixed but arbitrary n;

(ii) consider surfaces *in abstracto*, i.e. without reference to any surrounding space.

We take the latter course because surfaces do arise quite naturally which are not *given* as embedded in any Euclidean space. For example, the rotations R_2 of the plane about a fixed point are in (1–1) correspondence with the points of a circle and the *continuity* thus imparted to R_2 has an obvious significance. Similarly we can discuss the differential geometry of this space, but R_2 is not given as a subset of E_2 or any other Euclidean space. We give here a brief account of differentiable manifolds and refer the reader to Chevalley (1946) for a rigorous and more detailed treatment of the subject.

We assume that the reader is familiar with the notion of a Hausdorff topological space (cf. Patterson (1956), chapters ii, iii). Then, we define a *two-dimensional manifold* to be a Hausdorff topological space which is connected and has the property that each point has a neighbourhood homeomorphic to some open set in the Cartesian plane. As examples of two-dimensional manifolds we mention a sphere, a cylinder, a torus, and a Klein bottle. The torus may be represented by the rectangle shown in Fig. 15 where opposite pairs of edges are identified. The Klein bottle is represented in Fig. 16 where opposite pairs of edges are identified, one pair directly and the other pair in the opposite sense.

We now wish to introduce coordinate systems on the two-dimensional manifolds. We define *a system of differentiable coordinates in a two-dimensional manifold S* to be an indexed family $\{V_j, j \in J\}$ of open sets covering S, and for each j, a homeomorphism

$$\psi_j : E_j \to V_j.$$

where E_j is an open set in the Cartesian plane, such that the map

$$\psi_j^{-1}\psi_i : \psi_i^{-1}(V_i \cap V_j) \to \psi_j^{-1}(V_i \cap V_j), \qquad i, j \in J, \qquad (9.1)$$

is differentiable. If each map $\psi_j^{-1}\psi_i$ has continuous derivatives of order r, then the system is said to be of *class r*.

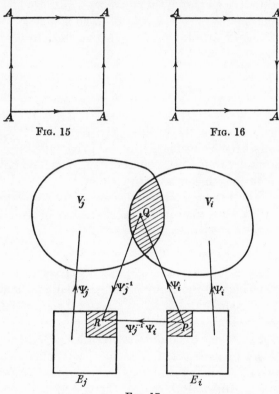

Fig. 15   Fig. 16

Fig. 17

The reader may find helpful the diagrammatic sketch of the maps ψ_i, ψ_j given in Fig. 17. The point P in E_i is mapped into Q by ψ_i, and Q is mapped into R by ψ_j^{-1}. The composite map $\psi_j^{-1}\psi_i$ maps the point P in the Cartesian plane into the point R in the Cartesian plane.

Suppose we have another system of differentiable coordinates of class r given by the indexed family $\{V'_k\}, \{\psi'_k\}$. We define these two systems to be *r-equivalent* if the composite families $\{V_j, V'_k\}$, $\{\psi_j, \psi'_k\}$ form a system of class r. It is easily verified that this is a proper equivalence relation, so the various systems fall into definite equivalence classes.

We can now define a *surface S of class r* to be *a two-dimensional manifold together with an r-equivalence class of systems of coordinates in S.*

The coordinates of Q, considered as an element of V_i, can be identified with the coordinates (u, v) of P in E_i. Alternatively, considered as an element of V_j the coordinates of Q can be identified with the coordinates (u', v') of R in E_j. The mapping $\psi_j^{-1}\psi_i$ gives a change of coordinates from (u, v) to (u', v') given by equations of the type

$$\left.\begin{array}{l} u' = g(u, v) \\ v' = h(u, v) \end{array}\right\}. \tag{9.2}$$

Let S be covered by a system of coordinates of class r. If $x \in V_i \cap V_j$, denote by $a_{ji}(x)$ the 2×2 matrix of first partial derivatives of the functions $\psi_j^{-1}\psi_i$ evaluated at $\psi_i^{-1}(x)$, i.e. the Jacobian matrix $\partial(u', v')/\partial(u, v)$ evaluated at P. [The suffixes i, j of the matrix $a_{ji}(x)$ refer to the neighbourhoods V_i, V_j and do not refer to the element in the jth row and ith column!]

If V_i, V_j, V_k are three open sets with a non-empty intersection and if $x \in V_i \cap V_j \cap V_k$, then the rule for the derivative of a function of a function leads to the equation

$$a_{kj}(x)\, a_{ji}(x) = a_{ki}(x), \quad x \in V_i \cap V_j \cap V_k. \tag{9.3}$$

Set $k = i$; then it follows from (9.3) that $a_{ji}(x)$ has an inverse, and hence $a_{ji}(x)$ belongs to the group of non-singular 2×2 matrices.

A system of coordinates is called *oriented* if the determinant of $a_{ji}(x)$ is positive for all i, j, and all $x \in V_i \cap V_j$. Suppose S admits two oriented systems of coordinates given by (V_i, ψ_i) and (V'_k, ψ'_k). Then the Jacobian matrices of $\psi_j'^{-1}\psi_i$ have determinants which are either positive for all i, j, and $x \in V_i \cap V'_j$ or negative for all i, j, and $x \in V_i \cap V'_j$. In the former case the two systems are said to be *positively related* and in the latter case *negatively related*. The oriented systems divide into two classes, those in the same class being positively related while two in different classes are negatively related. Each class is called an *orientation* of S. If S admits an oriented system it is said to be *orientable*. A sphere and a torus are examples of orientable manifolds, while the Klein bottle is a non-orientable manifold.

10. Triangulation

We have defined a surface as essentially a set of points carrying a special topological structure such that every point has a neighbourhood homeomorphic to an open set in the Cartesian plane. A convenient tool for the study of certain topological properties of a surface is *triangulation*, a process by which the surface is divided into a number of curvilinear triangles which are joined to each other in a certain manner. We refer the reader to textbooks on algebraic topology† for a rigorous and detailed account of triangulation, and here we merely recall that the essential idea is to divide the surface up into a family of curvilinear triangles which satisfy the following conditions:

T_1. No two triangles have a common interior point.

T_2. To each side of a triangle there correspond two and only two triangles with this common side; an exception is made so that the side of a triangle which lies on the boundary of the surface shall belong to only one triangle.

T_3. Any two triangles can be joined by a sequence of triangles such that each triangle in the sequence has one and only one side in common with the next one in the sequence.

T_4. All the triangles with a common vertex can be arranged in a definite order so that consecutive triangles have a common side passing through the vertex.

Such a partitioning of the surface is called a *triangulation*. Evidently if a surface can be triangulated, such a triangulation is by no means unique. However, there are many properties common to all possible triangulations of a given surface, and these in fact form the combinatorial topological properties of the surface. As an example, it is well known that if a surface can be partitioned by a finite set of triangles with F faces, E edges, and V vertices, then the number $\chi = F - E + V$ is the same for all possible triangulations. This number is thus a topological invariant of the surface, and is called the Euler–Poincaré characteristic.

It was proved by T. Rado (1925) and later by S. S. Cairns (1934) that every surface of the type considered in section 9 can be triangulated.

Triangulation gives an intuitive method of describing orientability or non-orientability of a surface. Suppose that the surface has been triangulated, and consider one particular triangle ABC, whose sides are considered oriented in the sense AB, BC, CA (see Fig. 18). This will induce an orientation in the neighbouring triangles, so that the orientation of a neighbouring triangle ABD will be given by BA, AD, DB. The side AB common to both triangles has orientation AB in the first triangle and orientation

---

† e.g. Wylie and Hilton (1960).

BA in the second. If this scheme of orientation can be induced in a consistent manner over the whole surface, then the surface is orientable. It can be proved that this definition is equivalent to the one given in section 9.

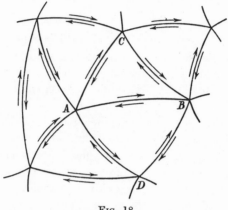

Fig. 18

If the surface is orientable and compact, another topological invariant can be defined called the *genus*, which is closely related to the Euler–Poincaré characteristic. The surface of a sphere is simply connected in the sense that every closed curve (homeomorphic image of a circle) on the surface can be deformed continuously into a point. (More precisely, a space is simply connected if its fundamental group consists of the identity.) On the other hand, the surface of an anchor ring is not simply connected, but it can be made into a simply connected region by two closed cuts. A double torus can be made into a simply connected region by four closed cuts. In general if it is necessary to introduce $2p$ simple cuts before a compact orientable surface is made into a simply connected region, the genus of the surface is defined to be p.

An alternative viewpoint is the following. If a closed curve is drawn on a sphere, the surface will be cut into two separate parts. In the case of an anchor ring it is possible to draw one closed curve without cutting the surface into more than one piece; but it is impossible to draw two non-intersecting curves on this surface without cutting it into more than one piece. The genus of a compact orientable surface is the maximum number of non-intersecting closed curves which can be drawn on the surface without cutting it into more than one piece. Again, genus could be defined for compact orientable surfaces by the statement that such a surface has genus p if it is homeomorphic to a sphere with p handles. For such a surface it can be shown that $\chi = 2(1-p)$. The fundamental theorem of the topology of compact surfaces states that *two compact surfaces are homeomorphic if and only if they have the same characteristic (or genus), and they are both orientable or both non-orientable.* The proof of this result may be found, for example, in Lefschetz (1949).

11. Two-dimensional Riemannian manifolds

Let V be any neighbourhood of a surface S of class $r > 1$ with the property that any point in V can be represented by two sets of coordinates (u, v) and (u', v'). Then the following relations hold between the differentials of the coordinates:

$$\begin{cases} du' = \dfrac{\partial u'}{\partial u}\, du + \dfrac{\partial u'}{\partial v}\, dv, \\[2mm] dv' = \dfrac{\partial v'}{\partial u}\, du + \dfrac{\partial v'}{\partial v}\, dv. \end{cases} \qquad \begin{cases} du = \dfrac{\partial u}{\partial u'}\, du' + \dfrac{\partial u}{\partial v'}\, dv', \\[2mm] dv = \dfrac{\partial v}{\partial u'}\, du' + \dfrac{\partial v}{\partial v'}\, dv'. \end{cases}$$

Suppose that there is associated with each point P belonging to V a *positive definite* quadratic differential form given in terms of coordinates (u, v) by

$$E\, du^2 + 2F\, du\, dv + G\, dv^2, \tag{11.1}$$

where E, F, G are functions of u and v of class $(r-1)$.

In terms of the coordinates (u', v') the differential form is

$$E'\, du'^2 + 2F'\, du'\, dv' + G'\, dv'^2, \tag{11.2}$$

where

$$\left. \begin{aligned} E' &= E\left(\frac{\partial u}{\partial u'}\right)^2 + 2F\,\frac{\partial u}{\partial u'}\,\frac{\partial v}{\partial u'} + G\left(\frac{\partial v}{\partial u'}\right)^2 \\[2mm] F' &= E\,\frac{\partial u}{\partial u'}\,\frac{\partial u}{\partial v'} + F\left(\frac{\partial u}{\partial u'}\,\frac{\partial v}{\partial v'} + \frac{\partial u}{\partial v'}\,\frac{\partial v}{\partial u'}\right) + G\,\frac{\partial v}{\partial u'}\,\frac{\partial v}{\partial v'} \\[2mm] G' &= E\left(\frac{\partial u}{\partial v'}\right)^2 + 2F\,\frac{\partial u}{\partial v'}\,\frac{\partial v}{\partial v'} + G\left(\frac{\partial v}{\partial v'}\right)^2 \end{aligned} \right\}. \tag{11.3}$$

Suppose now that there exists a positive definite quadratic differential form defined over each V_i of a coordinate system defined over S, with the property that if the form is defined over V_i by (11.1) and over V_j by (11.2), then in $V_i \cap V_j$ the coefficients of the two forms are related by (11.3). Then we shall say that such a surface S admits a *Riemannian structure*. S is called a *two-dimensional Riemannian manifold of class r*, and the quadratic differential form is called a *Riemannian metric*.

When the surface S is embedded in three-dimensional Euclidean space, the metric of the surrounding space induces a Riemannian metric given by (11.1) where E, F, G are the coefficients of the first form introduced in Chapter II. However, it was proved by N. E. Steenrod that when the surface S is defined intrinsically, without reference to the surrounding space, it nevertheless always admits globally a positive definite quadratic differential form. Hence any two-dimensional surface can be made into a two-dimensional Riemannian manifold. Moreover, even when the two-dimensional surface can be embedded in three-dimensional Euclidean space, the induced metric is not necessarily the same as the prescribed metric.

The local intrinsic theory of surfaces considered in Chapter II applies immediately to any one coordinate neighbourhood of a two-dimensional Riemannian manifold, provided that the first fundamental form is regarded as a prescribed Riemannian metric and not necessarily induced from the metric of a surrounding space. In particular, the intrinsic theory of geodesics and Gaussian curvature carries over without change.

The prescribed positive definite Riemannian metric can be used as in section 5 to define on S the structure of a metric space. The definition of completeness and the various ways of characterizing a complete surface described in section 6 carry over without change. Moreover, it can be proved that the topology of S regarded as a metric space is equivalent to the manifold topology by means of which S was defined.

The reader is warned against possible confusion in dealing with the structure of metric space induced on a submanifold. The Riemannian structure defined over the whole manifold defines a distance function over the whole manifold, and the restriction of this function to the submanifold gives this the structure of a metric space. Also, the Riemannian structure defined over the whole manifold induces a Riemannian structure over the submanifold and this gives rise to the structure of a metric space over the submanifold. It is important to realize that these two metric spaces which are induced on the submanifold may have quite different structures. For example, consider the two-dimensional Riemannian manifold formed by the torus in Fig. 19, with the prescribed Riemannian metric (dx^2+dy^2). Suppose that the ratio of the lengths of the sides AB, BC of the rectangle $ABCD$ is rational, and consider a straight line through A which makes an angle θ with AB such that $\tan\theta$ is irrational. Then it is easily verified that this geodesic is not closed, and hence there exist points on this curve whose 'distance' apart is as large as we please. On

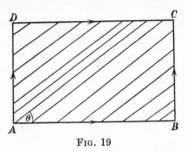

Fig. 19

the other hand, the 'distance' between any two points on the torus must be bounded.

Suppose we are given a Riemannian manifold S and a sub-manifold T. Then the Riemannian structure on S will induce a Riemannian structure on T. By means of the procedure in section 5 both S and T can be made into metric spaces. *It may happen that T is complete with respect to the distance function ρ_S associated with the Riemannian metric of S but not complete with respect to the distance function ρ_T associated with the Riemannian metric of T.*

12. The problem of metrization

In this problem one is given a differentiable surface of class r and considers whether a given quadratic differential form can be used to metrize the surface globally. The problem is difficult and is still not completely solved. Necessary conditions have been found which relate the topological and the differential invariants, but sufficient conditions are much more difficult to obtain.

As a first step towards the solution of this problem, we show that the Gauss–Bonnet theorem, which was proved in Chapter II as a local result, can be extended to apply to compact surfaces. Since the treatment in Chapter II involves only intrinsic properties, it is still valid for regions of compact surfaces which are simply connected.

Consider first a compact surface such as a sphere or ellipsoid. A simple closed curve Γ separates the surface into two simply connected regions I and II (Fig. 20). Apply the Gauss–Bonnet theorem to each region in turn to obtain

$$\int_{\Gamma} \kappa_g \, ds + \int_{I} K \, dS = 2\pi,$$

and

$$-\int_{\Gamma} \kappa_g \, ds + \int_{II} K \, dS = 2\pi,$$

since in the second case the curve is described in the reverse direction. Adding these equations we get $\int_{S} K \, dS = 4\pi$. It follows that the total curvature of a closed surface of the same connectivity

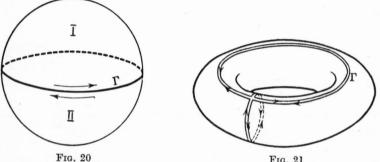

Fig. 20   Fig. 21

or genus as a sphere is 4π. Consider now an anchor ring, or rather a closed surface with the same connectivity as an anchor ring (Fig. 21). From section 10 it follows that this surface can be made into a simply connected region by means of two closed cuts, which can be so chosen that the region is bounded by four smooth arcs intersecting at angles of $\frac{1}{2}\pi$. Applying the Gauss–Bonnet theorem to this boundary, we get

$$\int \kappa_g \, ds + \int K \, dS + 2\pi = 2\pi,$$

from which

$$\int K \, dS = 0.$$

Thus the total curvature of a closed surface of the same genus as a torus is zero. A curve analogous to Γ in the case of a closed surface of genus p will have $4(2p-1)$ right angles, and in a similar manner we get

$$\int K \, dS = 4\pi(1-p).$$

Thus the total curvature of a closed surface of genus p is equal to $4\pi(1-p)$.

An alternative proof is the following. Let S be a compact orientable differentiable surface. It follows from section 10 that S can be triangulated into a finite number F of curvilinear triangles which can be oriented in a consistent manner over the whole surface. Apply the Gauss–Bonnet formula to a typical triangle T to get

$$\int_T \kappa_g \, ds + \sum \beta_i + \int_T K \, dS = 2\pi,$$

where the summation is taken over the exterior angles β_i of the triangle. Writing $\alpha_i = \pi - \beta_i$ for the interior angles, and adding the above formula for all triangles, we get

$$\int_S K \, dS + \sum (\pi - \alpha) = 2\pi F,$$

where the summation is now taken over all the triangles. The curvilinear integral vanishes since each edge of a triangle is traversed once in each direction. If V and E be the total number of vertices and edges in the triangulation, it can be seen that the term involving the summation sign can be written as $2E\pi - V2\pi$. The term π appears $2E$ times (twice for each edge); and the sum of all the interior angles at each vertex comes to 2π, making a total of $V2\pi$ for all the V vertices.

Hence
$$\int_S K \, dS = 2\pi(F - E + V)$$
$$= 2\pi\chi$$

where χ is the Euler–Poincaré characteristic. Comparing the two formulae for the total curvature we obtain the well-known relation $\chi = 2(1-p)$, referred to in section 10. The simple relation

$$\int_S K \, dS = 2\pi\chi$$

shows explicitly the relation between the differential geometric invariant K and the purely topological invariant of the underlying topological space. It follows immediately that a sphere cannot be metrized by a metric for which K is zero or negative. Similarly, a surface topologically equivalent to a torus can never be made into a differential geometric surface by means of a metric of positive curvature K.

An interesting differential geometric surface is the flat torus, i.e. a closed surface topologically equivalent to a torus but carrying a metric of the form $ds^2 = a \, d\theta^2 + c \, d\phi^2$, where a and c are constants.

This surface may be realized in Euclidean space of *four* dimensions by the equations

$$x_1 = \sqrt{a}\cos\theta, \qquad x_2 = \sqrt{a}\sin\theta, \qquad x_3 = \sqrt{c}\cos\phi,$$
$$x_4 = \sqrt{c}\sin\phi,$$

but it cannot be embedded in *three*-dimensional Euclidean space. It is evident that $K = 0$ and $\chi = 0$, in agreement with the Gauss–Bonnet formula.

So far in this section we have considered only compact surfaces. In an attempt to generalize the Gauss–Bonnet formula to complete but not compact surfaces, Cohn-Vossen (1935) has proved that

$$\int_S K\,dS \leqslant 2\pi\chi$$

provided that the left-hand integral exists as an improper integral. The proof is not brief and the reader is referred to Cohn-Vossen's paper for details.

13. The problem of continuation

In a sense the problem of continuation may be regarded as the inverse of the problem of metrization. Here a *surface element E* is given, i.e. a point P, a neighbourhood of P homeomorphic to a two-cell, and an *analytic* positive definite metric defined over the neighbourhood. The problem of continuation is to determine whether a given surface element can be continued to a complete surface.

In the special case when the metric has constant curvature, the problem has been considered by many mathematicians including Killing, Klein, Hopf, and others. A very good account of results obtained in this special case is given in a paper by Hopf (1925). In particular the main result is the following:

THEOREM 13.1. *To every constant K, there is one and only one simply connected, complete surface with constant curvature K.*

Alternatively, it may be said that any surface element E whose metric is of constant curvature K can be continued to a complete surface, and that any two such continuations must be isometric.

When, however, more general metrics are considered, the problem becomes very difficult and little is known. In a paper by Rinow (1932), the following uniqueness theorem is proved for general analytic metrics:

THEOREM 13.2. *Every surface element can be continued to at most one simply connected, complete surface, i.e. any two continuations must be isometric.*

An immediate corollary to this theorem is that a complete surface homeomorphic to a sphere cannot have the same geometry as one homeomorphic to a plane.

The main difference between Theorems 13.1 and 13.2 is that the former asserts existence and uniqueness, while the latter asserts only uniqueness.

There is no loss of generality in referring the metric of a surface element E to geodesic polar coordinates u, v where

$$ds^2 = du^2 + \lambda^2 dv^2.$$

It follows that information about continuability could be obtained solely from consideration of the behaviour of the analytic function λ. The following example due to Rinow shows the type of result which can be obtained.

THEOREM 13.3. *In order that the surface element, whose metric in geodesic polar coordinates is $ds^2 = du^2 + \lambda^2 dv^2$, can be continued to a complete surface, it is necessary that λ shall have no singularities at points with real v and positive u.*

An analogous result is

THEOREM 13.4. *A necessary condition for continuability is that the curvature K $(= -\lambda_{11}/\lambda)$ for any fixed v shall be an analytic function of u with no real singularities.*

As an example, consider the surface element with metric

$$ds^2 = du^2 + (u - u^3)^2 dv^2.$$

The curvature K is $6/(1 - u^2)$ which becomes infinite when $u = \pm 1$. It follows from Theorem 13.4 that this surface element cannot be continued to a complete surface.

The conditions in Theorems 13.3. and 13.4 are *necessary* but *not sufficient* for continuability. Examples are known of surface elements where λ is regular for all u and v, and which cannot be continued. The following positive result can be established.

THEOREM 13.5. *If, in addition, $\lambda \neq 0$ for all real v and positive u, then the surface element can be continued to a complete surface.*

For proofs of these theorems, the reader is referred to Rinow's paper quoted above. It should be noted that the metrics considered in this section are *analytic*, and no similar problem exists for metrics of class r.

14. Problems of embedding and rigidity

These two problems arise naturally from consideration of a differential geometric surface.

Embedding problem. Is a differential geometric surface isometric to a submanifold in a Euclidean space? If so, what is the least dimensional Euclidean space for which such embedding is possible?

Rigidity problem. Are two isometric submanifolds in a Euclidean space necessarily congruent or symmetric?

With regard to the first problem little is known. It remains unknown whether a differential geometric surface with constant curvature -1 can be embedded in a Euclidean space of sufficiently high dimension. It was proved in section 7 that the surface cannot be embedded in Euclidean space of three dimensions.

An example of a rigidity problem was dealt with in sections 3 and 4, in which the surface was defined by being embedded in Euclidean space of three dimensions. A generalization of these results is contained in the following theorem due to Cohn-Vossen.

THEOREM 14.1. *Two closed convex isometric surfaces embedded in Euclidean three-space are either congruent or symmetric.*

An interesting elementary proof of this result is given by S. Chern (1951).

15. Conclusion

The problems that have been discussed in this chapter can be easily generalized to manifolds of dimensions greater than two, i.e. to the so-called n-dimensional Riemannian manifolds. Little is known about the topology of these manifolds, and a complete set of conditions for the topological equivalence of two three-dimensional manifolds remains unknown. But, even in the case of two-dimensional differential geometric surfaces considered in this chapter, many problems remain unsolved though the topological characterization of these surfaces is fairly complete.

REFERENCES

BIEBERBACH, L., *Theorie der Differentialgleichungen*, Berlin (1923), p. 54.
—— *Acta Math.* **48** (1926) 319–27.
BLASCHKE, W., *Vorlesungen über Differentialgeometrie*, i, Berlin (1923–30), p. 206.

CAIRNS, S. S., *Ann. Math.* **35** (1934) 579.

CHERN, S., *Topics in Differential Geometry*, Institute for Advanced Study (1951).

CHEVALLEY, C., *Theory of Lie Groups*, vol. i, Princeton University Press (1946).

COHN–VOSSEN, S., *Compositio Math.* **2** (1935) 69–133.

DARBOUX, G., *Théorie Générale des Surfaces*, vol. iii, Gauthier–Villars (1896) § 629.

HOPF, H., 'Zum Clifford–Kleinschen Raumproblem', *Math. Ann.* **95** (1926) 313–39.

―― 'Differentialgeometrie und topologische Gestalt', *Jber. dtsch. Mat. Ver.* **41** (1932) 209–29.

―― and RINOW, W., 'Über den Begriff der vollständigen differential-geometrischen Fläche', *Comment. math. helvet.* **3** (1931) 209.

LEFSCHETZ, S., *Introduction to Topology*, Princeton University Press (1949).

MYERS, S. B., 'Riemannian manifolds in the large', *Duke Math. J.* **1** (1935) 39–49.

―― 'Connections between differential geometry and topology, I', ibid. 376–91.

―― 'Connections between differential geometry and topology, II', ibid. **2** (1936), 95–102.

―― 'Riemannian manifolds with positive mean curvature', ibid. **8** (1941) 401–4.

PATTERSON, E. M., *Topology*, Oliver and Boyd (1956).

PONTRJAGIN, L., *Topological Groups*, Princeton University Press (1946), § 46.

RADO, T., 'Über den Begriff der riemannschen Fläche', *Acta Szeged*, ii (1925).

DE RHAM, G., *Comment. math. helvet.* **26** (1952) 328–40.

RINOW, W., 'Über Zusammenhänge zwischen der Differentialgeometrie im Großen und im Kleinen', *Math. Z.* **35** (1932) 512–28.

VEBLEN, O., and WHITEHEAD, J. H. C., *Proc. Nat. Acad. Sci.* **17** (1931) 551.

WHITEHEAD, J. H. C., *Quart. J. Math.* **3** (1932) 33–42.

WYLIE, S., and HILTON, P., *Homology Theory*, Cambridge University Press (1960).

Part 2

DIFFERENTIAL GEOMETRY OF
n-DIMENSIONAL SPACE

V

TENSOR ALGEBRA

1. Vector spaces

THE elementary vector calculus used in Part 1 is not suitable for the study of the differential geometry of n-dimensional differentiable manifolds considered in Part 2, and it will be necessary to develop the more general *tensor calculus* which is concerned with the properties of fields of tensors defined over a differentiable manifold. A tensor is an algebraic construct associated with a basic abstract vector space, and the definition and properties of such tensors will be considered in the present chapter. In Chapter VI it will be shown that at each point P of a differentiable manifold there is a natural vector space—namely, the space of tangent vectors to the manifold at that point. The basic abstract vector space can be identified with this space of tangent vectors and the contents of the present chapter on tensor algebra thus has an immediate application to the study of differentiable manifolds. In this chapter certain properties of vector spaces will be stated without proof, and the reader is referred to such textbooks as that by P. Halmos (1953) for detailed proofs.

The reader is reminded that a vector space V over the field of real numbers consists of a set of elements (λ) called *vectors*, which may be added and subtracted in the usual way to give other vectors (addition is both commutative and associative). The set V contains a zero vector with the usual properties. Moreover, each vector λ of V can be multiplied by a real number a to give a vector $a\lambda$, the product satisfying the conditions

(i) $1\lambda = \lambda$, 　　　　　　　　(ii) $a(b\lambda) = ab(\lambda)$,

(iii) $(a+b)\lambda = a\lambda + b\lambda$, 　　　　(iv) $a(\lambda+\mu) = a\lambda + a\mu$.

A *vector subspace* of V is a subset of V which is itself a vector space under the induced laws of addition and multiplication by real numbers.

The vectors $\lambda_1, \lambda_2, ..., \lambda_p$ form a free system of order p when the relation

$$x^1\lambda_1 + x^2\lambda_2 + ... + x^p\lambda_p = 0$$

implies that all the coefficients x^i are zero. In this case the vectors are said to be *linearly independent*. Vectors which do not form a free system are said to be *linearly dependent*.

A given vector space V either admits free systems of arbitrarily large order, or else the order of the free systems is bounded and V is *finite dimensional*. In the latter case there exists some positive integer n with the property that there are free systems of order n but no free systems of order $(n+1)$, and the vector space is n-dimensional, i.e. dim $V = n$. All the vector spaces considered in this chapter will be finite dimensional. A free system of maximal order is called a *basis*. Standard theorems† state

THEOREM 1.1. *The number of elements in one basis of a finite dimensional vector space is the same as that in any basis.*

THEOREM 1.2. *A set of vectors is a basis of V if and only if* (i) *the vectors are linearly independent,* (ii) *every vector of V is expressible as a linear combination of these vectors.*

The following theorem will be used later and is stated here for convenience.

THEOREM 1.3. *When a free system of order r is given $(r < n)$, the system can be completed by means of $(n-r)$ other vectors to form a basis for V.*

Any n vectors forming a basis for V will be written $\mathbf{e}_1, \mathbf{e}_2, ..., \mathbf{e}_n$, and the set, i.e. the basis, will be written $(\mathbf{e}_i)$. As a consequence of Theorem 1.2 it follows that, relative to a basis $(\mathbf{e}_i)$, any vector λ is uniquely expressible in the form

$$\lambda = \sum_{i=1}^{n} \lambda^i \mathbf{e}_i.$$

Use will be made of the repeated suffix convention which states that a repeated suffix appearing once as a superscript and once as a subscript implies summation over the whole range of the suffix. Unless otherwise stated, this convention will be used throughout

† Cf. P. Halmos (1953).

the remainder of the book. The previous equation may now be written
$$\boldsymbol{\lambda} = \lambda^i \mathbf{e}_i. \tag{1.1}$$

The repeated suffix i is called a *dummy suffix*, as the symbol i has no special significance and could equally well have been replaced by j, k, etc. The numbers λ^i, where $i = 1, 2, ..., n$, are called the *components* of $\boldsymbol{\lambda}$ relative to the basis $(\mathbf{e}_i)$. It is evident that the vector $\boldsymbol{\lambda}$ is completely determined by specifying its components relative to any one basis. If μ^i are the components of another vector $\boldsymbol{\mu}$ relative to the same basis, then the components of the vector $\boldsymbol{\lambda} + \boldsymbol{\mu}$ are $\lambda^i + \mu^i$. Also if a is an arbitrary number, the components of the vector $a\boldsymbol{\lambda}$ are $a\lambda^i$.

Let the vectors $\mathbf{e}_{1'}, \mathbf{e}_{2'}, ..., \mathbf{e}_{n'}$ form another basis of V. Since each vector $\mathbf{e}_{i'}$ is uniquely expressible as a linear combination of the vectors $\mathbf{e}_i$, we have
$$\mathbf{e}_{i'} = p^i{}_{i'} \mathbf{e}_i, \tag{1.2}$$

where $(p^i{}_{i'})$ is an $n \times n$ matrix, non-singular because the vectors $\mathbf{e}_{i'}$ are linearly independent.

Similarly, the vector $\mathbf{e}_i$ is uniquely expressible in the form
$$\mathbf{e}_i = p^{i'}{}_i \mathbf{e}_{i'}, \tag{1.3}$$

where $(p^{i'}{}_i)$ is a non-singular $n \times n$ matrix. The matrix in (1.2) has primed lower suffixes and unprimed upper suffixes, whilst the matrix in (1.3) has primed upper suffixes and unprimed lower suffixes, so the notation prevents confusion between the two matrices.

Equations (1.2), (1.3) give
$$\mathbf{e}_i = p^{i'}{}_i p^j{}_{i'} \mathbf{e}_j.$$

The linear independence of the basis vectors implies that
$$p^j{}_{i'} p^{i'}{}_i = \delta^j{}_i, \tag{1.4}$$

where $\delta^j{}_i$, called the Kronecker delta, takes the value 1 if $i = j$ and is otherwise zero. The matrices $(p^i{}_{i'})$, $(p^{i'}{}_i)$ are therefore reciprocal, and we have immediately the relations
$$p^{j'}{}_i p^i{}_{i'} = \delta^{j'}{}_{i'}. \tag{1.5}$$

The components of a vector $\boldsymbol{\lambda}$ were defined relative to a basis $(\mathbf{e}_i)$, and a change of basis will induce a change of components. The law of transformation for the components of the vector $\boldsymbol{\lambda}$ will now be found when the basis is changed from $(\mathbf{e}_i)$ to $(\mathbf{e}_{i'})$ according to equation (1.2). If the vector $\boldsymbol{\lambda}$ has components λ^i relative to the basis $(\mathbf{e}_i)$, it is convenient to write $\lambda^{i'}$ for its components relative to the new basis $(\mathbf{e}_{i'})$, related to the former by (1.2). Then

$$\boldsymbol{\lambda} = \lambda^i \mathbf{e}_i = \lambda^{i'} \mathbf{e}_{i'}. \tag{1.6}$$

Equations (1.2), (1.6) give

$$\lambda^i \mathbf{e}_i = \lambda^{i'} p^i{}_{i'} \mathbf{e}_i,$$

from which, since the basis vectors $\mathbf{e}_i$ are linearly independent, it follows that

$$\lambda^i = p^i{}_{i'} \lambda^{i'}. \tag{1.7}$$

Similarly, substitute in (1.6) for $\mathbf{e}_i$ from (1.3) to get

$$\lambda^i p^{i'}{}_i \mathbf{e}_{i'} = \lambda^{i'} \mathbf{e}_{i'},$$

from which, since the basis vectors $\mathbf{e}_{i'}$ are linearly independent,

$$\lambda^{i'} = p^{i'}{}_i \lambda^i. \tag{1.8}$$

Then equation (1.8) expresses the new components in terms of the old components, while equation (1.7) expresses the old components in terms of the new.

The notation of using primed and unprimed suffixes has been used extensively by J. A. Schouten and it has the great advantage that it can be used as an aid to calculation. For example, change the dummy suffix i' of (1.7) to j' and multiply by $p^{i'}{}_i$ to get

$$\begin{aligned} p^{i'}{}_i \lambda^i &= p^{i'}{}_i (p^i{}_{j'} \lambda^{j'}) \\ &= \delta^{i'}{}_{j'} \lambda^{j'} \quad \text{from (1.5)} \\ &= \lambda^{i'}, \end{aligned}$$

which is equation (1.8). The reader will have noticed that in the previous calculations it was necessary to relabel dummy suffixes in order to avoid the same suffix appearing more than twice in the same expression. Failure to take care to relabel dummy suffixes often leads to errors in tensor calculations.

2. The dual space

In what follows the vectors of the n-dimensional vector space V of section 1 are called *contravariant vectors*. It will be shown that associated with V is another vector space, denoted by V^*, whose elements are called *covariant vectors*. Other vector spaces associated with V will appear later—in fact, tensors will appear as elements of these associated vector spaces. As far as possible the following convention will be used systematically. *Contravariant and covariant vectors and later, tensors, will be denoted by symbols* *printed in heavy type while real numbers will be denoted by symbols in ordinary type.*

A real-valued linear function μ over the space V will be called a *covariant vector*. If λ is a contravariant vector, the real number into which μ maps λ will be written as $\mu(\lambda)$. Then $\mu(\lambda)$ is the value taken by the function μ evaluated at the vector λ. By saying that μ is linear we mean that

$$\mu(a\lambda+b\nu) = a\mu(\lambda)+b\mu(\nu), \tag{2.1}$$

where λ, ν are any two contravariant vectors and a, b are any real numbers.

As an example of a covariant vector, consider the function μ which maps λ into λ^i, i.e. into its ith component with respect to a basis (e_i) of V. Evidently μ is a real-valued function over V and since the ith component of the vector $a\lambda+b\nu$ is $a\lambda^i+b\nu^i$, condition (2.1) is satisfied. Thus μ is a covariant vector.

It will now be shown that by suitably defining the sum of two covariant vectors and multiplication of a covariant vector by a real number, the set of all covariant vectors may be made into a vector space.

If α, β are any two covariant vectors, define their sum $\alpha+\beta$ to be the covariant vector γ such that

$$\gamma(\lambda) = \alpha(\lambda)+\beta(\lambda) \tag{2.2}$$

for all contravariant vectors λ, i.e. the real number $\gamma(\lambda)$ is just the sum of the real numbers $\alpha(\lambda)$, $\beta(\lambda)$. The product $a\alpha$ of a covariant vector α by the real number a is defined to be the covariant vector δ such that

$$\delta(\lambda) = a\alpha(\lambda), \tag{2.3}$$

for every contravariant vector λ. It is readily verified that with these laws of combination the covariant vectors form a vector space. This is called the *dual space of V* and is denoted by V^*.

It will now be shown that a basis (e_i) of V induces a unique basis in V^*. This is called the *dual basis* of (e_i) and is denoted by (e^i). Let λ^i be the components of the contravariant vector λ with respect to the basis (e_i), and let μ be any covariant vector. Then

$$\mu(\lambda) = \mu(\lambda^i e_i) = \lambda^i \mu(e_i).$$

Define n numbers μ_i by the relation

$$\mu_i = \mu(e_i).$$

Then

$$\mu(\lambda) = \lambda^i \mu_i. \tag{2.4}$$

It has already been seen that the real-valued linear function which maps a vector $\boldsymbol{\lambda}$ of V into its ith component with respect to a basis $(\mathbf{e}_i)$ is a covariant vector. Denote this vector by $\mathbf{e}^i$, so that

$$e^i(\boldsymbol{\lambda}) = \lambda^i,$$

and
$$e^i(\mathbf{e}_j) = \delta^i{}_j. \tag{2.5}$$

From (2.4) we have, for every contravariant vector $\boldsymbol{\lambda}$,

$$\mu(\boldsymbol{\lambda}) = \mu_i e^i(\boldsymbol{\lambda}),$$

and hence
$$\boldsymbol{\mu} = \mu_i \mathbf{e}^i. \tag{2.6}$$

Thus an arbitrary covariant vector $\boldsymbol{\mu}$ is expressible as a linear combination of the n vectors $\mathbf{e}^i$. Moreover, the relation

$$x_1 \mathbf{e}^1 + x_2 \mathbf{e}^2 + \ldots + x_n \mathbf{e}^n = \mathbf{0}$$

implies that for an *arbitrary* contravariant vector $\boldsymbol{\lambda}$ with components λ^i,

$$x_1 \lambda^1 + x_2 \lambda^2 + \ldots + x_n \lambda^n = 0.$$

Hence $x_1 = x_2 = \ldots = x_n = 0$, and the vectors $\mathbf{e}^i$ are linearly independent. It follows from Theorem 1.2 that the vectors $\mathbf{e}^i$ form a basis for V^*, and, as a corollary, it follows that $\dim V^* = n$.

A change of basis from $(\mathbf{e}_i)$ to $(\mathbf{e}_{i'})$ in V will induce a change of dual basis from $(\mathbf{e}^i)$ to $(\mathbf{e}^{i'})$ in V^*. The precise form of this change of dual basis will now be found. Let the transformation

$$\mathbf{e}_{j'} = p^j{}_{j'} \mathbf{e}_j$$

induce a transformation on the dual vectors given by

$$\mathbf{e}^{i'} = r^{i'}{}_i \mathbf{e}^i.$$

Then
$$\delta^{i'}{}_{j'} = e^{i'}(\mathbf{e}_{j'}) = r^{i'}{}_i e^i(p^j{}_{j'} \mathbf{e}_j) = r^{i'}{}_i p^i{}_{j'}.$$
Multiply by $p^{j'}{}_k$ to obtain

$$p^{i'}{}_k = r^{i'}{}_k,$$

and the required transformation for the dual vectors is therefore

$$\mathbf{e}^{i'} = p^{i'}{}_i \mathbf{e}^i. \tag{2.7}$$

It is well known (cf. Halmos (1953), p. 13) that all n-dimensional vector spaces are isomorphic, and so in particular V and V^* are isomorphic. However, certain vector spaces are related by a natural isomorphism, which is a much more strict relation. As an example we shall prove that there is a natural isomorphism relating V and V^{**}, the dual space of V^*.

Let $\boldsymbol{\lambda}$ have components λ^i with respect to the basis $(\mathbf{e}_i)$, and let $\boldsymbol{\mu}$ have components μ_i with respect to the dual basis $(\mathbf{e}^i)$. Then we have

$$\mu(\boldsymbol{\lambda}) = \lambda^i \mu_i. \tag{2.8}$$

When $\boldsymbol{\lambda}$ is regarded as fixed, this relation can be regarded as a mapping of an arbitrary covariant vector $\boldsymbol{\mu}$ into the real number $\lambda^i \mu_i$. Moreover, the form of (2.8) shows that this mapping is linear. Thus a given vector $\boldsymbol{\lambda}$ of V determines a real-valued linear function over the space V^*, i.e. $\boldsymbol{\lambda}$ determines a definite vector, say λ^{**}, of the vector space V^{**}. Equation (2.8) may then be written

$$\lambda^{**}(\boldsymbol{\mu}) = \mu(\boldsymbol{\lambda}). \tag{2.9}$$

Suppose now that two vectors $\boldsymbol{\lambda}$, $\boldsymbol{\nu}$ of V determine the same vector λ^{**} of V^{**}. Then from (2.9) it follows that for every covariant vector $\boldsymbol{\mu}$,
$$\mu(\boldsymbol{\lambda}) = \mu(\boldsymbol{\nu}).$$

In particular, by taking $\boldsymbol{\mu}$ to be the basis vector $\mathbf{e}^i$, we obtain $\lambda^i = \nu^i$ and hence $\boldsymbol{\lambda} = \boldsymbol{\nu}$. We have thus proved that the relation between $\boldsymbol{\lambda}$ and λ^{**} given by (2.9) is both linear and one-to-one. Let $\hat{V}$ denote the subvector space of V^{**} consisting of all vectors λ^{**} which arise from vectors of V by the relation (2.9). Evidently $\dim \hat{V} = n$. However, since $\dim V = n$ implies $\dim V^* = \cdot n$, it follows that $\dim V^{**} = n$ and the subspace $\hat{V}$ must indeed be the whole space V^{**}. Hence the relation (2.9) gives a natural isomorphism between V and V^{**}.

As a convention we agree to *identify* the two spaces V and V^{**} by regarding λ^{**} as the same vector as $\boldsymbol{\lambda}$. The dual basis of the basis $(\mathbf{e}^i)$ of V^* is then just the original basis $(\mathbf{e}_i)$ of V.

3. Tensor product of vector spaces

Let V, W be vector spaces with $\dim V = n$, $\dim W = m$. The reader is probably already familiar with the *direct sum* of V and W which is a vector space of dimensions $m+n$, and is usually denoted by $V \dotplus W$. The elements of $V \dotplus W$ are ordered pairs of vectors $(\boldsymbol{\alpha}, \boldsymbol{\lambda})$ where $\boldsymbol{\alpha}$ belongs to V and $\boldsymbol{\lambda}$ belongs to W. The operations of vector addition and multiplication by numbers are defined in $V \dotplus W$ as follows:
$$(\boldsymbol{\alpha}, \boldsymbol{\lambda}) + (\boldsymbol{\beta}, \boldsymbol{\mu}) = (\boldsymbol{\alpha}+\boldsymbol{\beta}, \boldsymbol{\lambda}+\boldsymbol{\mu}),$$
$$a(\boldsymbol{\alpha}, \boldsymbol{\lambda}) = (a\boldsymbol{\alpha}, a\boldsymbol{\lambda}).$$

We now show how to associate with the two vector spaces V, W another vector space called the *tensor product* of V and W. This space has dimensions mn and will be denoted by the symbol $V \otimes W$. Before defining $V \otimes W$ it will be convenient to define what is meant by a *real-valued bilinear function defined over the Cartesian product of two vector spaces*.

Let A, U be two vector spaces, and denote by $A \times U$ the Cartesian product of A and U, i.e. the set of all ordered pairs of vectors $(\boldsymbol{\alpha}, \boldsymbol{\lambda})$ where $\boldsymbol{\alpha}$ belongs to A and $\boldsymbol{\lambda}$ belongs to U. Although the elements of $A \times U$ are identical with those of the direct sum $A \dotplus U$ we prefer to use the notation $A \times U$ since the elements will be regarded only as members of a set and the vector structure will not be used. Then $\mathbf{T}$ is called a *real-valued bilinear function* over $A \times U$ if

$$T(a\boldsymbol{\alpha}+b\boldsymbol{\beta}, \boldsymbol{\mu}) = aT(\boldsymbol{\alpha}, \boldsymbol{\mu})+bT(\boldsymbol{\beta}, \boldsymbol{\mu}), \tag{3.1}$$

$$T(\boldsymbol{\alpha}, a\boldsymbol{\mu}+b\boldsymbol{\nu}) = aT(\boldsymbol{\alpha}, \boldsymbol{\mu})+bT(\boldsymbol{\alpha}, \boldsymbol{\nu}), \tag{3.2}$$

for all vectors $\boldsymbol{\alpha}$, $\boldsymbol{\beta}$ belonging to A, for all vectors $\boldsymbol{\mu}$, $\boldsymbol{\nu}$ belonging to U, and for all real numbers a, b.

Let us return to the two given spaces V, W and consider real-valued bilinear functions defined over the Cartesian product of the dual spaces, $V^* \times W^*$. Let $\mathbf{Q}$, $\mathbf{R}$ be two such functions. Then the sum $\mathbf{Q}+\mathbf{R}$ is defined to be that bilinear function $\mathbf{T}$ for which

$$T(\boldsymbol{\lambda}, \boldsymbol{\mu}) = Q(\boldsymbol{\lambda}, \boldsymbol{\mu})+R(\boldsymbol{\lambda}, \boldsymbol{\mu}), \tag{3.3}$$

while the product of $\mathbf{Q}$ with the real number a is defined to be the bilinear function $\mathbf{S}$ such that

$$S(\boldsymbol{\lambda}, \boldsymbol{\mu}) = aQ(\boldsymbol{\lambda}, \boldsymbol{\mu}). \tag{3.4}$$

It is readily verified that with the laws of composition (3.3), (3.4), the set of all such real-valued bilinear functions defined over $V^* \times W^*$ forms a vector space over the real numbers. This vector space is called the *tensor product* of the vector spaces V and W, and is denoted by $V \otimes W$. It should be noted that in defining the tensor product of V and W, the real-valued bilinear functions were defined over the Cartesian product of the dual spaces V^* and W^* and *not over the Cartesian product of the spaces V and W*.

Let $\mathbf{e}^i$ $(i = 1, 2,..., n)$ form the basis of V^* dual to the basis $(\mathbf{e}_i)$ of V, and let $\mathbf{e}^\alpha$ $(\alpha = 1, 2,..., m)$ form the basis of W^* dual to the basis $(\mathbf{e}_\alpha)$ of W. It will now be shown that these determine a unique basis of the tensor product $V \otimes W$.

Let $\boldsymbol{\lambda}$, $\boldsymbol{\mu}$ be arbitrary vectors of V^* and W^*, so that

$$\boldsymbol{\lambda} = \lambda_i \mathbf{e}^i, \qquad \boldsymbol{\mu} = \mu_\alpha \mathbf{e}^\alpha.$$

If $\mathbf{T}$ is any element of $V \otimes W$, then

$$T(\boldsymbol{\lambda}, \boldsymbol{\mu}) = T(\lambda_i \mathbf{e}^i, \mu_\alpha \mathbf{e}^\alpha) = \lambda_i \mu_\alpha T(\mathbf{e}^i, \mathbf{e}^\alpha).$$

Write $\qquad\qquad T^{i\alpha} = T(\mathbf{e}^i, \mathbf{e}^\alpha);$

then $\qquad\qquad T(\boldsymbol{\lambda}, \boldsymbol{\mu}) = \lambda_i \mu_\alpha T^{i\alpha}.$

By the previous definitions the function which maps the ordered pair $(\boldsymbol{\lambda}, \boldsymbol{\mu})$ into the real number $\lambda_i \mu_\alpha$ is a real-valued bilinear function which we denote by $\mathbf{e}_{i\alpha}$. Then $\mathbf{e}_{i\alpha}$ will map $(\mathbf{e}^j, \mathbf{e}^\beta)$ into the number $\delta^j{}_i \delta^\beta{}_\alpha$. Hence

$$\begin{aligned}
T(\boldsymbol{\lambda}, \boldsymbol{\mu}) &= T^{i\alpha} \delta^j{}_i \delta^\beta{}_\alpha \lambda_j \mu_\beta \\
&= T^{i\alpha} e_{i\alpha}(\lambda_j \mathbf{e}^j, \mu_\beta \mathbf{e}^\beta) \\
&= T^{i\alpha} e_{i\alpha}(\boldsymbol{\lambda}, \boldsymbol{\mu}).
\end{aligned}$$

Since this is true for arbitrary vectors $\boldsymbol{\lambda}$, $\boldsymbol{\mu}$, we have

$$\mathbf{T} = T^{i\alpha} \mathbf{e}_{i\alpha},$$

so any element of $V \otimes W$ is expressible as a linear combination of the vectors $\mathbf{e}_{i\alpha}$.

Suppose $\qquad x^{i\alpha} \mathbf{e}_{i\alpha} = \mathbf{0},$

for some coefficients $x^{i\alpha}$. Then for all vectors $\boldsymbol{\lambda}$, $\boldsymbol{\mu}$, we have

$$x^{i\alpha} e_{i\alpha}(\boldsymbol{\lambda}, \boldsymbol{\mu}) = 0,$$

and in particular $\qquad x^{i\alpha} e_{i\alpha}(\mathbf{e}^j, \mathbf{e}^\beta) = 0,$

i.e. $x^{j\beta} = 0$. Thus the vectors $\mathbf{e}_{i\alpha}$ are linearly independent. It follows from Theorem 1.2 that the vectors $\mathbf{e}_{i\alpha}$ form a basis, which will be written $(\mathbf{e}_{i\alpha})$. As a corollary, $\dim(V \otimes W) = mn$.

Let $\boldsymbol{\lambda}$ have components λ^i with respect to $(\mathbf{e}_i)$ and let $\boldsymbol{\mu}$ have components μ^α with respect to $(\mathbf{e}_\alpha)$. We define the tensor product of $\boldsymbol{\lambda}$ and $\boldsymbol{\mu}$, denoted by $\boldsymbol{\lambda} \otimes \boldsymbol{\mu}$, to be that element of $V \otimes W$ whose components with respect to the basis $(\mathbf{e}_{i\alpha})$ are $\lambda^i \mu^\alpha$, i.e.

$$\boldsymbol{\lambda} \otimes \boldsymbol{\mu} = \lambda^i \mu^\alpha \mathbf{e}_{i\alpha}. \tag{3.5}$$

Any element of $V \otimes W$ of the form $\boldsymbol{\lambda} \otimes \boldsymbol{\mu}$ is said to be *decomposable*. In particular, the basis vectors $\mathbf{e}_{i\alpha}$ are decomposable since we may write

$$\mathbf{e}_{i\alpha} = \mathbf{e}_i \otimes \mathbf{e}_\alpha. \tag{3.6}$$

Since any element of $V \otimes W$ is expressible as a linear combination of basis vectors, it follows that *it can be expressed as a linear combination of decomposable elements*.

In the previous section it was proved that the vector spaces V, V^{**} are naturally isomorphic. Further examples of naturally isomorphic vector spaces will now be considered.

Let $\mathbf{L}$ be a linear mapping of V^* into W, i.e. $\mathbf{L}$ is a linear function defined over V^* whose values are vectors in W. Then

$$\mathbf{L}(a\boldsymbol{\alpha}^* + b\boldsymbol{\beta}^*) = a\mathbf{L}(\boldsymbol{\alpha}^*) + b\mathbf{L}(\boldsymbol{\beta}^*), \tag{3.7}$$

for all vectors $\boldsymbol{\alpha}^*$, $\boldsymbol{\beta}^*$ of V^* and for all real numbers a, b. The *sum*

$L+M$ of two such linear functions L, M is defined to be the linear function N such that

$$N(\alpha^*) = L(\alpha^*) + M(\alpha^*) \tag{3.8}$$

for all vectors α^* of V^*. The product aL of L and the real number a is defined by the linear function P such that

$$P(\alpha^*) = aL(\alpha^*). \tag{3.9}$$

It is easily verified that with these laws of combination the set of all linear maps of V^* into W form a vector space, which we denote by $\mathscr{L}(V^*, W)$. *We shall now prove that the vector spaces $V \otimes W$ and $\mathscr{L}(V^*, W)$ are naturally isomorphic, so that a real-valued bilinear mapping of $V^* \times W^*$ determines and is determined by a linear mapping of V^* into W.*

Relative to bases (e_i) of V and (e_α) of W, any element T of $V \otimes W$ can be written in the form

$$T = T^{i\alpha} e_{i\alpha}, \tag{3.10}$$

where the coefficients $T^{i\alpha}$ are uniquely determined. Let λ^* in V^* have components λ_i relative to the basis (e^i), and let μ^α be the components of a vector μ of W relative to (e_α). Then the relation

$$\mu^\alpha = T^{i\alpha} \lambda_i \tag{3.11}$$

determines a unique linear mapping of V^* into W. Conversely, any linear mapping of V^* into W is uniquely determined relative to the bases (e^i), (e_α) by the coefficients $T^{i\alpha}$ in (3.11), and these coefficients determine a unique element of $V \otimes W$ given by (3.10). This correspondence between $V \otimes W$ and $\mathscr{L}(V^*, W)$ is easily seen to be a natural isomorphism. We leave as an exercise to the reader to prove that this isomorphism is natural, that is, independent of the particular bases in V or W. In later chapters we shall make use of this isomorphism. Sometimes it is more convenient to consider an element of $V \otimes W$ as a *linear* mapping of V^* into W rather than a *bilinear* mapping of $V^* \times W^*$ into the real numbers.

Let V, W, Y be three vector spaces with bases (e_i), (e_α), and (e_A). It will now be proved that the two vector spaces

$$(V \otimes W) \otimes Y, \qquad V \otimes (W \otimes Y)$$

are naturally isomorphic. From the previous results, it follows that any elements of $(V \otimes W) \otimes Y$ can be written in the form

$$T^{i\alpha A}(e_i \otimes e_\alpha) \otimes e_A,$$

and the coefficients $T^{i\alpha A}$ are uniquely determined. We make corresponding to this the vector in $V \otimes (W \otimes Y)$ given by

$$T^{i\alpha A}\mathbf{e}_i \otimes (\mathbf{e}_\alpha \otimes \mathbf{e}_A).$$

Conversely, any element in $V \otimes (W \otimes Y)$ determines a unique element in $(V \otimes W) \otimes Y$, and this correspondence is clearly a natural isomorphism. There will thus be no confusion if these two spaces are identified and the symbol $V \otimes W \otimes Y$ will be used to denote either space.

It will now be shown that the space $V \otimes W \otimes Y$ is naturally isomorphic to a third vector space whose elements are trilinear functions—a function of several variables is *multilinear* if it is linear in each variable when all the other variables are kept constant. In fact the set of all real-valued trilinear functions defined over the Cartesian product $V^* \times W^* \times Y^*$ form a vector space which is naturally isomorphic with $V \otimes W \otimes Y$. Let $\mathbf{S}, \mathbf{T}$ be two real-valued trilinear functions defined over $V^* \times W^* \times Y^*$. Then the sum of these two functions $\mathbf{S} + \mathbf{T}$ is defined as the trilinear function $\mathbf{U}$ such that

$$U(\boldsymbol{\lambda}, \boldsymbol{\mu}, \boldsymbol{\nu}) = S(\boldsymbol{\lambda}, \boldsymbol{\mu}, \boldsymbol{\nu}) + T(\boldsymbol{\lambda}, \boldsymbol{\mu}, \boldsymbol{\nu})$$

for all vectors $\boldsymbol{\lambda}, \boldsymbol{\mu}, \boldsymbol{\nu}$ belonging respectively to V^*, W^*, and Y^*. Also, the product $a\mathbf{S}$ of $\mathbf{S}$ and a real number a is defined as the trilinear function $\mathbf{R}$ such that

$$R(\boldsymbol{\lambda}, \boldsymbol{\mu}, \boldsymbol{\nu}) = aS(\boldsymbol{\lambda}, \boldsymbol{\mu}, \boldsymbol{\nu}).$$

It is readily verified that with these laws of composition the set of trilinear functions over $V^* \times W^* \times Y^*$ forms a vector space.

The trilinear function which maps the triple $(\boldsymbol{\lambda}, \boldsymbol{\mu}, \boldsymbol{\nu})$ into the product of the three components $\lambda_i \mu_\alpha \nu_A$ will be denoted by $\mathbf{e}_{i\alpha A}$. This function will map the triple $(\mathbf{e}^j, \mathbf{e}^\beta, \mathbf{e}^B)$ into the number $\delta_i^j \delta_\alpha^\beta \delta_A^B$. If $\mathbf{T}$ is any trilinear function, then

$$T(\boldsymbol{\lambda}, \boldsymbol{\mu}, \boldsymbol{\nu}) = T(\lambda_i \mathbf{e}^i, \mu_\alpha \mathbf{e}^\alpha, \nu_A \mathbf{e}^A)$$
$$= \lambda_i \mu_\alpha \nu_A T(\mathbf{e}^i, \mathbf{e}^\alpha, \mathbf{e}^A).$$

Write
$$T(\mathbf{e}^i, \mathbf{e}^\alpha, \mathbf{e}^A) = T^{i\alpha A};$$

then
$$T(\boldsymbol{\lambda}, \boldsymbol{\mu}, \boldsymbol{\nu}) = T^{i\alpha A} \lambda_i \mu_\alpha \nu_A$$
$$= T^{i\alpha A} \delta_i^j \lambda_j \delta_\alpha^\beta \mu_\beta \delta_A^B \nu_B$$
$$= T^{i\alpha A} e_{i\alpha A} (\lambda_j \mathbf{e}^j, \mu_\beta \mathbf{e}^\beta, \nu_B \mathbf{e}^B)$$
$$= T^{i\alpha A} e_{i\alpha A} (\boldsymbol{\lambda}, \boldsymbol{\mu}, \boldsymbol{\nu}).$$

Thus $\mathbf{T} = T^{i\alpha A}\mathbf{e}_{i\alpha A}$.

As in the case of the bilinear functions $\mathbf{e}_{i\alpha}$, it is easily seen that the vectors $\mathbf{e}_{i\alpha A}$ are linearly independent, and since any $\mathbf{T}$ is expressible as a linear combination of these functions they form a basis for the vector space.

The natural isomorphism between this vector space and the space $V \otimes W \otimes Y$ is easily obtained by making correspond to the element $T^{i\alpha A}\mathbf{e}_{i\alpha A}$ of the first space the element $T^{i\alpha A}\mathbf{e}_i \otimes \mathbf{e}_\alpha \otimes \mathbf{e}_A$ of the second space. The reader will easily complete details of the proof that this correspondence is a natural isomorphism.

In the same way it may be shown that the set of all real-valued multilinear functions defined over $U_1^* \times U_2^* \times ... \times U_p^*$ form a vector space which is naturally isomorphic to the tensor product

$$U_1 \otimes U_2 \otimes ... \otimes U_p.$$

EXERCISE 3.1. Prove that the spaces $V \otimes W$, $W \otimes V$ are naturally isomorphic, and hence prove that $V \otimes W$ is naturally isomorphic with the vector space of linear maps of W^* into V.

EXERCISE 3.2. Show that the space $(V \otimes W)^*$ is naturally isomorphic to the space $V^* \otimes W^*$.

The properties of tensor products of vector spaces described in the present section were given in order to introduce the concept of a *tensor*. We are now in a position to define tensors and to study their algebraic properties.

A tensor is an element of the vector space formed by taking repeated tensor products of spaces taken from the pair V, V^*.

If we write V^1 for the space of contravariant vectors V, then the elements of the tensor product of V^1 taken r times, denoted by V^r, are called *contravariant tensors of order r*. Similarly, writing V_1 for the space of covariant vectors V^*, the elements of the tensor product of V_1 taken s times, denoted by V_s, are called *covariant tensors of order s*. An element of $V^r \otimes V_s$, called a *tensor of order $r+s$, contravariant of order r and covariant of order s*, will be referred to as a *tensor of type (r,s)*. Contravariant vectors are thus tensors of type $(1,0)$, and covariant vectors are tensors of type $(0,1)$. It will be convenient to consider scalars (real numbers) as tensors of type $(0,0)$. Sometimes $V^r \otimes V_s$ will be denoted by V_s^r.

By making use of the natural isomorphism between tensors and real-valued multilinear functions, we could have *defined* a tensor of type (r, s) to be a real-valued multilinear function over the Cartesian product

$$(V_1 \times V_1 \times \ldots \times V_1) \times (V^1 \times V^1 \times \ldots \times V^1),$$

where the first bracket contains r factors and the second bracket s factors. The reader should be careful to note that a tensor of type (r, s) can be regarded either as an element of the *tensor product* of r spaces V^1 and s spaces V_1, or else as a real-valued multilinear function over the *Cartesian product* of s spaces V^1 and r spaces V_1.

Components of tensors

A given basis $(\mathbf{e}_i)$ of V^1 determines a basis $(\mathbf{e}_{i_1 i_2})$ of V^2, and since any element $\mathbf{T}$ of V^2 can be written uniquely in the form

$$\mathbf{T} = T^{i_1 i_2} \mathbf{e}_{i_1 i_2}, \tag{3.12}$$

the numbers $T^{i_1 i_2}$ are called the *components of* $\mathbf{T}$ *relative to the basis* $(\mathbf{e}_i)$ *of* V^1.

The basis $(\mathbf{e}_i)$ of V^1 induces a unique dual basis $(\mathbf{e}^j)$ of V_1 and hence a unique basis of V_1^1 denoted by $(\mathbf{e}_{i_1}^{j_1})$. Since any element of V_1^1 can be written uniquely in the form

$$\mathbf{T} = T_{j_1}^{i_1} \mathbf{e}_{i_1}^{j_1}, \tag{3.13}$$

the numbers $T_{j_1}^{i_1}$ are called the *components of* $\mathbf{T}$ *relative to the basis* $(\mathbf{e}_i)$ *of* V^1. Similarly, the basis $(\mathbf{e}_i)$ of V^1 determines a basis $(\mathbf{e}^{j_1 j_2})$ in V_2, and since any element of V_2 can be written uniquely in the form

$$\mathbf{T} = T_{j_1 j_2} \mathbf{e}^{j_1 j_2}, \tag{3.14}$$

the numbers $T_{j_1 j_2}$ are called the *components of* $\mathbf{T}$ *relative to the basis* $(\mathbf{e}_i)$ *of* V^1. More generally, a basis $(\mathbf{e}_i)$ determines uniquely a basis $(\mathbf{e}_{i_1 i_2 \ldots i_r}^{j_1 j_2 \ldots j_s})$ in V_s^r, and since any element of V_s^r is uniquely expressible in the form

$$\mathbf{T} = T_{j_1 j_2 \ldots j_s}^{i_1 i_2 \ldots i_r} \mathbf{e}_{i_1 i_2 \ldots i_r}^{j_1 j_2 \ldots j_s}, \tag{3.15}$$

the numbers $T_{j_1 j_2 \ldots j_s}^{i_1 i_2 \ldots i_r}$ are called the *components of* $\mathbf{T}$ *relative to the basis* $(\mathbf{e}_i)$ *of* V^1. Evidently a tensor $\mathbf{T}$ is determined by its components relative to any one basis.

Two tensors of the same type (r, s) may be added or multiplied by scalars to give tensors of the same type. Relative to a basis, the components of the sum of two tensors are just the sum of the corresponding components of each tensor.

Two tensors of different types cannot be added together since they are elements of different vector spaces. However, two tensors $\mathbf{R}, \mathbf{S}$ of types $(r, s), (p, q)$ can always be multiplied to form a tensor

$\mathbf{R} \otimes \mathbf{S}$ of type $(r+p, s+q)$, defined as follows. If $\mathbf{R}, \mathbf{S}$ have components $R^{i_1 i_2 \ldots i_r}_{j_1 j_2 \ldots j_s}$, $S^{i_{r+1} \ldots i_{r+p}}_{j_{s+1} \ldots j_{s+q}}$ respectively with respect to the basis $(\mathbf{e}_i)$ of V^1, the tensor $\mathbf{T} = \mathbf{R} \otimes \mathbf{S}$ is defined by the components

$$T^{i_1 \ldots i_{r+p}}_{j_1 \ldots j_{s+q}} = R^{i_1 \ldots i_r}_{j_1 \ldots j_s} S^{i_{r+1} \ldots i_{r+p}}_{j_{s+1} \ldots j_{s+q}}. \tag{3.16}$$

Although $\mathbf{R} \otimes \mathbf{S}$ has been defined relative to a particular basis $(\mathbf{e}_i)$ of V^1, it will be shown in the next section that this product remains invariant when the basis is changed and hence the product is independent of the particular basis $(\mathbf{e}_i)$.

4. Transformation formulae

It has been shown in section 1 that a change of basis of the vector space V given by

$$\mathbf{e}_{i'} = p^i{}_{i'} \mathbf{e}_i \tag{4.1}$$

induces the transformation of components of a vector $\boldsymbol{\lambda}$ given by

$$\lambda^{i'} = p^{i'}{}_i \lambda^i. \tag{4.2}$$

The corresponding transformation for the components of a vector $\boldsymbol{\mu}$ of the dual space V^* will now be obtained. Writing $\mu_i, \mu_{i'}$ for the components of $\boldsymbol{\mu}$ relative to bases $(\mathbf{e}_i)$, $(\mathbf{e}_{i'})$, then

$$\mu(\boldsymbol{\lambda}) = \mu_i \lambda^i = \mu_{i'} \lambda^{i'} = \mu_{i'} p^{i'}{}_i \lambda^i, \quad \text{from (4.2)}.$$

The components λ^i of an arbitrary vector of V must therefore satisfy the condition

$$\lambda^i(\mu_i - \mu_{i'} p^{i'}{}_i) = 0,$$

from which $\mu_i = \mu_{i'} p^{i'}{}_i.$

This is equivalent to $\mu_{i'} = p^i{}_{i'} \mu_i, \tag{4.3}$

which is the required law of transformation.

Let $(\mathbf{e}_i)$, $(\mathbf{e}_\alpha)$ be bases for the spaces V, W, and let $(\mathbf{e}_{i\alpha})$ be the corresponding basis of $V \otimes W$. Let the basis of V be changed according to (4.1), and let a change of basis of W be given by

$$\mathbf{e}_{\alpha'} = p^\alpha{}_{\alpha'} \mathbf{e}_\alpha. \tag{4.4}$$

The change of basis in V and W will induce a change of basis in $V \otimes W$ given by

$$\mathbf{e}_{i'\alpha'} = p^i{}_{i'} p^\alpha{}_{\alpha'} \mathbf{e}_{i\alpha}. \tag{4.5}$$

If $T^{i\alpha}$ are the components of an element $\mathbf{T}$ of $V \otimes W$ relative to the basis $(\mathbf{e}_{i\alpha})$, it is convenient to write $T^{i'\alpha'}$ for the components relative to the basis $(\mathbf{e}_{i'\alpha'})$. Then

$$\mathbf{T} = T^{i\alpha} \mathbf{e}_{i\alpha} = T^{i'\alpha'} \mathbf{e}_{i'\alpha'} = T^{i'\alpha'} p^i{}_{i'} p^\alpha{}_\alpha \mathbf{e}_{i\alpha}, \quad \text{from (4.5)}.$$

Since the basis vectors $\mathbf{e}_{i\alpha}$ are linearly independent, it follows that

$$T^{i\alpha} = p^i{}_{i'} p^\alpha{}_{\alpha'} T^{i'\alpha'}, \tag{4.6}$$

which is equivalent to

$$T^{i'\alpha'} = p^{i'}{}_i p^{\alpha'}{}_\alpha T^{i\alpha}. \tag{4.7}$$

It will be seen that if the basis in V is changed but the basis of W remains unaltered, then the law of transformation (4.6) reduces to

$$T^{i\alpha} = p^i{}_{i'} T^{i'\alpha},$$

or alternatively $\qquad T^{i'\alpha} = p^{i'}{}_i T^{i\alpha}, \tag{4.8}$

which is of the same type as the law of transformation (4.2) for the components of a vector of V. Similarly, if the basis of V is un-changed but the basis of W is changed, the components of $\mathbf{T}$ transform like the components of a vector of W.

For the remainder of this section we shall be concerned with the law of transformation of components of tensors of type (r, s) induced by a change of basis in V^1 given by (4.1). The laws of transformation for components of tensors of type $(1, 0)$, $(0, 1)$ are given by (4.2), (4.3) respectively. The corresponding laws of transformation for tensors of type $(2, 0)$, $(1,1)$, and $(0, 2)$ will now be obtained.

By taking $V = V^1$, $W = V^1$, the law of transformation for the components of contravariant tensors of order 2 is seen from (4.7) to be

$$T^{i'j'} = p^{i'}{}_i p^{j'}{}_j T^{ij}. \tag{4.9}$$

Similarly, by taking $V = V^1$, $W = V_1$, the law of transformation for the components of tensors of type $(1, 1)$ is

$$T^{i'}_{j'} = p^{i'}{}_i p^j{}_{j'} T^i_j. \tag{4.10}$$

By taking $V = V_1$, $W = V_1$, components of covariant tensors of order 2 are seen to transform according to the law

$$T_{i'j'} = p^i{}_{i'} p^j{}_{j'} T_{ij}. \tag{4.11}$$

More generally, a change of basis of V^1 given by (4.1) induces a change of basis in V^r_s given by

$$\mathbf{e}^{j_1'j_2'\ldots j_r'}_{i_1'i_2'\ldots i_r'} = p^{j_1'}_{j_1} p^{j_2'}_{j_2} \ldots p^{j_r'}_{j_s} p^{i_1}_{i_1'} p^{i_2}_{i_2'} \ldots p^{i_r}_{i_{r'}} \mathbf{e}^{j_1 j_2 \ldots j_s}_{i_1 i_2 \ldots i_r}. \tag{4.12}$$

Since $\qquad \mathbf{T} = T^{i_1 i_2 \ldots i_r}_{j_1 j_2 \ldots j_s} \mathbf{e}^{j_1 j_2 \ldots j_s}_{i_1 i_2 \ldots i_r} = T^{i_1' i_2' \ldots i_r'}_{j_1' j_2' \ldots j_s'} \mathbf{e}^{j_1' j_2' \ldots j_r'}_{i_1' i_2' \ldots i_r'},$

we have from (4.12),

$$(T^{i_1 i_2 \ldots i_r}_{j_1 j_2 \ldots j_s} - T^{i_1' i_2' \ldots i_r'}_{j_1' j_2' \ldots j_s'} p^{j_1'}_{j_1} p^{j_2'}_{j_2} \ldots p^{j_r'}_{j_s} p^{i_1}_{i_1'} p^{i_2}_{i_2'} \ldots p^{i_r}_{i_{r'}}) \mathbf{e}^{j_1 j_2 \ldots j_s}_{i_1 i_2 \ldots i_r} = 0.$$

Since the vectors $e^{j_1 j_2 \ldots j_s}_{i_1 i_2 \ldots i_r}$ are linearly independent, this implies

$$T^{i_1 i_2 \ldots i_r}_{j_1 j_2 \ldots j_s} = p^{i_1}_{i_{1'}} p^{i_2}_{i_{2'}} \ldots p^{i_r}_{i_{r'}} p^{j_{1'}}_{j_1} p^{j_{2'}}_{j_2} \ldots p^{j_{s'}}_{j_s} T^{i_{1'} i_{2'} \ldots i_{r'}}_{j_{1'} j_{2'} \ldots j_{s'}},$$

which is equivalent to

$$T^{i_{1'} i_{2'} \ldots i_{r'}}_{j_{1'} j_{2'} \ldots j_{s'}} = p^{i_{1'}}_{i_1} p^{i_{2'}}_{i_2} \ldots p^{i_{r'}}_{i_r} p^{j_1}_{j_{1'}} p^{j_2}_{j_{2'}} \ldots p^{j_s}_{j_{s'}} T^{i_1 i_2 \ldots i_r}_{j_1 j_2 \ldots j_s}. \tag{4.13}$$

In particular, the components of a contravariant tensor of order r transform according to the law

$$T^{i_{1'} i_{2'} \ldots i_{r'}} = p^{i_{1'}}_{i_1} p^{i_{2'}}_{i_2} \ldots p^{i_{r'}}_{i_r} T^{i_1 i_2 \ldots i_r}, \tag{4.14}$$

while the components of a covariant tensor of order s transform according to the law

$$T_{j_{1'} j_{2'} \ldots j_{s'}} = p^{j_1}_{j_{1'}} p^{j_2}_{j_{2'}} \ldots p^{j_s}_{j_{s'}} T_{j_1 j_2 \ldots j_s}. \tag{4.15}$$

It has been shown that a tensor **T** of type (r, s) has associated with each basis (e_i) of V^1, n^{r+s} components $T^{i_1 \ldots i_r}_{j_1 \ldots j_s}$ which necessarily transform according to (4.13) when the basis of V^r_s is changed according to (4.12). We now remark that *these conditions are also sufficient for the numbers $T^{i_1 \ldots i_r}_{j_1 \ldots j_s}$ to be components of a tensor*. For if (4.13) is satisfied, then the vector defined by

$$\mathbf{T} = T^{i_1 i_2 \ldots i_r}_{j_1 j_2 \ldots j_s} e^{j_1 j_2 \ldots j_s}_{i_1 i_2 \ldots i_r},$$

is independent of the particular basis, and the sets of numbers $(T^{i_1 \ldots i_r}_{j_1 \ldots j_s})$ are components of the same tensor **T** with respect to different bases.

The following theorem can also be used to decide whether a given system of numbers $T^{i_1 \ldots i_r}_{j_1 \ldots j_s}$ associated with each basis of V^1 can be regarded as the components of a tensor of type (r, s).

THEOREM 4.1. *In order that n^{r+s} numbers $T^{i_1 i_2 \ldots i_r}_{j_1 j_2 \ldots j_s}$ associated with each basis of V^1 can be regarded as the components of a tensor **T** of type (r, s), it is necessary and sufficient that for any r covariant vectors $\alpha, \beta, \ldots, \gamma$, and any s contravariant vectors $\lambda, \mu, \ldots, \nu$, the expression*

$$T^{i_1 i_2 \ldots i_r}_{j_1 j_2 \ldots j_s} \lambda^{j_1} \mu^{j_2} \ldots \nu^{j_s} \alpha_{i_1} \beta_{i_2} \ldots \gamma_{i_r} \tag{4.16}$$

shall be invariant under a change of basis of V^1.

We first prove that the invariance of (4.16) is a necessary condition that the numbers $T^{i_1 i_2 \ldots i_r}_{j_1 j_2 \ldots j_s}$ are components of a tensor of type (r, s). When a change of basis (4.1) is made in V^1, the components of **T** transform according to (4.13) while the components of the contravariant and covariant vectors transform according to (4.2) and (4.3) respectively. Hence

$$T_{j_{1'}j_{2'}\ldots j_{s'}}^{i_{1'}i_{2'}\ldots i_{r'}} \lambda^{j_{1'}}\mu^{j_{2'}}\ldots\nu^{j_{s'}}\alpha_{i_{1'}}\beta_{i_{2'}}\ldots\gamma_{i_{r'}}$$

$$= T_{j_1 j_2\ldots j_s}^{i_1 i_2\ldots i_r} p_{i_1}^{i_{1'}} p_{i_2}^{i_{2'}}\ldots p_{i_r}^{i_{r'}} p_{j_{1'}}^{j_1} p_{j_{2'}}^{j_2}\ldots p_{j_{s'}}^{j_s} p_{k_1}^{j_{1'}} \lambda^{k_1} p_{k_2}^{j_{2'}} \mu^{k_2}\ldots$$

$$p_{k_s}^{j_{s'}} \nu^{k_s} p_{i_{1'}}^{h_1} \alpha_{h_1} p_{i_{2'}}^{h_2} \beta_{h_2}\ldots p_{i_{r'}}^{h_r} \gamma_{h_r}$$

$$= T_{j_1 j_2\ldots j_s}^{i_1 i_2\ldots i_r} (p_{i_1}^{i_{1'}} p_{i_{1'}}^{h_1})(p_{i_2}^{i_{2'}} p_{i_{2'}}^{h_2})\ldots(p_{i_r}^{i_{r'}} p_{i_{r'}}^{h_r})(p_{j_{1'}}^{j_1} p_{k_1}^{j_{1'}})(p_{j_{2'}}^{j_2} p_{k_2}^{j_{2'}})\ldots$$

$$(p_{j_{s'}}^{j_s} p_{k_s}^{j_{s'}})\lambda^{k_1}\mu^{k_2}\ldots\nu^{k_s}\alpha_{h_1}\beta_{h_2}\ldots\gamma_{h_r}$$

$$= T_{j_1 j_2\ldots j_s}^{i_1 i_2\ldots i_r} \lambda^{j_1}\mu^{j_2}\ldots\nu^{j_s}\alpha_{i_1}\beta_{i_2}\ldots\gamma_{i_r},$$

since $p_{i_1}^{i_{1'}} p_{i_{1'}}^{h_1} = \delta_{i_1}^{h_1}$, etc. This proves the invariance of (4.16) to be a necessary condition.

To prove the condition sufficient we proceed as follows. By hypothesis,

$$T_{j_{1'}j_{2'}\ldots j_{s'}}^{i_{1'}i_{2'}\ldots i_{r'}} \lambda^{j_{1'}}\mu^{j_{2'}}\ldots\nu^{j_{s'}}\alpha_{i_{1'}}\beta_{i_{2'}}\ldots\gamma_{i_{r'}} = T_{j_1 j_2\ldots j_s}^{i_1 i_2\ldots i_r} \lambda^{j_1}\mu^{j_2}\ldots\nu^{j_s}\alpha_{i_1}\beta_{i_2}\ldots\gamma_{i_r}.$$

Making use of equations (4.2), (4.3), the left-hand member can be written

$$T_{j_{1'}j_{2'}\ldots j_{s'}}^{i_{1'}i_{2'}\ldots i_{r'}} p_{j_1}^{j_{1'}} p_{j_2}^{j_{2'}}\ldots p_{j_s}^{j_{s'}} p_{i_{1'}}^{i_1} p_{i_{2'}}^{i_2}\ldots p_{i_{r'}}^{i_r} \lambda^{j_1}\mu^{j_2}\ldots\nu^{j_s}\alpha_{i_1}\beta_{i_2}\ldots\gamma_{i_r}.$$

We thus have

$$(T_{j_{1'}j_{2'}\ldots j_{s'}}^{i_{1'}i_{2'}\ldots i_{r'}} p_{j_1}^{j_{1'}} p_{j_2}^{j_{2'}}\ldots p_{j_s}^{j_{s'}} p_{i_{1'}}^{i_1} p_{i_{2'}}^{i_2}\ldots p_{i_{r'}}^{i_r} - T_{j_1 j_2\ldots j_s}^{i_1 i_2\ldots i_r})\lambda^{j_1}\mu^{j_2}\ldots\nu^{j_s}\alpha_{i_1}\beta_{i_2}\ldots\gamma_{i_r} = 0,$$

satisfied for all arbitrary numbers $\lambda^{j_1}, \mu^{j_2},\ldots$, etc. Hence

$$T_{j_1 j_2\ldots j_s}^{i_1 i_2\ldots i_r} = T_{j_{2'}j_{2'}\ldots j_{s'}}^{i_{1'}i_{2'}\ldots i_{r'}} p_{j_1}^{j_{1'}} p_{j_2}^{j_{2'}}\ldots p_{j_s}^{j_{s'}} p_{i_{1'}}^{i_1} p_{i_{2'}}^{i_2}\ldots p_{i_{r'}}^{i_r},$$

which is equivalent to

$$T_{j_{1'}j_{2'}\ldots j_{s'}}^{i_{1'}i_{2'}\ldots i_{r'}} = p_{i_1}^{i_{1'}} p_{i_2}^{i_{2'}}\ldots p_{i_r}^{i_{r'}} p_{j_{1'}}^{j_1} p_{j_{2'}}^{j_2}\ldots p_{j_{s'}}^{j_s} T_{j_1 j_2\ldots j_s}^{i_1 i_2\ldots i_r}.$$

The previous result now shows that the numbers $T_{j_1 j_2\ldots j_s}^{i_1 i_2\ldots i_r}$ are components of a tensor **T** of type (r, s), and the proof of the theorem is thus complete.

The previous theorem can be generalized to the following:

THEOREM 4.2. *In order that a set of $n^{(r+R)+(s+S)}$ numbers*

$$T_{j_1\ldots j_s j_{s+1}\ldots j_{s+S}}^{i_1\ldots i_r i_{r+1}\ldots i_{r+R}}$$

corresponding to each basis of V_{s+S}^{r+R} shall be the components of a tensor of type $(r+R, s+S)$, it is necessary and sufficient that for r arbitrary covariant vectors **α**, **β**, ..., **γ** *and s arbitrary contravariant vectors* **λ**, **μ**,...,**ν**, *the numbers*

$$T_{j_1\ldots j_s j_{s+1}\ldots j_{s+S}}^{i_1\ldots i_r i_{r+1}\ldots i_{r+R}} \lambda^{j_1}\mu^{j_2}\ldots\nu^{j_s}\alpha_{i_1}\beta_{i_2}\ldots\gamma_{i_r}.$$

shall be components of a tensor of type (R, S).

In order not to complicate matters unduly, we shall prove this theorem only in the special case $r = 1$, $s = 2$, $R = 1$, $S = 3$, but it will be seen that the same method of proof can be used to prove the general theorem. We have to show that in order that a set of n^7 numbers $T^{i_1 i_2}_{j_1 j_2 j_3 j_4 j_5}$ corresponding to each basis of V_5^2 shall be components of a tensor of type $(2, 5)$ it is necessary and sufficient that for an arbitrary covariant vector $\boldsymbol{\alpha}$ and two arbitrary contravariant vectors $\boldsymbol{\lambda}, \boldsymbol{\mu}$, the numbers

$$T^{i_1 i_2}_{j_1 j_2 j_3 j_4 j_5} \lambda^{j_1} \mu^{j_2} \alpha_{i_1}$$

shall be components of a tensor of type $(1, 3)$.

Let $\boldsymbol{\beta}$ be an arbitrary covariant vector, and let $\boldsymbol{\nu}, \boldsymbol{\sigma}, \boldsymbol{\tau}$ be arbitrary contravariant vectors. Then a necessary and sufficient condition that $T^{i_1 i_2}_{j_1 j_2 j_3 j_4 j_5}$ shall be the components of a tensor of type $(2, 5)$ is that the expression

$$T^{i_1 i_2}_{j_1 j_2 j_3 j_4 j_5} \lambda^{j_1} \mu^{j_2} \nu^{j_3} \sigma^{j_4} \tau^{j_5} \alpha_{i_1} \beta_{i_2}$$

shall be invariant over a change of basis. However, this is just a necessary and sufficient condition that

$$T^{i_1 i_2}_{j_1 j_2 j_3 j_4 j_5} \lambda^{j_1} \mu^{j_2} \alpha_{i_1}$$

should be the components of a tensor of type (1.3), and the theorem is therefore proved.

The tensor product $\mathbf{R} \otimes \mathbf{S}$ of a tensor $\mathbf{R}$ of type (r, s) with a tensor $\mathbf{S}$ of type (p, q) was defined in (3.16) by reference to a particular basis $(\mathbf{e}_i)$ of V^1. When the basis is changed to $(\mathbf{e}_{i'})$ we have

$$R^{i_{1'} \dots i_{r'}}_{j_{1'} \dots j_{s'}} = p^{i_{1'}}_{i_1} \dots p^{i_{r'}}_{i_r} p^{j_1}_{j_{1'}} \dots p^{j_s}_{j_{s'}} R^{i_1 \dots i_r}_{j_1 \dots j_s},$$

$$S^{i_{r'+1'} \dots i_{r'+p'}}_{j_{s'+1'} \dots j_{s'+q'}} = p^{i_{r'+1'}}_{i_{r+1}} \dots p^{i_{r'+p'}}_{i_{r+p}} p^{j_{s+1}}_{j_{s'+1'}} \dots p^{j_{s+q}}_{j_{s'+q'}} S^{i_{r+1} \dots i_{r+p}}_{j_{s+1} \dots j_{s+q}},$$

$$T^{i_{1'} \dots i_{r'+p'}}_{j_{1'} \dots j_{s'+q'}} = p^{i_{1'}}_{i_1} \dots p^{i_{r'+p'}}_{i_{r+p}} p^{j_1}_{j_{1'}} \dots p^{j_{s+q}}_{j_{s'+q'}} T^{i_1 \dots i_{r+p}}_{j_1 \dots j_{s+q}}.$$

Hence $$T^{i_{1'} \dots i_{r'+p'}}_{j_{1'} \dots j_{s'+q'}} = R^{i_{1'} \dots i_{r'}}_{j_{1'} \dots j_{s'}} S^{i_{r'+1'} \dots i_{r'+p'}}_{j_{s'+1'} \dots j_{s'+q'}},$$

and the relation (3.16) is seen to be invariant over a change of basis. Thus although reference was made to a basis in defining $\mathbf{R} \otimes \mathbf{S}$, the product is independent of the particular basis chosen.

5. Contraction

This is an operation which when applied to a tensor of type (r, s) $(r, s \geqslant 1)$ yields a tensor of type $(r-1, s-1)$. Although the process of contraction can be defined without reference to a basis, we shall not introduce the idea in this way as in many respects it is simpler to regard contraction as an operation applied to the components of a tensor relative to some basis.

Let **T** have components $T^{i_1 i_2 \ldots i_r}_{j_1 j_2 \ldots j_s}$ with reference to a basis $(\mathbf{e}_i)$. Suppose some upper suffix, say i_p, is identified with some lower suffix, say j_q, and, in accordance with the summation convention, the terms are summed over the range 1 to n of this repeated suffix. The n^{r+s} components of **T** will give rise to n^{r+s-2} numbers $S^{i_1 \ldots i_r}_{j_1 \ldots j_s}$ defined by

$$S^{i_1 \ldots i_{p-1} i_{p+1} \ldots i_r}_{j_1 \ldots j_{q-1} j_{q+1} \ldots j_s} = T^{i_1 i_2 \ldots i_{p-1} i p i_{p+1} \ldots i_r}_{j_1 j_2 \ldots j_{q-1} i p j_{q+1} \ldots j_s}. \tag{5.1}$$

We shall now prove that these numbers are the components of a tensor **S** of type $(r-1, s-1)$, obtained from **T** by *contraction*. We do this by showing that a change of basis induces the appropriate tensor law for the coefficients of **S**.

A change of basis $\mathbf{e}_{i'} = p^i_{i'} \mathbf{e}_i$ induces the transformation

$$T^{i_{1'} i_{2'} \ldots i_{p'-1'} i_{p'} i_{p'+1'} \ldots i_{r'}}_{j_{1'} j_{2'} \ldots j_{q'-1'} j_{q'} j_{q'+1'} \ldots j_{s'}}$$
$$= p^{i_{1'}}_{i_1} p^{i_{2'}}_{i_2} \ldots p^{i_{p'-1'}}_{i_{p-1}} p^{i_{p'}}_{i_p} p^{i_{p'+1'}}_{i_{p+1}} \ldots p^{i_{r'}}_{i_r} p^{j_1}_{j_{1'}} p^{j_2}_{j_{2'}} \ldots$$
$$p^{j_{q-1}}_{j_{q'-1'}} p^{j_q}_{j_{q'}} p^{j_{q+1}}_{j_{q'+1'}} \ldots p^{j_s}_{j_{s'}} T^{i_1 i_2 \ldots i_{p-1} i p i_{p+1} \ldots i_r}_{j_1 j_2 \ldots j_{q-1} j q j_{q+1} \ldots j_s}. \tag{5.2}$$

Identification of the suffixes i_p, j_q introduces on the right-hand side the symbol $p^{i_{p'}}_{i_p} p^{i_p}_{j_{q'}} = \delta^{i_{p'}}_{j_{q'}}$, and equation (5.2) becomes

$$S^{i_{1'} \ldots i_{p'-1'} i_{p'+1'} \ldots i_{r'}}_{j_{1'} \ldots j_{q'-1'} j_{q'+1'} \ldots j_{s'}}$$
$$= p^{i_{1'}}_{i_1} \ldots p^{i_{p'-1'}}_{i_{p-1}} p^{i_{p'+1'}}_{i_{p+1}} \ldots p^{i_{r'}}_{i_r} p^{j_1}_{j_{1'}} \ldots p^{j_{q-1}}_{j_{q'-1'}} p^{j_{q+1}}_{j_{q'+1'}} \ldots p^{j_s}_{j_{s'}} S^{i_1 \ldots i_{p-1} i_{p+1} \ldots i_r}_{j_1 \ldots j_{q-1} j_{q+1} \ldots j_s}, \tag{5.3}$$

showing that these numbers are the components of a tensor of type $(r-1, s-1)$. Evidently, different tensors **S** arise from the same tensor **T** by identifying different pairs of suffixes. The particular tensor **S** defined by (5.1) arises by contracting the pth contravariant suffix with the qth covariant suffix.

We shall now give an intrinsic definition to the process of contraction. It is simple to define this process for *decomposable tensors*, i.e. tensors which are expressible as repeated tensor products of contravariant and covariant vectors. Suppose a decomposable tensor of type (r, s) is expressed as the tensor product of r contravariant vectors $\boldsymbol{\lambda}, \boldsymbol{\mu}, \boldsymbol{\nu}, \ldots, \boldsymbol{\sigma}$ and s covariant vectors $\boldsymbol{\alpha}, \boldsymbol{\beta}, \boldsymbol{\gamma}, \ldots, \boldsymbol{\epsilon}$, i.e.

$$\mathbf{T} = \boldsymbol{\lambda} \otimes \boldsymbol{\mu} \otimes \boldsymbol{\nu} \otimes \ldots \otimes \boldsymbol{\sigma} \otimes \boldsymbol{\alpha} \otimes \boldsymbol{\beta} \otimes \boldsymbol{\gamma} \otimes \ldots \otimes \boldsymbol{\epsilon}. \tag{5.4}$$

Then, for example, the tensor obtained by contracting the second contravariant vector with the first covariant vector is defined by

$$\mathbf{S} = \alpha(\boldsymbol{\mu}) \boldsymbol{\lambda} \otimes \boldsymbol{\nu} \otimes \ldots \otimes \boldsymbol{\sigma} \otimes \boldsymbol{\beta} \otimes \boldsymbol{\gamma} \otimes \ldots \otimes \boldsymbol{\epsilon}. \tag{5.5}$$

Equation (5.5) evidently defines a tensor of type $(r-1, s-1)$ so *we do not have the task of proving that* **S** *is a tensor*. Since any arbitrary tensor of V^r_s is expressible as a linear combination of decomposable tensors and since contraction is a linear operation, it follows that our intrinsic definition of contraction can be extended to apply to arbitrary tensors. In this case, contraction with respect to the second contravariant vector and the first covariant vector corresponds to contraction with respect to the second contravariant suffix and the first covariant suffix.

We have seen that there are four fundamental operations associated with tensors:

(i) addition of tensors of type (r, s) giving a tensor of type (r, s);

(ii) multiplication of a tensor of type (r, s) by a real number giving a tensor of type (r, s);

(iii) the tensor product of tensors of types (r, s), (r', s') giving a tensor of type $(r+r', s+s')$;

(iv) contraction which associates with a tensor of type (r, s) another tensor of type $(r-1, s-1)$.

Tensor algebra is concerned with the properties of tensors with reference to these four laws.

6. Special tensors

The Kronecker tensor

A special tensor which plays an important role in tensor algebra is the Kronecker tensor **K** of type $(1, 1)$, which may be defined by specifying its components with respect to a basis $(\mathbf{e}_i)$ as the Kronecker delta δ^i_j. The components of **K** relative to the basis $(\mathbf{e}_{i'})$ are given by

$$\delta^{i'}_{j'} = p^{i'}_i p^j_{j'} \delta^i_j = p^{i'}_i p^i_{j'},$$

from which it follows that **K** has the same components with respect to any basis. If we regard a tensor of type $(1, 1)$ as a real-valued bilinear function defined over $V_1 \times V^1$, we can define **K** as that real-valued bilinear function which maps the pair $(\boldsymbol{\alpha}, \boldsymbol{\lambda})$ into the real number $\alpha(\boldsymbol{\lambda})$.

Again, if we regard a tensor of type $(1, 1)$ as a linear mapping of the space V^1 into V^1, **K** can be defined as the tensor which gives the identity mapping, i.e. **K** maps each contravariant vector into itself.

Symmetric tensors

A tensor **T** of type $(2, 0)$ is said to be *symmetric* if

$$T(\boldsymbol{\alpha}, \boldsymbol{\beta}) = T(\boldsymbol{\beta}, \boldsymbol{\alpha})$$

for all covariant vectors $\boldsymbol{\alpha}$, $\boldsymbol{\beta}$.

In particular, by taking $\boldsymbol{\alpha} = \mathbf{e}^i$, $\boldsymbol{\beta} = \mathbf{e}^j$ it follows that the components of a symmetric tensor of type $(2, 0)$ satisfy the symmetry relations $T^{ij} = T^{ji}$. Alternatively, a symmetric tensor of type $(2, 0)$ could have been *defined* as one whose components relative to any basis satisfied the conditions $T^{ij} = T^{ji}$. Since

$$T^{i'j'} = p^{i'}{}_i p^{j'}{}_j T^{ij},$$

the relations $T^{ij} = T^{ji}$ imply the relations $T^{i'j'} = T^{j'i'}$, and the condition for symmetry is seen to be invariant over a change of basis.

More generally, a tensor $\mathbf{T}$ of type $(r, 0)$ is symmetric in the pth and qth place if

$$T(\boldsymbol{\alpha}_1, \boldsymbol{\alpha}_2, ..., \boldsymbol{\alpha}_p, ..., \boldsymbol{\alpha}_q, ..., \boldsymbol{\alpha}_r) = T(\boldsymbol{\alpha}_1, \boldsymbol{\alpha}_2, ..., \boldsymbol{\alpha}_q, ..., \boldsymbol{\alpha}_p, ..., \boldsymbol{\alpha}_r) \quad (6.1)$$

for all covariant vectors $\boldsymbol{\alpha}_1, \boldsymbol{\alpha}_2, ..., \boldsymbol{\alpha}_r$. By taking the basis vectors $\mathbf{e}^i$ for the vectors $\boldsymbol{\alpha}$, it follows that the components of $\mathbf{T}$ satisfy the symmetry conditions

$$T^{i_1 i_2 ... i_p ... i_q ... i_r} = T^{i_1 i_2 ... i_q ... i_p ... i_r}. \quad (6.2)$$

We leave the reader to verify that these conditions remain invariant over a change of basis, so that the symmetry of $\mathbf{T}$ could have been *defined* by requiring (6.2) to hold relative to any one basis.

So far we have considered only symmetric contravariant tensors, but evidently symmetric covariant tensors are defined in a similar manner. For example, a symmetric tensor $\mathbf{T}$ of type $(0, 2)$ is characterized either by the condition

$$T(\boldsymbol{\lambda}, \boldsymbol{\mu}) = T(\boldsymbol{\mu}, \boldsymbol{\lambda})$$

for all contravariant vectors $\boldsymbol{\lambda}, \boldsymbol{\mu}$, or by the restrictions

$$T_{ij} = T_{ji}$$

on the components relative to any one basis.

A *completely symmetric tensor*, or briefly, a *symmetric tensor*, is one which is symmetric with respect to *all* pairs of suffixes. Such tensors must be either of type $(r, 0)$ or $(0, r)$, i.e. contravariant or covariant tensors.

The property of symmetry with respect to a pair of contravariant suffixes or covariant suffixes may evidently apply to the components of a tensor of type (r, s). The reader is left to verify that such symmetry conditions are invariant over a change of basis.

Skew symmetric tensors

A tensor of type $(2, 0)$ is skew symmetric if

$$T(\alpha, \beta) = -T(\beta, \alpha)$$

for all covariant vectors α, β. Alternatively, it is sufficient to require the condition $T^{ij} = -T^{ji}$ to hold relative to any one basis. Tensors of type (r, s) whose components are skew symmetric in pairs of contravariant or covariant suffixes may be defined as in the symmetric case, with an obvious modification. Covariant tensors which are skew symmetric for all pairs of suffixes will play an important role in a later part of the book.

The use of components

The reader will have noticed that special tensors may be defined either in an invariant manner, e.g. by equation (6.1), or else by imposing restrictions on the components relative to a basis, e.g. equation (6.2). In the latter case it is necessary to verify that the conditions imposed are invariant over a change of basis, a task which is quite pointless if the special tensor is defined without reference to a basis. There are a number of intrinsic operations, e.g. contraction, which seem natural when applied to the components of a tensor relative to a basis but which nevertheless can be defined intrinsically. We do not wish to restrict ourselves to the policy of defining every intrinsic operation without making use of a basis. Élie Cartan, in a memorable passage in the preface to the first edition of *Leçons sur la géométrie des espaces de Riemann* (1928) warns the reader that formal tensor calculations involving suffixes may often conceal a simple geometric fact: 'Les services éminents qu'a rendus et que rendra encore le Calcul différentiel absolu de Ricci et Levi-Civita ne doivent pas nous empêcher d'éviter les calculs trop exclusivement formels où les débauches d'indices masquent une realité géométrique souvent très simple.' However, Cartan himself did not hesitate to use tensor components when more convenient to do so, and we propose to follow this course. It is a fact that many new results in differential geometry have been first discovered by 'formal suffix manipulation', and only subsequently have they been reobtained in an invariant, though probably more significant manner. The reader is advised to acquire skill both in using tensors without reference to a basis, and also in suffix manipulation when working with components. These two methods should be regarded as complementary to one another.

7. Inner product

The reader is already familiar with the idea of scalar or inner product of two vectors in three-dimensional Euclidean space. In many applications it will be necessary to perform an analogous operation in the fundamental vector space V^1, and in this section we show that a suitable scalar product may be defined by means of a tensor $\mathfrak{g}$ of type $(0, 2)$ which is non-singular, i.e. its components form a non-singular matrix.

A suitable inner product will be given by a function ϕ which maps a pair of contravariant vectors $\boldsymbol{\lambda}, \boldsymbol{\mu}$ to a real number, such that

  (i) ϕ is bilinear;
  (ii) ϕ is symmetric, i.e. $\phi(\boldsymbol{\lambda}, \boldsymbol{\mu}) = \phi(\boldsymbol{\mu}, \boldsymbol{\lambda})$;
  (iii) the conditions $\phi(\boldsymbol{\lambda}, \mathbf{x}) = 0$ for all $\mathbf{x}$ must imply $\boldsymbol{\lambda} = \mathbf{0}$.

Conditions (i) and (ii) show that ϕ is a symmetric tensor $\mathfrak{g}$ of type $(0, 2)$. If we interpret $\mathfrak{g}$ as a linear mapping from V^1 into V_1, condition (iii) implies that this mapping must be one-one, for if $\boldsymbol{\lambda}$ and $\boldsymbol{\mu}$ both map into the same vector in V_1, then the vector $\boldsymbol{\lambda} - \boldsymbol{\mu}$ will map into $\mathbf{0}$ and this implies $\boldsymbol{\lambda} = \boldsymbol{\mu}$. Hence condition (iii) ensures that the tensor $\mathfrak{g}$ is non-singular.

An alternative method of interpreting (iii) will now be given. Let $(\mathbf{e}_i)$ be a basis in V^1, and let $\mathfrak{g}, \boldsymbol{\lambda}, \mathbf{x}$ have components g_{ij}, λ^i, x^j relative to this basis. Then condition (iii) states that if the relation $g_{ij}\lambda^i x^j = 0$ holds for arbitrary numbers x^j, then all the numbers λ^i must be zero. Alternatively, the relations $g_{ij}\lambda^i = 0$ imply that all the numbers λ^i are zero, and this is the case if and only if the matrix (g_{ij}) is non-singular. Thus the tensor $\mathfrak{g}$ is non-singular.

Since $\mathfrak{g}$ is non-singular, there exists a reciprocal tensor $\mathfrak{g}^{-1}$ whose components g^{ij} relative to the basis $(\mathbf{e}_i)$ are symmetric and satisfy the conditions
$$g_{ij} g^{jk} = \delta^k{}_i.$$

The tensor $\mathfrak{g}$ is called the *metric tensor* associated with the inner product ϕ. Since $\mathfrak{g}$ is a reversible linear mapping of V^1 onto V_1, a metric tensor establishes an isomorphism between V^1 and V_1.

8. Associated tensors

A metric tensor $\mathfrak{g}$ associates with a tensor of type (r, s), other tensors of type $(r-1, s+1)$, $(r-2, s+2)$, $(r-3, s+3)$,..., etc. For example, consider a tensor $\mathbf{T}$ of type $(2, 1)$ whose components relative to a basis $(\mathbf{e}_i)$ are $T^{ij}{}_k$. From section 4, it follows that the numbers $T_i{}^j{}_k$ defined by

$$T_i{}^j{}_k = g_{il} T^{lj}{}_k \tag{8.1}$$

are components of a tensor of type $(1, 2)$, while the numbers T_{ijk} defined by

$$T_{ijk} = g_{jh} T_i{}^h{}_k \tag{8.2}$$

are components of a tensor of type $(0, 3)$.

The tensor with components $T_i{}^j{}_k$ is said to be obtained from the tensor with components $T^{ij}{}_k$ 'by *lowering the suffix i*'.

If equation (8.1) is multiplied by the components g^{hi} of $\mathring{g}^{-1}$, we get

$$g^{hi} T_i{}^j{}_k = \delta_l^h T^{lj}{}_k = T^{hj}{}_k,$$

i.e.

$$T^{ij}{}_k = g^{ih} T_h{}^j{}_k. \tag{8.3}$$

Thus the original tensor with components $T^{ij}{}_k$ may be recovered from the tensor with components $T_i{}^j{}_k$ by means of the reciprocal metric tensor. This process is called *raising the suffix i by means of the reciprocal metric tensor*. All the tensors which can be obtained from a given tensor $\mathbf{T}$ by means of raising or lowering suffixes are called *associated tensors*.

The process of obtaining associated tensors can be described in an invariant manner by noting that the metric tensor $\mathring{g}$ establishes an isomorphism between the spaces V_s^r and V_{s+1}^{r-1}, while the tensor $\mathring{g}^{-1}$ gives the inverse of this isomorphism.

EXERCISE 8.1. The components of ϕ with respect to a basis of a two-dimensional vector space are $g_{\alpha\beta}$, and g is the determinant $|g_{\alpha\beta}|$. The symbols $e_{\alpha\beta}$, $e^{\alpha\beta}$ are defined to take values 1 when $\alpha = 1$, $\beta = 2$; -1 when $\alpha = 2$, $\beta = 1$; 0 when $\alpha = \beta$. Numbers $\epsilon_{\alpha\beta}$, $\epsilon^{\alpha\beta}$ are defined by the equations

$$\epsilon_{\alpha\beta} = e_{\alpha\beta} \sqrt{g}, \qquad \epsilon^{\alpha\beta} = e^{\alpha\beta}/\sqrt{g}.$$

Prove that the numbers $\epsilon_{\alpha\beta}$ transform like the components of a skew symmetric covariant tensor of order 2, while $\epsilon^{\alpha\beta}$ transform like the components of a skew symmetric contravariant tensor of order 2, provided that the matrix corresponding to the change of basis has positive determinant.

9. Exterior algebra

The theory of exterior algebras was introduced by Grassmann in order to study algebraically certain geometrical problems on linear varieties in projective space. However, this theory remained mainly unknown until Élie Cartan discovered it again and applied it to the study of differentials and their multiple integrals over a differentiable manifold. The theory of exterior algebras has very wide applications, including the theory of determinants, the theory of linear varieties in projective space, and the theory of differential forms. We shall give a brief account in this chapter of only those properties of exterior algebra which will be used later. For a more detailed treatment the reader is referred to works by Bourbaki (1948), Cartan (1945), Goursat (1922), and Chevalley (1946).

The Cartan calculus of exterior forms which will be used later in the book depends essentially on the exterior algebra of skew symmetric covariant tensors. From the fundamental space V^1 are derived the vector spaces of covariant tensors of order $0, 1, ..., n$, and the direct sum of these $(n+1)$ spaces consists of elements of the exterior or Grassmann algebra. The elements of this direct sum can be added and multiplied by real numbers because they are elements of a vector space, but this structure alone will not permit the multiplication of vectors *by vectors*. The vector space is made into an algebra by defining a multiplication operation, called *wedge*, denoted by $\wedge$, which enables vectors to be multiplied by vectors, analogous to the operation of vector product in elementary vector algebra of three-dimensional vector space.

We first show how to obtain a skew symmetric covariant tensor of type $(0, r)$ from a given general tensor of type $(0, r)$. Let the symbol $e^{i_1...i_r}$ take values $+1$ if $(i_1...i_r)$ is an even permutation of the numbers $(1, 2, ..., r)$, -1 if the permutation is odd, and zero otherwise. Let $\mathbf{f}$ be the given tensor of type $(0, r)$, and define $\mathrm{Sk}\,\mathbf{f}$ (pronounced *skew-f*) by the equation

$$\mathrm{Sk}f(\boldsymbol{\lambda}_1, \boldsymbol{\lambda}_2, ..., \boldsymbol{\lambda}_r) = \frac{1}{r!}\,e^{i_1...i_r}f(\boldsymbol{\lambda}_{i_1}, ..., \boldsymbol{\lambda}_{i_r}),$$

where $\boldsymbol{\lambda}_1, \boldsymbol{\lambda}_2, ..., \boldsymbol{\lambda}_r$ are r arbitrary contravariant vectors, and the summation is taken over all suffixes.

In a similar way a skew symmetric contravariant tensor of type $(r, 0)$ can be obtained from a general tensor of type $(r, 0)$. Such skew symmetric tensors are sometimes called *multivectors of order r*. We shall find, however, that skew symmetric *covariant* tensors will play a more important role than multivectors.

Let A_r be the vector space of skew symmetric covariant tensors of order r, with $r \geqslant 2$. Denote by A_1 the vector space of covariant vectors, and by A_0 the one-dimensional vector space which is isomorphic with the real numbers.

If we denote by A the direct sum of the vector spaces A_r so that
$$A = A_0 \dotplus A_1 \dotplus A_2 \dotplus ... \dotplus A_n,$$

then A is a vector space of dimensions 2^n. We make A into an algebra by defining a multiplication $\wedge$ on its elements. We require this multiplication to be distributive, i.e. we must have

$$\mathbf{f} \wedge (\mathbf{g}_1 + \mathbf{g}_2) = \mathbf{f} \wedge \mathbf{g}_1 + \mathbf{f} \wedge \mathbf{g}_2,$$

and
$$(\mathbf{f}_1 + \mathbf{f}_2) \wedge \mathbf{g} = \mathbf{f}_1 \wedge \mathbf{g} + \mathbf{f}_2 \wedge \mathbf{g},$$

where $\mathbf{f}_1, \mathbf{f}_2, \mathbf{g}_1, \mathbf{g}_2, \mathbf{f}, \mathbf{g}$ belong to A.

If $\mathbf{f}$ belongs to A_r and $\mathbf{g}$ belongs to A_s, we define $\mathbf{f} \wedge \mathbf{g}$ by the rule

$$\mathbf{f} \wedge \mathbf{g} = \mathrm{Sk}(\mathbf{f} \otimes \mathbf{g}).$$

This means we take the tensor product of $\mathbf{f}$ and $\mathbf{g}$, and obtain the corresponding skew symmetric tensor. Since any element of A is expressible as a sum of elements each of which lies in some A_r, the rule we have just given determines the wedge product of any two elements of A.

From the definition of multiplication it follows that if $\mathbf{f} \in A_r$, $\mathbf{g} \in A_s$, then
$$\mathbf{f} \wedge \mathbf{g} = (-)^{rs} \mathbf{g} \wedge \mathbf{f}.$$

If $(\mathbf{e}^i)$ is a basis of V_1, now identified with A_1, then a basis of the algebra A is formed by

$$1, \mathbf{e}^i, \mathbf{e}^i \wedge \mathbf{e}^j \, (i < j), \mathbf{e}^i \wedge \mathbf{e}^j \wedge \mathbf{e}^k \, (i < j < k),..., \mathbf{e}^1 \wedge \mathbf{e}^2 \wedge \mathbf{e}^3 \wedge ... \wedge \mathbf{e}^n.$$

Many results on linear dependence can be expressed in a concise form by means of the wedge operator, as is illustrated by the following examples.

(1) *The elements* $\boldsymbol{\omega}^1, \boldsymbol{\omega}^2,..., \boldsymbol{\omega}^r$ *of* V_1 *are linearly dependent if and only if*
$$\boldsymbol{\omega}^1 \wedge \boldsymbol{\omega}^2 \wedge ... \wedge \boldsymbol{\omega}^r = 0.$$

(2) *Let* $\boldsymbol{\omega}^1, \boldsymbol{\omega}^2,..., \boldsymbol{\omega}^r$ *be linearly independent elements of* V_1 *and suppose there are* r *elements* $\boldsymbol{\pi}^1, \boldsymbol{\pi}^2,..., \boldsymbol{\pi}^r$ *of* V_1 *such that*

$$\sum_{s=1}^{r} \boldsymbol{\pi}^s \wedge \boldsymbol{\omega}^s = 0.$$

Then there exists a symmetric matrix $(a^s{}_t)$ *such that*

$$\boldsymbol{\pi}^s = a^s{}_t \boldsymbol{\omega}^t, \qquad s = 1, 2,..., r.$$

REFERENCES

BOURBAKI, N., *Algèbre*, Chapter III, Actualités scientifiques et industrielles, No. 1044, Hermann (1948).

CARTAN, É., *Les Systèmes différentiels extérieurs et leurs applications géométriques*, Actualités scientifiques et industrielles, No. 994, Hermann (1945).

—— *Leçons sur la géométrie des espaces de Riemann*, Gauthier–Villars (1928; 2nd edition 1946).

CHEVALLEY, C., *Theory of Lie Groups*, vol. i, Princeton University Press (1946), Chapter V.

GOURSAT, E., *Leçons sur le problème de Pfaff*, Hermann (1922).

HALMOS, P., *Finite Dimensional Vector Spaces*, Princeton University Press (1953).

MISCELLANEOUS EXERCISES V

1. If a_{rs}, b_{rs} are the components of two symmetric covariant tensors such that
$$a_{rs}b_{tu} - a_{ru}b_{st} + a_{st}b_{ru} - a_{tu}b_{rs} = 0,$$
show that $a_{rs} = kb_{rs}$ for some real number k.

2. If $a_{mn}x^m x^n = b_{mn}x^m x^n$ for arbitrary values of x^r, show that
$$a_{mn} + a_{nm} = b_{mn} + b_{nm}.$$
Hence if a_{mn}, b_{mn} are symmetric prove that $a_{mn} = b_{mn}$.

3. If a_s^r satisfies the relations $a_s^r a_t^s = \delta_t^r$, show that (i) $\|a_s^r\| = 1$; or (ii) $\|a_s^r\| = -1$, in which case $\|a_s^r + \delta_s^r\| = 0$ if $\|a_s^r\|$ has odd order.

4. If $a_{\alpha\beta}, b_\alpha$ are respectively components of a symmetric covariant tensor and a covariant vector which satisfy the relation
$$a_{\beta\gamma}b_\alpha + a_{\gamma\alpha}b_\beta + a_{\alpha\beta}b_\gamma = 0$$
for $\alpha, \beta, \gamma = 1, 2, ..., n$, prove that either $a_{\alpha\beta} = 0$ or $b_\alpha = 0$.

VI

TENSOR CALCULUS

1. Differentiable manifolds

TWO-DIMENSIONAL surfaces were first defined in Chapter II with reference to the three-dimensional Euclidean space in which they were immersed, but it was seen in Chapter IV that they can be defined intrinsically as two-dimensional differentiable manifolds without reference to a containing Euclidean space. We now define intrinsically n-dimensional differentiable manifolds, and the reader will see that these are natural generalizations of the two-dimensional manifolds of Chapter IV. The essential feature of a differentiable manifold is that it is covered by a set of coordinate neighbourhoods, each having the same number of coordinates, with the property that two different systems of coordinates in a common region are related by a differentiable transformation of class not less than 1. We now give a more formal definition.

An n-dimensional manifold is a Hausdorff topological space which is connected and has the property that each point has a neighbourhood homeomorphic to some open set in Cartesian n-space.

A system S of differentiable coordinates in an n-dimensional manifold X is an indexed family $\{V_j, j \in J\}$ of open sets covering X, and, for each j, a homeomorphism

$$\psi_j : E_j \to V_j$$

where E_j is an open set in Cartesian n-space, such that the map

$$\psi_j^{-1}\psi_i : \psi_i^{-1}(V_i \cap V_j) \to \psi_j^{-1}(V_i \cap V_j), \quad i, j \in J \qquad (1.1)$$

is differentiable. If each such map has continuous derivatives of order r, then S is said to be of *class r*. If S is of class r for every positive integer r, then S is said to be of *class ∞*. If every map $\psi_j^{-1}\psi_i$ is analytic in the sense that it is expressible as a convergent multiple power series in the n variables, then S is *analytic* or *of class ω*.

Two systems of coordinates S, S' in X of class r are said to be *r-equivalent* if the composite families $\{V_j, V_k'\}$, $\{\psi_j, \psi_k'\}$ form a system of class r. It is easily verified that this defines a proper equivalence relation, and the various systems of coordinates separate into disjoint equivalence classes. We can now make the formal definition:

A differentiable n-manifold X of class r is an n-manifold X, together with an r-equivalence class of systems of coordinates in X.†

The open sets V_j are called *coordinate neighbourhoods*, while the coordinate systems valid in V_j are called *local coordinate systems*. Let P be a point of X which lies in the overlap of two coordinate neighbourhoods, say V_j and $V_{j'}$. Then the homeomorphism ψ_j gives P the coordinates $x^1, x^2, ..., x^n$, while the homeomorphism $\psi_{j'}$ gives P the coordinates $x^{1'}, x^{2'}, ..., x^{n'}$. These two sets of local coordinates are related by equations of the type

$$x^{i'} = f^{i'}(x^1, x^2, ..., x^n), \tag{1.2}$$

where the functions $f^{i'}$ are of class r, and the Jacobian $\partial x^{i'}/\partial x^i$ is non-zero. The transformation (1.2) is called *a transformation of local coordinates*.

The simplest example of a differentiable n-manifold is Cartesian n-space itself. Here one neighbourhood V covers the whole space and we take for ψ the identity mapping. As a second example, denote by L_n the group of non-singular real $n \times n$ matrices with positive determinant. The elements of a matrix can be regarded as the coordinates of a point in n^2-dimensional Cartesian space and then L_n is an open subset. Again, one neighbourhood $V = L_n$ covers the whole space and the identity map ψ defines an analytic system of coordinates S in L_n. Then L_n together with the ω-equivalence class is an analytic differentiable manifold.

Let S be a system of coordinates of class r in an n-manifold X. If $x \in V_i \cap V_j$, denote by $(a_{ji}(x))$ the $n \times n$ Jacobian matrix of first partial derivatives of the functions (1.1) evaluated at $\psi_i^{-1}(x)$. Then, just as in Chapter IV, we have

$$a_{kj}(x)a_{ji}(x) = a_{ki}(x), \qquad x \in V_i \cap V_j \cap V_k. \tag{1.3}$$

Write $k = i$; then it follows that $a_{ji}(x)$ possesses an inverse matrix, and so is non-singular.

A system S of coordinates is called *oriented* if the determinant $|a_{ij}(x)|$ is positive for all i, j and for all $x = V_i \cap V_j$. A manifold X

---

† A natural question which arises is whether a given topological manifold can possess at most one differentiable structure. Until the appearance of a paper by John Milnor (1956), it was thought probable that this was the case, but Milnor has shown that the 7-dimensional sphere possesses several inequivalent differentiable structures.

which admits an oriented system of coordinates is called an *orientable* manifold.

It will be seen that the surfaces considered in Part 1 of this book are differentiable manifolds of two dimensions, the local coordinate systems being given by the parameters (u, v). The change of parameters $(u, v) \rightarrow (u', v')$ corresponds to a change from one system of coordinates to another.

2. Tangent vectors

In the previous section differentiable manifolds were defined intrinsically, i.e. without reference to a containing Euclidean space of higher dimensions. We now have the problem of giving an intrinsic definition to a tangent vector at a point P on the manifold, for clearly a tangent vector can no longer be considered as the limiting position of a secant PQ as Q tends to P. One solution to this problem, suggested by a geometrical approach, is to consider a tangent vector to the manifold at P as a tangent vector to a curve which lies on the manifold and passes through P. By considering all curves passing through P we obtain a vector space of n dimensions which is the space of tangent vectors to the manifold at P. This vector space is identified with the abstract vector space V^1 considered in Chapter V, and the tangent vectors are thus contravariant vectors. By means of the tensor algebra of V^1 we can associate with each point P of the manifold a space of tensors of type (r, s) and a Grassmann algebra.

Let X be an n-dimensional differentiable manifold of class r, with a system of differentiable coordinates $\{V_j, \psi_j\}$. Let P be a point of X which lies in some neighbourhood V_j where the points are given coordinates by means of the function ψ_j. Each point in V_j may be represented in this way by its coordinates $x^1, x^2, ..., x^n$. A curve γ which lies on X and passes through P is given by a continuous function ϕ which maps the interval $0 \leqslant t \leqslant 1$ of the real line into X such that $\phi(0) = P$. The coordinates x^i of those points of γ which lie in V_j can then be written as functions of t in the form

$$x^i = \phi^i(t) \qquad (i = 1, 2, ..., n). \qquad (2.1)$$

If each of the functions ϕ^i is of class s $(s \leqslant r)$, then we say that γ is of class s. Clearly the class s of γ is independent of the particular system of coordinates (x^i) chosen to describe the points of X near P.

A tangent vector to γ at P is defined as that vector λ determined by the ordered set of n numbers $\dot{x}_0^i$, $i = 1, 2, ..., n$, where $\dot{x}_0^i$ denotes the function dx^i/dt evaluated when $t = 0$.

Each curve γ through P determines a definite vector λ; conversely each ordered set of n numbers $\dot{x}_0^i$, $i = 1, 2, ..., n$, is a tangent vector of the curve given by

$$x^i = \dot{x}_0^i\, t. \tag{2.2}$$

Hence all tangent vectors corresponding to all curves passing through P form an n-dimensional vector space, denoted by V^1, called *the space of tangent vectors to X at P*.

Consider the curve γ_i through P with the property that all coordinates of points on γ_i are constant except the ith coordinate, i.e. the curve represented by

$$\begin{cases} x^i = x_0^i + t, \\ x^j = x_0^j, \quad j \neq i. \end{cases}$$

Denoting by $\mathbf{e}_i$ the tangent vector to γ_i at P, we see that the n numbers corresponding to $\mathbf{e}_i$ are all zero, except for the number 1 in the ith place. Thus we may write

$$\mathbf{e}_1 = (1, 0, 0, ..., 0), \quad \mathbf{e}_2 = (0, 1, 0, ..., 0), \quad ..., \quad \mathbf{e}_n = (0, 0, 0, ..., 1),$$

and the tangent vectors $\mathbf{e}_i$ form a basis $(\mathbf{e}_i)$ for V^1 naturally associated with the coordinate system (x^i). For this reason $(\mathbf{e}_i)$ will be called the *natural basis* associated with the coordinate system.

The definition of tangent vector to X at P which we have given is not very satisfactory for two reasons. In the first place we defined tangent vectors at P to curves on X and only subsequently the space of all such vectors was identified with the space of tangent vectors to X at P. A more direct method avoiding the use of curves would have been preferable. Secondly, our definition has made use of one particular coordinate system (x^i). If we choose another system of local coordinates $(x^{i'})$ so that points near P have coordinates $x^{1'}, x^{2'}, ..., x^{n'}$, then the tangent vector λ is determined by the set of numbers $(\dot{x}_0^{i'})$. The two sets of coordinates (x^i), $(x^{i'})$ are related by equations of the type

$$x^{i'} = f^{i'}(x^1, x^2, ..., x^n), \tag{2.3}$$

while the two sets $(\dot{x}_0^i)$, $(\dot{x}_0^{i'})$ are related by

$$\left(\frac{dx^{i'}}{dt}\right)_0 = \left(\frac{\partial x^{i'}}{\partial x^i}\right)_0 \left(\frac{dx^i}{dt}\right)_0,$$

i.e. $$\dot{x}_0^{i'} = p^{i'}{}_i \dot{x}_0^i, \tag{2.4}$$

where we have written $(p^{i'}{}_i)$ for the Jacobian matrix $(\partial x^{i'}/\partial x^i)$ evaluated at P. Thus the same tangent vector $\boldsymbol{\lambda}$ is represented either by the components $(\dot{x}^i)$ relative to the coordinate system (x^i), or by the components $(\dot{x}^{i'})$ relative to the coordinate system $(x^{i'})$.

The second objection to our definition of tangent vector, viz. the dependence on a particular coordinate system, can be removed as follows. We define a *representation of a tangent vector at P, relative to a coordinate system* (x^i), *by the triple* $(P, x^i, \dot{x}^i)$. The representation $(Q, x^{i'}, \dot{x}^{i'})$ is said to be *equivalent* to $(P, x^i, \dot{x}^i)$ if $P = Q$ and if $(\dot{x}^i)$, $(\dot{x}^{i'})$ are related by (2.4) when $(x^i), (x^{i'})$ are related by (2.3). It is easily verified that this is a proper equivalence relation giving rise to equivalence classes. A tangent vector at P could then be *defined* as an equivalence class of triples.

We now give a second definition of tangent vector which is more abstract than the first but which is more convenient for our purpose. The essential idea is to regard a tangent vector at P as an operator which maps a *real-valued differentiable function at P* into a real number. We define first what we mean by a real-valued differentiable function of class s at P. Such a function maps a neighbourhood of P into the real numbers, and when expressed in terms of some local coordinate system valid in a neighbourhood of P the function is of class s $(s \leqslant r)$ in these coordinates. Since the transformation from one system of local coordinates to another is of class r, it is evidently sufficient for the function to be of class s relative to any one local coordinate system. The set of all real-valued differentiable functions of class s at P, denoted by $\mathscr{D}^s(P)$, is evidently a linear space since if the functions f, g belong to $\mathscr{D}^s(P)$ so does the function $af+bg$, for arbitrary real numbers a and b. We can now give our second definition of tangent vector.

A *tangent vector* $\boldsymbol{\lambda}$ at P is a linear mapping of $\mathscr{D}^r(P)$ into the real numbers which satisfies the following differentiation property:

If $g_1, g_2, ..., g_k$ are functions belonging to $\mathscr{D}^r(P)$, and if f is any function of the variables $g_1, g_2, ..., g_k$ of class r in each variable g_j, then

$$\lambda\{f(g_1, g_2, ..., g_k)\} = \sum_{j=1}^{k} f_j(P)\lambda(g_j) \tag{2.5}$$

where $f_j(P) = \partial f/\partial g_j$ evaluated at P.

By saying that $\boldsymbol{\lambda}$ is a linear mapping of $\mathscr{D}^r(P)$ we mean that

$$\lambda(af+bg) = a\lambda(f)+b\lambda(g), \tag{2.6}$$

for all functions f, g belonging to $\mathscr{D}^r(P)$ and all real numbers a, b. This definition of tangent vector may appear at first sight to be peculiarly lacking in geometrical significance, but, as we shall now show, entities so defined do in fact possess all the properties that one would expect tangent vectors to have.

By taking for f in (2.5) the particular function for which $f(g_1, g_2) = g_1 g_2$, it follows from (2.5) that

$$\lambda(g_1 g_2) = g_2(P)\lambda(g_1)+g_1(P)\lambda(g_2). \tag{2.7}$$

The sum $\boldsymbol{\nu}$ of two tangent vectors $\boldsymbol{\lambda}$ and $\boldsymbol{\mu}$ at P is defined by

$$\nu(f) = \lambda(f)+\mu(f)$$

for every f belonging to $\mathscr{D}^r(P)$, while the product $\boldsymbol{\tau}$ of a vector $\boldsymbol{\lambda}$ and a real number a, denoted by $a\boldsymbol{\lambda}$, is defined by

$$\tau(f) = a\lambda(f).$$

With these laws of combination the tangent vectors at P form a vector space, and we now show that the differentiation property implies that this space is n-dimensional.

Let (x^i) be a system of local coordinates at P. Then any function f belonging to $\mathscr{D}^r(P)$ can be expressed in the form $f(x^1, x^2, ..., x^n)$, and the n functions $g_j = x^j$ also belong to $\mathscr{D}^r(P)$. The mappings $\mathbf{e}_j$, $j = 1, 2, ..., n$, defined by

$$e_j(f) = (\partial f/\partial x^j)_P$$

evidently satisfy conditions (2.5), (2.6), and are thus tangent vectors at P. Let $\boldsymbol{\lambda}$ be an arbitrary tangent vector at P, and let n numbers λ^i be defined by

$$\lambda^i = \lambda(x^i). \tag{2.8}$$

Then, from (2.5), for any function $f(x^1, x^2, ..., x^n)$ belonging to $\mathscr{D}^r(P)$ we have

$$\lambda(f) = \sum_{j=1}^{n} (\partial f/\partial x^j)_0 \lambda(x^j) = \sum_{j=1}^{n} \lambda^j e_j(f),$$

i.e.

$$\boldsymbol{\lambda} = \sum_{j=1}^{n} \lambda^j \mathbf{e}_j = \lambda^j \mathbf{e}_j \tag{2.9}$$

using the repeated suffix convention. Thus any vector $\boldsymbol{\lambda}$ is expressible as a linear combination of the n vectors $\mathbf{e}_j$. Suppose now that there are numbers a^j such that $a^j\mathbf{e}_j = 0$. Operating on the function x^i we have
$$a^j e_j(x^i) = a^j \delta_j^i = a^i = 0,$$
and the n vectors $\mathbf{e}_j$ are therefore linearly independent. Thus the set of n vectors $\mathbf{e}_i$ form a basis $(\mathbf{e}_i)$ naturally associated with the coordinate system. Therefore we have proved that *the space of tangent vectors at P is n-dimensional*. Sometimes it is convenient to write

$$\mathbf{e}_i = \frac{\partial}{\partial x^i},$$

and hence
$$\boldsymbol{\lambda} = \lambda^i \frac{\partial}{\partial x^i},$$

showing that a tangent vector can be regarded as a differential operator.

We now show that a tangent vector to a curve through P is a tangent vector according to our second definition. Let the curve have equations $x^i = \phi^i(t)$ in some coordinate neighbourhood containing P. A function $f(x^1, x^2, ..., x^n)$ belonging to $\mathscr{D}^s(P)$ can be expressed at points on the curve near P as a function of the single variable t. Consider now the mapping $\boldsymbol{\lambda}$ which is defined by

$$\lambda(f) = \left[\frac{d}{dt}\{f(\phi^1(t), \phi^2(t), ..., \phi^n(t))\}\right]_{t=0}.$$

This is a tangent vector, and its components relative to the natural basis are
$$\lambda(x^i) = \lambda^i = \left[\frac{d}{dt}\phi^i(t)\right]_{t=0} = \dot{x}_0^i.$$

These, however, are the components of the tangent vector to γ according to our first definition.

It has been shown that a consequence of the differentiation condition (2.5) is (2.7), i.e.
$$\lambda(g_1 g_2) = g_2(P)\lambda(g_1) + g_1(P)\lambda(g_2).$$

An operator with this property is called a *differentiation*, and such operators play an important role in other branches of mathematics. It will be shown in Appendix VI. 1 that in order to define a tangent vector on manifolds of class ∞ or ω, it is sufficient to impose condition (2.7) instead of condition (2.5). *Thus, for these manifolds, a tangent vector is a linear mapping of $\mathscr{D}^\infty(P)$ (or $\mathscr{D}^\omega(P)$) into the real numbers which is also a differentiation.*

3. Affine tensors and tensorial forms

It has been shown that at each point x_0 on an n-dimensional differentiable manifold X there is an n-dimensional vector space T_{x_0}. By identifying the space T_{x_0} with the vector space V^1 considered in Chapter V, section 3, there is determined at each point the space of affine tensors of all types and also the Grassmann algebra of forms. In what follows we shall omit the adjective 'affine' as applied to these tensors.

In differential geometry one is concerned not so much with vectors and tensors at isolated points as with vector and tensor fields defined over X. A law which associates with each point P of a coordinate neighbourhood U a tangent vector $\boldsymbol{\lambda}$ at P gives rise to a vector field in U. Relative to the natural basis $(\mathbf{e}_i)$ associated with a local coordinate system around P, we have

$$\boldsymbol{\lambda} = \lambda^i \mathbf{e}_i \tag{3.1}$$

and if the components λ^i are functions of (x^i) of class s $(s \leqslant r-1)$, we say that the vector field is of class s.

Similarly, we can define tensor fields of class s over U by requiring that, relative to the natural basis in any one coordinate system, the components $T^{i_1 \ldots i_r}_{j_1 \ldots j_r}$ are functions of class s.

The bracket operation. Let $\boldsymbol{\lambda}$ denote a field of tangent vectors of class s on a differentiable manifold of class r. In a coordinate neighbourhood of X we can write

$$\boldsymbol{\lambda} = \lambda^i \frac{\partial}{\partial x^i}$$

so that $\boldsymbol{\lambda}$ can be regarded as a differential operator. If $\boldsymbol{\mu}$ is a similar field of tangent vectors then the equation

$$(\boldsymbol{\lambda}\boldsymbol{\mu} - \boldsymbol{\mu}\boldsymbol{\lambda})(f) = \boldsymbol{\nu}(f) \tag{3.2}$$

for all $f \in \mathscr{D}^r(x)$ defines a tangent vector $\boldsymbol{\nu}$ denoted by $[\boldsymbol{\lambda}, \boldsymbol{\mu}]$ of class $(s-1)$. The components of $\boldsymbol{\nu}$ are easily seen to be given by

$$\nu^i = \lambda^j \partial_j \mu^i - \mu^j \partial_j \lambda^i, \tag{3.3}$$

where $\partial_j = \partial/\partial x^j$.

We have already seen that at each point P of a neighbourhood V_i there is a natural basis $(\mathbf{e}_i)$ associated with the coordinates (x^i) defined over V_i. Consider now points P which lie in the overlap of coordinate neighbourhoods V_i, V_j, the latter carrying a system of coordinates (x^i). Then there will be two natural bases for the tangent space at P, viz. $(\mathbf{e}_i)$ and $(\mathbf{e}_{i'})$. Since

$$\frac{\partial}{\partial x^i} = \frac{\partial x^{i'}}{\partial x^i} \frac{\partial}{\partial x^{i'}},$$

we have $$\mathbf{e}_i = p^{i'}{}_i \mathbf{e}_{i'}, \tag{3.4}$$

where $$p^{i'}{}_i = \frac{\partial x^{i'}}{\partial x^i}. \tag{3.5}$$

Also $$\frac{\partial}{\partial x^{i'}} = \frac{\partial x^i}{\partial x^{i'}} \frac{\partial}{\partial x^i},$$

so that $$\mathbf{e}_{i'} = p^i{}_{i'} \mathbf{e}_i, \tag{3.6}$$

where $$p^i{}_{i'} = \frac{\partial x^i}{\partial x^{i'}}. \tag{3.7}$$

From (3.5), (3.7) it follows that

$$p^i{}_{i'} p^{i'}{}_j = \delta^i_j. \tag{3.8}$$

Thus the transformation formulae obtained in Chapter V, section 4, give the change of components of a tensor field when the coordinates are changed, provided that $(p^{i'}{}_i)$, $(p^i{}_{i'})$ are the Jacobian matrices associated with the transformation of coordinates.

A *system of frames* over some coordinate neighbourhood V_i is formed by a system of n vector fields over V_i with the property that at each point x of V_i the n vectors are linearly independent. The natural basis vectors over V_i form such a system, called the system of *natural frames*. A change of coordinates gives rise to a change of natural frames. Sometimes we shall consider a system of frames which are not the natural frames of any system of coordinates.

Since at each point $x \in V_i$ a basis of tangent vectors determines a dual basis of the space of covariant vectors, a system of frames over V_i determines a system of n linearly independent 1-forms over V_i, called a *system of co-frames*. The natural basis of the space of covariant vectors is the set of differentials (dx^i). Thus a basis for the Grassmann algebra $\mathscr{A}$ is given by

$$1,\ dx^i,\ dx^i \wedge dx^j\ (i < j),\ dx^i \wedge dx^j \wedge dx^k\ (i < j < k),$$

$$\ldots,\ dx^1 \wedge dx^2 \wedge \ldots \wedge dx^n. \tag{3.9}$$

Relative to this basis a q-form is represented by the expression

$$A_{i_1 i_2 \ldots i_q} dx^{i_1} \wedge dx^{i_2} \wedge \ldots \wedge dx^{i_q} \tag{3.10}$$

where the coefficients form a covariant tensor of order q, skew symmetric in all suffixes. If these coefficients are functions of (x^i) of class s, then we say that there is defined over U a field of q-forms of class s, or more briefly, a differential q-form.

A *tensorial form* at x of type (k, l) and degree q is defined as a linear mapping $\mathbf{\Phi}_x$ of the space of multivectors of order q at x into the space of tensors of type (k, l) at x.

A *tensorial differential form* $\mathbf{\Phi}$ of type (k, l) and degree q is defined over X by a law which associates with each point $x \in X$ such a mapping $\mathbf{\Phi}_x$. In particular, a tensorial differential form of type $(0, 0)$ and degree q is just a q-form. The components of $\mathbf{\Phi}$ referred to a basis $(\mathbf{\omega}^i)$ of T_x^* defined over the coordinate neighbourhood U are given by

$$\Phi^{i\ldots}_{jk\ldots} = A^{i\ldots}_{jk\ldots\alpha_1\alpha_2\ldots\alpha_q}\mathbf{\omega}^{\alpha_1} \wedge \mathbf{\omega}^{\alpha_2} \wedge \ldots \wedge \mathbf{\omega}^{\alpha_q}. \tag{3.11}$$

The coefficients $A^{i\ldots}_{jk\ldots\alpha_1\alpha_2\ldots\alpha_q}$ when restricted to the point $x \in U$ are the components of a tensor at x of type $(k, l+q)$.

In terms of a new basis $(\tilde{\mathbf{\omega}}^a)$ defined over U and related to the previous basis by equations

$$\tilde{\mathbf{\omega}}^a = u^a_i\, \mathbf{\omega}^i, \tag{3.12}$$

the new components of $\mathbf{\Phi}$ are given by

$$\tilde{\Phi}^{a\ldots}_{bc\ldots} = u^a_{\ i}\, v^j_{\ b}\, v^k_{\ c} \ldots \Phi^{i\ldots}_{jk\ldots}, \tag{3.13}$$

where

$$u^a_j\, v^j_b = \delta^a_b.$$

The new coefficients are given by

$$\tilde{A}^{a\ldots}_{bc\ldots\beta_1\beta_2\ldots\beta_q} = u^a_{\ i}v^j_{\ b}v^k_{\ c}\ldots v^{\alpha_1}_{\ \beta_1}\ldots v^{\alpha_q}_{\ \beta_q} A^{i\ldots}_{jk\ldots\alpha_1\ldots\alpha_q}. \tag{3.14}$$

Alternatively, a tensorial form $\mathbf{\Phi}_x$ at a point x of type (k, l) and degree q can be regarded as an element of the tensor product of the vector space of tensors of type (k, l) at x with the vector space of q-forms at x. In this way a tensorial differential form $\mathbf{\Phi}$ of type (k, l) and degree q appears either as a differential q-form with tensor coefficients of type (k, l), or, alternatively, as a tensor field of type (k, l) whose coefficients are differential q-forms.

The class of a tensorial differential form $\mathbf{\Phi}$ is defined to be the class of its components when referred to the natural basis associated with some system of coordinates.

We now introduce an operation called *exterior differentiation* which maps differential p-forms into differential $(p+1)$-forms. This operation, denoted by d, can be defined by the following properties:

When $\boldsymbol{\omega}$ is a p-form, $\boldsymbol{\eta}$ a q-form,

(1) $d(\boldsymbol{\omega}+\boldsymbol{\eta}) = d\boldsymbol{\omega}+d\boldsymbol{\eta}$;

(2) $d(\boldsymbol{\omega}\wedge\boldsymbol{\eta}) = d\boldsymbol{\omega}\wedge\boldsymbol{\eta}+(-1)^p\boldsymbol{\omega}\wedge d\boldsymbol{\eta}$;

(3)   if f is a scalar (0-form) and $\boldsymbol{\lambda}$ an arbitrary vector, then $df(\boldsymbol{\lambda}) = \boldsymbol{\lambda}(f)$, i.e. d is the ordinary differential of f;

(4)   for any scalar f, $d(df) = 0$.

If, relative to a natural basis, we have

$$\eta = A_{i_1 i_2 \ldots i_q} \, dx^{i_1} \wedge dx^{i_2} \wedge \ldots \wedge dx^{i_q},$$

then from conditions (2) and (3) we see that

$$d\eta = dA_{i_1 i_2 \ldots i_q} \wedge dx^{i_1} \wedge dx^{i_2} \wedge \ldots \wedge dx^{i_q},$$

where $dA_{i_1 i_2 \ldots i_q}$ is the ordinary differential which can be written as

$$\partial_{q+1} A_{i_1 i_2 \ldots i_q} \, dx^{q+1}.$$

Thus the operator d exists and is uniquely determined.

Since any 1-form is expressible as

$$\boldsymbol{\omega} = A^i df_i \tag{3.15}$$

for some functions A^i, f_i, it follows from (2) and (4) that

$$d\boldsymbol{\omega} = dA^i \wedge df_i \tag{3.16}$$

and
$$d(d\boldsymbol{\omega}) = 0. \tag{3.17}$$

Moreover, since any p-form is expressible as a linear combination of exterior products of 1-forms, it follows from (3.17) together with (1) and (2) that for any p-form $\boldsymbol{\eta}$,

$$d(d\boldsymbol{\eta}) = 0. \tag{3.18}$$

The relation between bracket operations on vector fields and the exterior differentiation operator is shown by the following identity satisfied by arbitrary vector fields $\boldsymbol{\lambda}$, $\boldsymbol{\mu}$ and an arbitrary 1-form $\boldsymbol{\omega}$:

$$\boldsymbol{\lambda}(\omega(\boldsymbol{\mu}))-\boldsymbol{\mu}(\omega(\boldsymbol{\lambda})) = \omega([\boldsymbol{\lambda},\boldsymbol{\mu}])+2\,d\omega(\boldsymbol{\lambda},\boldsymbol{\mu}). \tag{3.19}$$

The verification of this identity is left as an exercise for the reader.

The effect of a mapping. Let M, M' be two differentiable manifolds such that $\dim M = n$ and $\dim M' = n'$. Let ϕ be a mapping from M into M' so that $x \in M$ is mapped into $x' \in M'$, i.e. $x' = \phi(x)$. The mapping ϕ is *admissible* if, for any $f' \in \mathscr{D}^r(x')$, the composite map $f' \circ \phi \in \mathscr{D}^r(x)$. The differential map, denoted by $d\phi$, maps the tangent space T_x at x into the tangent space $T_{x'}$ at x', and $d\phi$ is defined by the relation

$$\lambda'(f') = \lambda(f' \circ \phi), \tag{3.20}$$

where $\lambda \in T_x$ and $\lambda' = d\phi(\lambda)$.

In order to obtain (3.20) in coordinate form, write

$$\lambda^i = \lambda(x^i), \quad \lambda^{i'} = \lambda'(x^{i'}),$$

where i runs from 1 to n, and i' runs from $1'$ to n'. By $x^{i'} \circ \phi$ is meant the function $x^{i'}$ expressed as a function of (x^i); in classical notation we could write $x^{i'} = x^{i'}(x)$. Then

$$\lambda^{i'} = \lambda(x^{i'} \circ \phi) = \lambda(x^{i'}(x)) = \frac{\partial x^{i'}}{\partial x^i} \lambda^i,$$

and so

$$\lambda^{i'} = \frac{\partial x^{i'}}{\partial x^i} \lambda^i, \tag{3.21}$$

which gives the required relation.

Similarly, the mapping $d\phi$ induces a map of the space of contravariant tensors of order p at x into the space of contravariant tensors of the same order at x'. In terms of coordinates the relation is

$$T^{i'j'...l'} = \frac{\partial x^{i'}}{\partial x^i} \frac{\partial x^{j'}}{\partial x^j} \cdots \frac{\partial x^{l'}}{\partial x^l} T^{ij...l}. \tag{3.22}$$

The map ϕ also induces a *dual map*, denoted by $\delta\phi$, which sends the dual space $T_{x'}^*$ into T_x^*, i.e. $\delta\phi$ maps a *covariant vector at x'* into a *covariant vector at x*. The map $\delta\phi$ is defined by the equation

$$\{\delta\phi(\mu')\}\lambda = \mu'\{d\phi(\lambda)\} \tag{3.23}$$

for all $\lambda \in T_x$.

In order to clarify the meaning of (3.23), consider a covariant vector μ' at x'. The image under $\delta\phi$, written $\delta\phi(\mu')$, will be specified when we know the real number into which it maps an arbitrary contravariant vector λ at x. The right-hand member of (3.23) merely gives a rule for the calculation of this real number, viz. it is precisely the real number obtained by operating on the image of λ under $d\phi$ with the given linear function μ'.

In terms of coordinates, equation (3.23) may be written

$$\mu_i = \frac{\partial x^{i'}}{\partial x^i} \mu_{i'} \tag{3.24}$$

where μ_i, $\mu_{i'}$ are the components of μ and μ' at x and x' respectively.

The mapping $\delta\phi$ induces a map of covariant tensors at x' into covariant tensors at x, given in coordinates by

$$T_{ij...l} = \frac{\partial x^{i'}}{\partial x^i}\frac{\partial x^{j'}}{\partial x^j}...\frac{\partial x^{l'}}{\partial x^l}T_{i'j'...l'}. \tag{3.25}$$

The effect of $\delta\phi$ on exterior differential forms will now be considered, and it will be shown that $\delta\phi$ *maps a p-form at x' into a p-form at x.* Let $\boldsymbol{\eta}$ be a p-form at x' defined locally by

$$\boldsymbol{\eta} = \frac{1}{p!}a_{i'j'...l'}dx^{i'} \wedge dx^{j'} \wedge ... \wedge dx^{l'}. \tag{3.26}$$

From (3.25) it follows that the image of the skew-symmetric covariant tensor associated with $\boldsymbol{\eta}$ at x' will have components

$$(\delta\eta)_{ij...l} = \frac{\partial x^{i'}}{\partial x^i}\frac{\partial x^{j'}}{\partial x^j}...\frac{\partial x^{l'}}{\partial x^l}a_{i'j'...l'}. \tag{3.27}$$

Thus

$$\delta\boldsymbol{\eta} = \frac{1}{p!}a_{i'j'...l'}\frac{\partial x^{i'}}{\partial x^i}\frac{\partial x^{j'}}{\partial x^j}...\frac{\partial x^{l'}}{\partial x^l}dx^i \wedge dx^j \wedge ... \wedge dx^l$$

$$= \frac{1}{p!}a_{i'j'...l'}(x)dx^{i'}(x) \wedge dx^{j'}(x) \wedge ... \wedge dx^{l'}(x). \tag{3.28}$$

Thus the local representation of $\delta\boldsymbol{\eta}$ is obtained from the local representation of $\boldsymbol{\eta}$ by expressing the coordinates $(x^{i'})$ as functions of the coordinates of (x^i). From this it follows immediately that if $\boldsymbol{\eta}$, $\boldsymbol{\omega}$ are any differential forms on M', then

$$\delta(\boldsymbol{\omega} \wedge \boldsymbol{\eta}) = \delta\boldsymbol{\omega} \wedge \delta\boldsymbol{\eta}, \tag{3.29}$$

and

$$\delta(d\boldsymbol{\omega}) = d(\delta\boldsymbol{\omega}). \tag{3.30}$$

The last equation shows that the operators d and δ are commutative.

4. Connexions

It has been seen that at each point P of a differentiable manifold X there is an n-dimensional tangent space T_P. Hence the tangent spaces T_P, T_Q at points P and Q are isomorphic since they are both n-dimensional. However, in order to obtain a definite isomorphism relating T_P and T_Q it is necessary to introduce some additional structure on the manifold called a *connexion*. This 'connects' the tangent spaces at different points on the manifold. Such a connexion would be obtained, for example, by a non-singular linear mapping of a basis $(\mathbf{e}_P)$ of T_P onto a basis $(\mathbf{e}_Q)$ of T_Q. The most suitable tool for developing this line of approach is the theory of fibre-bundles,† a theory which lies outside the scope of this book. However, the following treatment which is based on more elementary considerations does have the merit of leading directly to the essential features of a connexion.

Let γ be a differentiable curve lying on X and joining P to Q. We require a definite isomorphism ϕ_{QP} which maps T_P onto T_Q such that ϕ depends only on the points P, Q and the curve γ joining P to Q. In addition we require ϕ to satisfy the two conditions:

(1) if R lies on γ between P and Q, then we require

$$\phi_{QP} = \phi_{QR} \circ \phi_{RP},$$

i.e. the isomorphism $T_P \to T_Q$ must be identical with the successively applied isomorphisms $T_P \to T_R$ and $T_R \to T_Q$;

(2) if ϕ_{PQ} denotes the isomorphism $T_Q \to T_P$, then we require ϕ_{PQ} to be the inverse of the isomorphism ϕ_{QP}.

Suppose now that the points P, Q lie in a coordinate neighbourhood U which also contains the curve γ. Then γ can be represented by equations
$$x^i = f^i(t), \quad 0 \leqslant t \leqslant 1,$$
where $f(0) = P, f(1) = Q.$

Let $\boldsymbol{\lambda}_0$ be a tangent vector belonging to T_P. We wish to find a function ϕ_t at each point of parameter t on γ such that the mapping

$$\phi_t: T_0 \to T_t \tag{4.1}$$

is an isomorphism of the tangent space at P onto the tangent space at the point of parameter t. We require therefore

$$\phi_t(a\boldsymbol{\lambda}+b\boldsymbol{\mu}) = a\phi_t(\boldsymbol{\lambda})+b\phi_t(\boldsymbol{\mu}), \tag{4.2}$$

for any vectors $\boldsymbol{\lambda}$, $\boldsymbol{\mu}$ belonging to T_P and for all real numbers a, b.

Relative to a basis $(\mathbf{e}_i)$ of T_P, ϕ_t may be represented by a non-singular matrix and $\boldsymbol{\lambda}$ by a column vector. With this interpretation equation (4.1) may be written in the matrix form

$$\boldsymbol{\lambda}_t = \phi_t \boldsymbol{\lambda}_0. \tag{4.3}$$

Let us now assume that the elements of ϕ_t considered as functions of t are of class s, where $s \geqslant 1$. Then, on differentiating (4.3) and writing $\dot{\phi}$ for $d\phi/dt$, we obtain

$$\dot{\boldsymbol{\lambda}}_t = \dot{\phi}\boldsymbol{\lambda}_0 = (\dot{\phi}\phi^{-1})\boldsymbol{\lambda}_t.$$

Writing $-\omega$ for the matrix $(\dot{\phi}\phi^{-1})$ we have

$$\dot{\boldsymbol{\lambda}}_t+\omega\boldsymbol{\lambda}_t = \mathbf{0}, \tag{4.4}$$

† Ehresmann (1950).

where ω now depends upon γ and the parameter t. *Suppose now that ω is restricted so that it depends only upon $x(t)$ and $\dot{x}(t)$, i.e. we have $\omega = \omega(x, \dot{x})$.* Then equation (4.4) gives

$$\dot{\boldsymbol{\lambda}} + \omega(x, \dot{x})\boldsymbol{\lambda} = 0. \tag{4.5}$$

However, a standard result in the theory of linear differential equations of the first degree ensures that there is a unique solution which assumes the initial value $\boldsymbol{\lambda}_0$ when $t = 0$ (existence theorem, Appendix I. 1). Thus a given vector $\boldsymbol{\lambda}_0$ at P gives rise to a vector field $\boldsymbol{\lambda}$ along γ. It is necessary, however, that the isomorphism ϕ_t shall be independent of the particular parametric representation of γ, i.e. equation (4.5) must be invariant over a change of parameter t. From the form of equation (4.5) this will be possible if $\omega(x, \dot{x})$ is homogeneous of degree one in $\dot{x}$. Thus we see that an isomorphic mapping of the tangent spaces at all points along a curve γ will be obtained by attaching to each point of γ a matrix ω whose elements are homogeneous of degree one in $\dot{x}$.

So far we have been concerned only with a neighbourhood U and a definite system of coordinates (x^i) over U. We now consider the effect on ω of a change of coordinates $(x^i) \to (x^{i'})$. Evidently we require (4.5) to be invariant over such a change of coordinates. We have

$$\dot{\lambda}^{i'} = \frac{d}{dt}(p^{i'}_i \lambda^i) = p^{i'}_{ij}\dot{x}^j\lambda^i + p^{i'}_i\frac{d\lambda^i}{dt},$$

where $\qquad\qquad p^{i'}_{ij} = \partial^2 x^{i'}/\partial x^i \partial x^j.$

Hence $\qquad\qquad -\omega^{i'}_{j'}\lambda^{j'} = p^{i'}_{ij}\dot{x}^j\lambda^i - p^{i'}_i\omega^i_j\lambda^j,$

i.e. $\qquad\qquad [\omega^{i'}_{j'}p^{j'}_k + p^{i'}_{kj}\dot{x}^j - p^{i'}_i\omega^i_k]\lambda^k = 0.$

Changing suffixes we obtain

$$\omega^{i'}_{j'}p^{j'}_j = \omega^i_j p^{i'}_i - p^{i'}_{ji}\dot{x}^i. \tag{4.6}$$

We can now say that *a connexion is defined along the curve γ relative to the coordinate system (x^i) by a matrix (ω^i_j). The components of the matrix $(\omega^{i'}_{j'})$ which determine the same connexion relative to the coordinate system $(x^{i'})$ are given by equations (4.6).*

In this book we shall be concerned only with the particular case† when ω *is linear in $\dot{x}$*, and ω is then said to be *an affine connexion.*

If $\qquad\qquad\qquad \omega^i_j = L^i_{jk}\dot{x}^k, \tag{4.7}$

then L^i_{jk} are called the *connexion coefficients.*

---

† The more general case when $\omega(x, \dot{x})$ is homogeneous of degree 1 in $\dot{x}$ (*but not linear in $\dot{x}$*) leads to 'Finsler Geometry'.

From (4.6) it is readily deduced that the law of transformation of connexion coefficients associated with a change of coordinates is

$$L^{i'}_{j'k'} p^{j'}_j p^{k'}_k = L^i_{jk} p^{i'}_i - p^{i'}_{jk}, \qquad (4.8)$$

from which it follows that *the connexion coefficients are not components of a tensor.*

An arbitrary set of n^3 *real-valued functions* L^i_{jk}, *of class* > 1, *defined in some coordinate system over a neighbourhood* U, *determines a unique affine connexion.* The components of the connexion relative to another system of coordinates defined over U are then given by (4.8). We are now in a position to give a more satisfactory definition of an affine connexion over a manifold M.

Let P be a point of M, let $U(x)$ be a coordinate neighbourhood of P, and let (L^i_{jk}) $(i, j, k = 1, 2,..., n)$ be an ordered set of n^3 real numbers. Then we define the triple $(P, U(x), L^i_{jk})$ to be *a representation of an affine connexion at* P. Two such triples $(P, U(x), L^i_{jk})$, $(P', U'(x'), L^{i'}_{j'k'})$ are said to be equivalent if $P = P'$ and if L^i_{jk}, $L^{i'}_{j'k'}$ are related by equations (4.8), the derivatives being evaluated at P. We leave the reader to check that this is a proper equivalence relation, so the triples $(P, U(x), L^i_{jk})$ fall into mutually exclusive equivalence classes. *An equivalence class is an affine connexion at* P.

If we have an affine connexion at each point P such that the components L^i_{jk} are functions of class s in some local coordinate system about P, then we say that the manifold M admits an affine connexion of class s. We shall see in the next chapter that every manifold of class r admits an affine connexion of class $r-2$.

From (4.8) it follows that if L^i_{jk}, $\bar{L}^i_{jk}$ are the components of two affine connexions defined over some coordinate neighbourhood, then the functions

$$a^i_{jk} = L^i_{jk} - \bar{L}^i_{jk} \qquad (4.9)$$

are components of a tensor field. Alternatively, if L^i_{jk} are components of an affine connexion, and if a^i_{jk} are the components of a tensor field, then

$$\bar{L}^i_{jk} = L^i_{jk} + a^i_{jk} \qquad (4.10)$$

are components of another connexion. Moreover, the components of *every* affine connexion $\bar{L}^i_{jk}$ can be written in this form.

If in one coordinate system the connexion coefficients satisfy the symmetry relation

$$L^i_{jk} = L^i_{kj}, \qquad (4.11)$$

it follows from (4.8) that a similar relation holds for any other coordinate system. In this case the connexion is sometimes said to be *symmetric*.

From the given set of connexion coefficients L^i_{kj} two other connexions may be obtained, namely one with coefficients L^i_{kj}, and the other a symmetric connexion given by coefficients Γ^i_{jk} where

$$\Gamma^i_{jk} = \tfrac{1}{2}(L^i_{jk} + L^i_{kj}).\qquad(4.12)$$

It is readily verified that if (L^i_{jk}), (L^i_{kj}) satisfy (4.8), then so does (Γ^i_{jk}). The tensor with components T^i_{jk} defined by

$$T^i_{jk} = \tfrac{1}{2}(L^i_{jk} - L^i_{kj})\qquad(4.13)$$

is called the *torsion tensor of the connexion*. The torsion tensor of a symmetric connexion is evidently zero, and a symmetric connexion is sometimes called *torsionless* or *free from torsion*. We prefer this terminology because we shall see later that when the connexion coefficients of a torsionless affine connexion are given with respect to an arbitrary frame (as distinct from the natural frame), then the corresponding coefficients are no longer symmetric.

5. Covariant differentiation

Let L^i_{jk} be the components of an affine connexion of class $\geqslant 1$ defined over some coordinate neighbourhood U of a differentiable manifold M. Let γ be an arc lying in U given parametrically by equations $x^i = x^i(t)$, where $0 \leqslant t \leqslant 1$. Let P be the point on γ with parameter 0, and let λ^i_0 be the components of a tangent vector $\boldsymbol{\lambda}_0$ to M at P.

Consider the set of linear differential equations

$$\frac{d\lambda^i}{dt} + L^i_{jk}\lambda^j\frac{dx^k}{dt} = 0,\qquad(5.1)$$

where the connexion coefficients are expressed as functions of t by means of the equations of the curve γ. From the existence theorem, Appendix I. 1, it follows that (5.1) admits a unique solution $\lambda^i(t)$ which assumes prescribed values λ^i_0 when $t = 0$. The functions $\lambda^i(t)$ therefore determine a vector field $\boldsymbol{\lambda}_t$ along γ, said to be generated from $\boldsymbol{\lambda}_0$ at P by *parallel transport along the arc γ*. The vector $\boldsymbol{\lambda}_{t_1}$ at the point of parameter t_1 is said to be *parallel* to the vector $\boldsymbol{\lambda}_0$, and the set of such vectors forms a *parallel vector field along γ*.

Let $\mu_i(t)$ be the components of a given field of covariant vectors $\boldsymbol{\mu}_t$ defined along γ. Let $\lambda^i(t)$ be the components of an *arbitrary* parallel field of contravariant vectors λ_t along γ. Then, by contraction, $\lambda^i \mu_i$ is a scalar function of t defined along γ. We have

$$\frac{d}{dt}(\lambda^i \mu_i) = \frac{d\lambda^i}{dt}\mu_i + \lambda^i \frac{d\mu_i}{dt}$$

$$= \frac{d\mu_i}{dt}\lambda^i - \mu_i L_{jk}^i \lambda^j \frac{dx^k}{dt} \qquad \text{(from (5.1)),}$$

$$= \lambda^j \left(\frac{d\mu_j}{dt} - L_{jk}^i \mu_i \frac{dx^k}{dt} \right).$$

Since the left-hand side is an invariant, so is the right-hand side. Since λ_t is arbitrary, it follows from Chapter V, Theorem 4.1, that, on writing

$$\frac{D\mu_j}{dt} = \frac{d\mu_j}{dt} - \mu_i L_{jk}^i \frac{dx^k}{dt}, \qquad (5.2)$$

the numbers $D\mu_j/dt$ are components of a covariant vector field defined along γ.

The covariant vector determined by these components is called the *intrinsic derivative* of μ_t along γ, and is denoted by $D\mu_t/dt$. When

$$\frac{D\mu_j}{dt} = 0 \qquad (5.3)$$

we say that the *covariant vector field* μ_t *is parallel along* γ.

Let $\lambda^i(t)$ be the components of a given field of contravariant tangent vectors λ_t defined along γ, and let $\mu_i(t)$ be the components of an *arbitrary* field of parallel covariant vectors μ_t. Then, by contraction, we obtain a field of scalars $\lambda^i \mu_i$ defined along γ, and by differentiation we get

$$\frac{d}{dt}(\lambda^i \mu_i) = \frac{d\lambda^i}{dt}\mu_i + \lambda^i \frac{d\mu_i}{dt}$$

$$= \left(\frac{d\lambda^i}{dt} + L_{jk}^i \lambda^j \frac{dx^k}{dt} \right) \mu_i \qquad \text{from (5.3).}$$

Since the left-hand member is an invariant, it follows that

$$\frac{D\lambda^i}{dt} = \frac{d\lambda^i}{dt} + L_{jk}^i \lambda^j \frac{dx^k}{dt} \qquad (5.4)$$

are the components of a vector called the *intrinsic derivative of* λ_t *along* γ, denoted by $D\lambda_t/dt$.

It follows from (5.1) that the contravariant vector field λ_t is parallel along γ when

$$\frac{D\lambda^i}{dt} = 0. \qquad (5.5)$$

Consider now a field of tensors $\mathbf{A}_t$ of type $(2, 1)$ defined along γ. Let μ_i, ν_i be components of two *arbitrary* parallel covariant vector fields along γ and let λ^k be components of an *arbitrary* parallel contravariant vector field along γ. Then by contraction we obtain a scalar field

$$a^{ij}{}_k \mu_i \nu_j \lambda^k.$$

We have

$$\frac{d}{dt}(a^{ij}{}_k \mu_i \nu_j \lambda^k)$$

$$= \frac{da^{ij}{}_k}{dt} \mu_i \nu_j \lambda^k + a^{ij}{}_k \frac{d\mu_i}{dt} \nu_j \lambda^k + a^{ij}{}_k \mu_i \frac{d\nu_j}{dt} \lambda^k + a^{ij}{}_k \mu_i \nu_j \frac{d\lambda^k}{dt}.$$

Substituting from (5.3), (5.5) we get

$$\frac{d}{dt}(a^{ij}{}_k \mu_i \nu_j \lambda^k) = \frac{da^{ij}{}_k}{dt} \mu_i \nu_j \lambda^k + a^{ij}{}_k \mu_p \nu_j \lambda^k L^p_{il} \frac{dx^l}{dt} +$$

$$+ a^{ij}{}_k \mu_i \lambda^k \nu_p L^p_{jl} \frac{dx^l}{dt} - a^{ij}{}_k \mu_i \nu_j \lambda^p L^k_{pl} \frac{dx^l}{dt}$$

$$= \left[\frac{da^{ij}{}_k}{dt} + (a^{pj}{}_k L^i_{pl} + a^{ip}{}_k L^j_{pl} - a^{ij}{}_p L^p_{kl}) \frac{dx^l}{dt} \right] \mu_i \nu_j \lambda^k.$$

Since the components μ_i, ν_j, λ^k are arbitrary, it follows that

$$\frac{D}{dt} a^{ij}{}_k = \frac{da^{ij}{}_k}{dt} + (L^i_{pl} a^{pj}{}_k + L^j_{pl} a^{ip}{}_k - L^p_{kl} a^{ij}{}_p) \frac{dx^l}{dt} \qquad (5.6)$$

are the components of a tensor of type $(2, 1)$ called *the intrinsic derivative of $\mathbf{A}_t$ along γ*, denoted by $D\mathbf{A}_t/dt$.

A similar argument applied to a field of tensors $\mathbf{T}_t$ of type (m, p), defined along γ gives rise to the intrinsic derivative $D\mathbf{T}_t/dt$ whose components are

$$\frac{DT^{r_1 \ldots r_m}_{s_1 \ldots s_p}}{dt} = \frac{dT^{r_1 \ldots r_m}_{s_1 \ldots s_p}}{dt} + \sum_{\alpha=1}^{m} L^{r_\alpha}_{ji} T^{r_1 \ldots r_{\alpha-1} j r_{\alpha+1} \ldots r_m}_{s_1 \ldots s_p} \frac{dx^i}{dt} -$$

$$- \sum_{\beta=1}^{p} L^k_{s_\beta i} T^{r_1 \ldots r_m}_{s_1 \ldots s_{\beta-1} k s_{\beta+1} \ldots s_p} \frac{dx^i}{dt}. \qquad (5.7)$$

The intrinsic derivative of a tensor field of type (m, p) is thus seen to be a tensor field of type (m, p). Moreover, the intrinsic derivative is equal to the ordinary derivative together with a number of terms involving the connexion coefficients, each term corresponding to a suffix of the given tensor field. In particular, the intrinsic derivative of a scalar field ϕ_t defined along γ coincides with the ordinary derivative.

So far we have assumed that the field of tensors **T** is defined along a given curve γ. Suppose now that $(T^{r_1 \ldots r_m}_{s_1 \ldots s_p})$ are the components of a tensor field **T** of class $\geqslant 1$ *defined over some coordinate neighbourhood* U. Let P be any point of U and let γ be a curve lying in U passing through P. Then there is defined along γ a field of tensors **T** and the corresponding field of tensors obtained by intrinsic differentiation given by (5.7).

Since
$$\frac{dT^{r_1 \ldots r_m}_{s_1 \ldots s_p}}{dt} = \frac{\partial T^{r_1 \ldots r_m}_{s_1 \ldots s_p}}{\partial x^i} \frac{dx^i}{dt},$$

the right-hand member of (5.7) can be written

$$\left[\frac{\partial T^{r_1 \ldots r_m}_{s_1 \ldots s_p}}{\partial x^i} + \sum_{\alpha=1}^{m} L^{r_\alpha}_{ji} T^{r_1 \ldots r_{\alpha-1} j r_{\alpha+1} \ldots r_m}_{s_1 \ldots s_p} - \sum_{\beta=1}^{p} L^{k}_{s_\beta i} T^{r_1 \ldots r_m}_{s_1 \ldots s_{\beta-1} k s_{\beta+1} \ldots s_p} \right] \frac{dx^i}{dt}.$$

These are the components of a tensor at P, and this is the case for all curves γ at P, i.e. for all directions dx^i/dt. It follows that the expression in brackets gives the components of a tensor at P of type $(m, p+1)$, called the *covariant derivative* of the given tensor field at P with respect to the coordinate x^i. The components of this tensor are denoted either by

$$D_i \, T^{r_1 \ldots r_m}_{s_1 \ldots s_p}$$

or by
$$T^{r_1 \ldots r_m}_{s_1 \ldots s_p, i}.$$

Thus

$$D_i \, T^{r_1 \ldots r_m}_{s_1 \ldots s_p} = T^{r_1 \ldots r_m}_{s_1 \ldots s_p, i} = \frac{\partial}{\partial x^i} T^{r_1 \ldots r_m}_{s_1 \ldots s_p} + \sum_{\alpha=1}^{m} L^{r_\alpha}_{ji} \, T^{r_1 \ldots r_{\alpha-1} j r_{\alpha+1} \ldots r_m}_{s_1 \ldots s_p} -$$

$$- \sum_{\beta=1}^{p} L^{k}_{s_\beta i} T^{r_1 \ldots r_m}_{s_1 \ldots s_{\beta-1} k s_{\beta+1} \ldots s_p}. \qquad (5.8)$$

Covariant differentiation thus maps tensor fields of class r and type (m, p) into tensor fields of class $(r-1)$ and type $(m, p+1)$. It should be noticed that ordinary partial differentiation does not map a tensor field into a tensor field, as the functions

$$\frac{\partial}{\partial x^i} \, T^{r_1 \ldots r_m}_{s_1 \ldots s_p}$$

are not the components of a tensor field. It is for this reason that covariant differentiation plays the same role in tensor calculus as partial differentiation in 'ordinary' calculus. However, in the case of a field of scalars, covariant differentiation is identical with partial differentiation.

From (5.8) it follows that covariant differentiation is a linear operation, i.e.

$$D_i(aS^{r_1...r_m}_{s_1...s_p} + bT^{r_1...r_m}_{s_1...s_p}) = aD_i\,S^{r_1...r_m}_{s_1...s_p} + bD_i\,T^{r_1...r_m}_{s_1...s_p}, \qquad (5.9)$$

where **S**, **T** are tensor fields and a, b are real numbers.

Moreover,

$$D_i(S^{r_1...r_m}_{s_1...s_p}\,T^{t_1...t_h}_{u_1...u_k}) = (D_i\,S^{r_1...r_m}_{s_1...s_p})T^{t_1...t_h}_{u_1...u_k} + S^{r_1...r_m}_{s_1...s_p}(D_i\,T^{t_1...t_h}_{u_1...u_k}). \qquad (5.10)$$

We note that the covariant derivative of the Kronecker tensor is identically zero. This follows from the relation

$$\delta^i{}_{j,k} = L^i_{\alpha k}\delta^\alpha{}_j - L^\beta_{jk}\delta^i{}_\beta = L^i_{jk} - L^i_{jk} = 0. \qquad (5.11)$$

We now prove that the two operations of contraction and covariant differentiation commute. In order not to complicate matters we shall illustrate this property for a tensor field **A** of type $(2, 1)$ and leave the reader to verify that a similar proof applies to a general field of type (m, p).

The covariant derivative has components $a^{ij}{}_{k,l}$ where

$$a^{ij}{}_{k,l} = \frac{\partial a^{ij}{}_k}{\partial x^l} + L^i_{sl}a^{sj}{}_k + L^j_{sl}a^{is}{}_k - L^s_{kl}a^{ij}{}_s.$$

Contracting the suffixes i and k gives

$$\frac{\partial a^{ij}{}_i}{\partial x^l} + L^i_{sl}a^{sj}{}_i + L^j_{sl}a^{is}{}_i - L^s_{il}a^{ij}{}_s. \qquad (5.12)$$

Now the covariant derivative of the contracted tensor field $a^{ij}{}_i$ is

$$\frac{\partial a^{ij}{}_i}{\partial x^l} + L^j_{sl}a^{is}{}_i. \qquad (5.13)$$

Expressions (5.12), (5.13) differ by

$$L^i_{sl}a^{sj}{}_i - L^s_{il}a^{ij}{}_s,$$

which is easily seen to be zero on interchanging the suffixes i and s in the last term. This completes the proof that the two operations are commutative. It follows that there will be no ambiguity in using the notation

$$a^{ij}{}_{i,l}$$

to denote the result of performing both operations on the tensor field with components $a^{ij}{}_k$.

Sometimes it is convenient to use the *covariant differential* or *absolute differential* of a tensor field **T** instead of the covariant derivative. This is denoted by $D\mathbf{T}$ and defined by

$$DT^{r_1...r_m}_{s_1...s_p} = (D_i\,T^{r_1...r_m}_{s_1...s_p})\,dx^i. \qquad (5.14)$$

Hence the absolute differential of a tensor field of type (m, p) is seen to be a tensorial 1-form of type (m, p). As we shall see in section 7, the absolute differential of a tensor field can be defined without explicit reference to a coordinate system whereas the covariant derivative necessarily involves the use of coordinates. For certain investigations the absolute differential is more suitable.

Non-commutativity of covariant differentiations

It is well known that if ϕ is a real-valued function of the n variables $x^1, x^2, ..., x^n$ of sufficiently high class, then

$$\partial_l(\partial_k \phi) = \partial_k(\partial_l \phi)$$

where $\partial_k = \partial/\partial x^k$, i.e. partial differentiation with respect to x^k followed by partial differentiation with respect to x^l gives the same result as differentiation with respect to x^l followed by differentiation with respect to x^k. Briefly we may say that the *partial* differentiations are commutative. We now show that, on the contrary, *covariant* differentiations are not commutative.

Let **A** be a tensor field of type (m, p) whose components referred to the natural basis are $a^{r_1...r_m}_{s_1...s_p}$. The tensor obtained by covariant differentiation with respect to x^k will have components $a^{r_1...r_m}_{s_1...s_p,k}$, while the tensor obtained from this latter tensor by covariant differentiation with respect to x^l will have components $(a^{r_1...r_m}_{s_1...s_p,k})_{,l}$ which will be denoted by the simpler notation

$$a^{r_1...r_m}_{s_1...s_p,kl}.$$

Consider first the special case when **A** is a scalar. Then

$$\phi_{,k} = \partial_k \phi,$$

$$\phi_{,kl} = \partial_l(\partial_k \phi) - L^h_{kl} \phi_{,h},$$

so that
$$\phi_{,kl} - \phi_{,lk} = (L^h_{lk} - L^h_{kl})\phi_{,h} = -2\phi_{,h} T^h_{kl}, \tag{5.15}$$

where T^h_{kl} are the components of the torsion tensor of the connexion, defined by (4.13). From (5.15) it follows that successive covariant differentiations of a scalar are commutative only when the connexion has zero torsion.

Consider next the case when **A** is a contravariant vector field $\boldsymbol{\lambda}$. Then

$$\lambda^i_{,k} = \partial_k \lambda^i + L^i_{hk} \lambda^h,$$

$$\lambda^i_{,kl} = \partial_l(\partial_k \lambda^i + L^i_{hk} \lambda^h) + L^i_{hl} \lambda^h_{,k} - L^h_{kl} \lambda^i_{,h}$$

$$= \partial_l(\partial_k \lambda^i) + (\lambda^h \partial_l L^i_{hk} + L^i_{hk} \partial_l \lambda^h) + L^i_{hl} \lambda^h_{,k} - L^h_{kl} \lambda^i_{,h}$$

$$= \partial_l(\partial_k \lambda^i) + \lambda^h \partial_l L^i_{hk} + L^i_{hk}(\lambda^h_{,l} - L^h_{ml} \lambda^m) + L^i_{hl} \lambda^h_{,k} - L^h_{kl} \lambda^i_{,h}$$

$$= \partial_l(\partial_k \lambda^i) + \lambda^h(\partial_l L^i_{hk} - L^i_{mk} L^m_{hl}) + (L^i_{hk} \lambda^h_{,l} + L^i_{hl} \lambda^h_{,k}) - L^h_{kl} \lambda^i_{,h}.$$

Since the first and third terms of the right-hand member are symmetric in suffixes k and l we have

$$\lambda^i{}_{,kl}-\lambda^i{}_{,lk}$$
$$= -\lambda^h(\partial_k L^i_{hl}-\partial_l L^i_{hk}+L^i_{mk}L^m_{hl}-L^i_{ml}L^m_{hk})-2\lambda^i{}_{,h}T^h_{kl}. \quad (5.16)$$

Writing $\qquad L^i_{jkl} = \partial_k L^i_{jl}-\partial_l L^i_{jk}+L^h_{jl}L^i_{hk}-L^h_{jk}L^i_{hl}, \qquad (5.17)$

the above equation becomes

$$\lambda^i{}_{,kl}-\lambda^i{}_{,lk} = -\lambda^h L^i_{hkl}-2\lambda^i{}_{,h}T^h_{kl}. \quad (5.18)$$

Since (5.18) is satisfied by an arbitrary vector field $\boldsymbol{\lambda}$, it follows from the form of equation (5.18) that the coefficients L^i_{jkl} defined by (5.17) are components of a tensor, called the *curvature tensor of the connexion*. The use of the word 'curvature' for this tensor will be justified in the next chapter. From (5.18) it is seen that, *even when the connexion has zero torsion, successive covariant differentiations of a contravariant vector field are not commutative except when the curvature tensor is zero.*

Taking **A** to be a covariant vector field μ, it may be shown that

$$\mu_{i,kl}-\mu_{i,lk} = \mu_h L^h_{ikl}-2\mu_{i,h}T^h_{kl}. \quad (5.19)$$

Also, if **A** is a tensor field of type $(0, 2)$, in a similar manner we find

$$a_{ij,kl}-a_{ij,lk} = a_{hj}L^h_{ikl}+a_{ih}L^h_{jkl}-2a_{ij,h}T^h_{kl}. \quad (5.20)$$

More generally, if **A** is a tensor field of type (m, p), we find

$$a^{r_1\ldots r_m}_{s_1\ldots s_p,kl}-a^{r_1\ldots r_m}_{s_1\ldots s_p,lk}$$
$$= \sum_{\alpha=1}^{p} a^{r_1\ldots r_m}_{s_1\ldots s_{\alpha-1}hs_{\alpha+1}\ldots s_p}L^h_{s_\alpha kl}- \sum_{\beta=1}^{m} a^{r_1\ldots r_{\beta-1}hr_{\beta+1}\ldots r_m}_{s_1\ldots s_p}L^{r_\beta}_{hkl}-2a^{r_1\ldots r_m}_{s_1\ldots s_p,h}T^h_{kl},$$
$$(5.21)$$

relations which were first discovered by Ricci.

We have already constructed the torsion tensor and the curvature tensor from the connexion coefficients L^i_{jk}. We now obtain other tensors determined by these coefficients. Writing Γ^i_{jk} for the components of the associated symmetric connexion so that

$$\Gamma^i_{jk} = \tfrac{1}{2}(L^i_{jk}+L^i_{kj}),$$
then $\qquad\qquad L^i_{jk} = \Gamma^i_{jk}+T^i_{jk}.$

Substituting for L^i_{jk} in (5.17) we find

$$L^i_{jkl} = B^i_{jkl}+\Omega^i_{jkl} \quad (5.22)$$

where $\qquad B^i_{jkl} = \partial_k \Gamma^i_{jl}-\partial_l \Gamma^i_{jk}+\Gamma^h_{jl}\Gamma^i_{hk}-\Gamma^h_{jk}\Gamma^i_{hl}, \quad (5.23)$

and $\qquad \Omega^i_{jkl} = T^i_{jl,k}-T^i_{jk,l}+T^i_{hl}T^h_{jk}-T^i_{hk}T^h_{jl}-2T^i_{jh}T^h_{kl}. \quad (5.24)$

Other tensors may be obtained from these tensors by contraction. For example, write

$$B_{jk} = B^i_{jki} \tag{5.25}$$

$$b_{jk} = \tfrac{1}{2}(B_{jk} + B_{kj}), \tag{5.26}$$

$$\beta_{jk} = \tfrac{1}{2}(B_{jk} - B_{kj}), \tag{5.27}$$

so that b_{jk}, β_{jk} are respectively the symmetric and skew-symmetric parts of B_{jk}. It follows from (5.23) that

$$b_{jk} = \tfrac{1}{2}\left(\frac{\partial \Gamma^h_{hj}}{\partial x^k} + \frac{\partial \Gamma^h_{hk}}{\partial x^j}\right) - \frac{\partial \Gamma^h_{jk}}{\partial x^h} + \Gamma^h_{ji}\,\Gamma^i_{hk} - \Gamma^h_{jk}\,\Gamma^i_{hi}, \tag{5.28}$$

$$\beta_{jk} = \tfrac{1}{2}\left(\frac{\partial \Gamma^h_{hj}}{\partial x^k} - \frac{\partial \Gamma^h_{hk}}{\partial x^j}\right). \tag{5.29}$$

Another tensor with components S_{kl}, defined from B^i_{jkl} by contracting suffixes i and j,

$$S_{kl} = B^i_{ikl} \tag{5.30}$$

is not essentially a new tensor since it is easily verified that

$$S_{kl} = 2\beta_{lk}. \tag{5.31}$$

6. Connexions over submanifolds

Let M be a differentiable manifold of m dimensions and M' be a differentiable manifold of n dimensions embedded in M, both manifolds being of class ∞. We assume that M' is regularly embedded in M in the sense that if I is the identity mapping of M' into M, then its differential dI is of class ∞ and also one to one. Each point P of M' has associated with it two different tangent spaces T_P, T'_P, of dimensions m and n respectively. T_P is the tangent space to M at P, T'_P the tangent space to M' at P: T'_P is a subspace of T_P. Taking T_P as our fundamental vector space we can construct affine tensors of various orders at P. Similarly, T'_P gives rise to a class of affine tensors at P.

Let $\mathscr{T}^p_q$ be the vector space of tensors of type (p,q) based on the fundamental space T_P, and let $\mathscr{T}'^r_s$ be the vector space of tensors of type (r,s) based on T'_P. Form the tensor product of these two spaces, denoted by $\mathscr{T}^{p,r}_{q,s}$, so that

$$\mathscr{T}^{p,r}_{q,s} = \mathscr{T}^p_q \otimes \mathscr{T}'^r_s. \tag{6.1}$$

An element of $\mathscr{T}^{p,r}_{q,s}$ will be called a *double tensor*.

Let $T^{p,r}_{q,s}$ be a double tensor field defined over M' of class ∞, i.e. its components referred to some coordinate neighbourhoods X, U of M, M' respectively are of class ∞. Suppose now that M carries an affine connexion L which induces an affine connexion L' over M'. Then we shall show that the process of covariant differentiation already defined for ordinary tensor fields over M' can be extended in a natural manner to apply to double tensor fields over M' of type $\mathcal{T}_{q,s}^{p,r}$.

It will be convenient to use Roman suffixes $i, j, k,...$ for the range $1, 2,..., m$, and Greek suffixes $\alpha, \beta, \gamma,...$ for the range $1, 2,..., n$. In order to illustrate the process, consider a double tensor field of type $\mathcal{T}_{0,2}^{1,0}$. Relative to the natural basis defined over $X \times U$, such a tensor field will have components $T_{\alpha\beta}^i$. Let γ be any curve on M' with parameter t, so that γ is specified by two sets of parametric equations $x^i = x^i(t)$, and $u^\alpha = u^\alpha(t)$. Let A_i be the components of an arbitrary field of covariant vectors of M defined along γ which is parallel with respect to the connexion L. Let B^α, C^β be two arbitrary fields of contravariant vectors of M' defined along γ which are parallel with respect to the connexion L'. Then by contraction of tensors it follows that

$$T_{\alpha\beta}^i A_i B^\alpha C^\beta$$

is a scalar field defined along γ, and hence

$$\frac{d}{dt}(T_{\alpha\beta}^i A_i B^\alpha C^\beta) = \frac{dT_{\alpha\beta}^i}{dt} A_i B^\alpha C^\beta + T_{\alpha\beta}^i \frac{dA_i}{dt} B^\alpha C^\beta +$$

$$+ T_{\alpha\beta}^i A_i \frac{dB^\alpha}{dt} C^\beta + T_{\alpha\beta}^i A_i B^\alpha \frac{dC^\beta}{dt}$$

$$= \left\{ \frac{dT_{\alpha\beta}^i}{dt} + L_{jk}^i T_{\alpha\beta}^j \frac{dx^k}{dt} - L_{\alpha\delta}'^\gamma T_{\gamma\beta}^i \frac{du^\delta}{dt} - L_{\beta\delta}'^\gamma T_{\alpha\gamma}^i \frac{du^\delta}{dt} \right\} \dot{A}_i B^\alpha C^\beta,$$

where use has been made of the conditions that (A_i), (B^α), (C^β) form parallel fields.

It follows that

$$\frac{dT_{\alpha\beta}^i}{dt} + L_{jk}^i T_{\alpha\beta}^j \frac{dx^k}{dt} - L_{\alpha\delta}'^\gamma T_{\gamma\beta}^i \frac{du^\delta}{dt} - L_{\beta\delta}'^\gamma T_{\alpha\gamma}^i \frac{du^\delta}{dt} \qquad (6.2)$$

are the components of a tensor, called the intrinsic derivative of $(T_{\alpha\beta}^i)$ along γ.

By varying the curve γ passing through P we obtain the covariant derivative for the tensor field $T_{\alpha\beta}^i$ defined over U by

$$T_{\alpha\beta,\gamma}^i = \frac{\partial T_{\alpha\beta}^i}{\partial u^\gamma} + L_{jk}^i T_{\alpha\beta}^j x_\gamma^k - L_{\alpha\gamma}'^\delta T_{\delta\beta}^i - L_{\beta\gamma}'^\delta T_{\alpha\delta}^i, \qquad (6.3)$$

where

$$x_\gamma^k = \partial x^k / \partial u^\gamma. \qquad (6.4)$$

The appropriate generalization of equations (6.2) and (6.3) to give formulae for the intrinsic and covariant derivative of a general double tensor field defined over M' is obvious. We note that the formula for derivation of a double tensor reduces to the usual formula when the tensor has all suffixes from the same range. Use will be made of fields of double tensors in Chapter VIII.

7. Absolute derivation of tensorial forms

It was seen in section 4 that an affine connexion is determined in a system of coordinates by a set of n^3 functions L^i_{jk} which obey the transformation law (4.8). In this section an alternative definition is given which is more appropriate for use with the calculus of differential forms. The classical theory of tensor calculus and the more recent tensor calculus of forms should be regarded as complementary tools. The classical calculus appears to be particularly suitable for investigating properties of special types of differential geometric structures, while the form calculus seems to be particularly suitable for development of the more general theory. It is firmly believed that the reader should master both types of calculus and should be able to pass easily from one to the other.

We write
$$\omega^i_j = L^i_{jk} dx^k$$
so that (ω^i_j) is a matrix of 1-forms. If in terms of an admissible coordinate system (x') we write
$$\omega^{i'}_{j'} = L^{i'}_{j'k'} dx^{k'},$$
it follows from (4.8) that
$$\omega^{i'}_{k'} = p^{i'}_i \omega^i_j p^j_{k'} - p^j_{k'} dp^{i'}_j, \tag{7.1}$$
where
$$dx^{i'} = p^{i'}_i dx^i. \tag{7.2}$$

This suggests that, relative to any basis (ω^i), an affine connexion can be *defined by a matrix* (ω^i_j) *whose elements are 1-forms*. Relative to a new basis $(\tilde{\omega}^a)$, where
$$\tilde{\omega}^a = u^a_i \omega^i, \qquad \omega^i = v^i_a \tilde{\omega}^a, \tag{7.3}$$
the connexion is given by $(\tilde{\omega}^a_b)$, where
$$\tilde{\omega}^a_b = u^a_i w^i_j v^j_b - du^a_i v^i_b. \tag{7.4}$$

In the particular case where (ω^i), $(\tilde{\omega}^a)$ are the natural bases associated with the coordinate systems (x), (x'), then equations (7.3), (7.4) reduce to (7.2), (7.1) respectively.

The absolute differential of a vector field (λ^i) is given by
$$D\lambda^i = d\lambda^i + \omega^i_j \lambda^j.$$

Similarly, for a covariant field (μ_i) we have

$$D\mu_i = d\mu_i - \omega^j{}_i \mu_j,$$

and, in general,

$$DT^{r_1...r_m}_{s_1...s_p} = dT^{r_1...r_m}_{s_1...s_p} + \sum_{\alpha=1}^{m} \omega^{r_\alpha}{}_j \, T^{r_1...r_{\alpha-1}jr_{\alpha+1}...r_m}_{s_1...s_p} -$$

$$- \sum_{\beta=1}^{p} \omega^k{}_{s_\beta} \, T^{r_1...r_m}_{s_1...s_{\beta-1}ks_{\beta+1}...s_p}. \qquad (7.5)$$

The operator D thus maps a field of tensors of type (m, p) into a tensorial 1-form of the same type. The operator D can be extended to operate on tensorial forms as follows. Let Φ be a tensorial q-form which, relative to the basis (ω^i), may be written

$$\Phi^{i...}_{jk..} = A^{i...}_{jk..\alpha_1...\alpha_q} \omega^{\alpha_1} \wedge \omega^{\alpha_2} \wedge ... \wedge \omega^{\alpha_q}.$$

It will be proved that the forms $D\Phi^{i...}_{jk..}$ defined by

$$D\Phi^{i...}_{jk..} = d\Phi^{i...}_{jk..} + \omega^i{}_s \wedge \Phi^{s...}_{jk..} - \omega^h{}_j \wedge \Phi^{i...}_{hk..} - \omega^l{}_k \wedge \Phi^{i...}_{jl..} + ... \quad (7.6)$$

are components relative to (ω^i) of a tensorial $(q+1)$-form called the *covariant differential* of Φ with respect to the connexion $(\omega^i{}_j)$, and this will be denoted by $D\Phi$.

Let $\tilde{\Phi}^{a...}_{bc..}$ be the components of Φ corresponding to the basis $(\tilde{\omega}^a)$, which is related to (ω^i) by equation (7.3). Define forms $D\tilde{\Phi}^{a...}_{bc..}$ by the equations

$$D\tilde{\Phi}^{a...}_{bc..} = d\tilde{\Phi}^{a...}_{bc..} + \tilde{\omega}^a{}_f \wedge \tilde{\Phi}^{f...}_{bc..} - \tilde{\omega}^g{}_b \wedge \tilde{\Phi}^{a...}_{gc..} - \tilde{\omega}^h{}_c \wedge \tilde{\Phi}^{a...}_{bh..} + ... \quad (7.7)$$

Then equations (7.4), (7.6), (7.7) together give the relation

$$D\tilde{\Phi}^{a...}_{bc..} = u^a{}_i v^j{}_b v^k{}_c ... D\Phi^{i...}_{jk..},$$

from which it follows that $D\Phi$ is a tensorial form. Hence the covariant differential $D\Phi$ of a tensorial q-form Φ is a tensorial $(q+1)$-form of the same type.

The vectorial 1-form whose components relative to the basis (ω^i) are $\omega^1, \omega^2, ..., \omega^n$ gives rise to a vectorial 2-form with components $\Omega^i = D\omega^i$, called the *torsion form* of the connexion. Since $(\omega^i \wedge \omega^j)$ is a basis for the space of 2-forms, we have

$$\Omega^i = T^i_{jk} \omega^j \wedge \omega^k, \qquad (7.8)$$

where T^i_{jk} is a tensor skew symmetric in j and k, called the *torsion tensor*.

In terms of a natural basis we have

$$\Omega^i = D\omega^i = \omega^i{}_j \wedge \omega^j = L^i_{jk} dx^k \wedge dx^j, \qquad (7.9)$$

and

$$2T^i_{jk} = L^i_{jk} - L^i_{kj},$$

so that this is consistent with our previous definition of torsion
tensor. It follows that a symmetric connexion has zero torsion and
conversely.

It will now be proved that the 2-forms defined by

$$\Omega^i{}_j = d\omega^i{}_j + \omega^i{}_s \wedge \omega^s{}_j$$

are components with respect to (ω^i) of a tensorial form, called the
curvature form of the connexion. In terms of the basis $(\tilde{\omega}^a)$, the
connexion $(\tilde{\omega}^a{}_b)$ is given by (7.4) and the corresponding forms
$\tilde{\Omega}^l{}_k$ are defined by

$$\tilde{\Omega}^l{}_k = d\tilde{\omega}^l{}_k + \tilde{\omega}^l{}_h \wedge \tilde{\omega}^h{}_k. \tag{7.10}$$

Then from (7.3), (7.4), (7.10) it follows that

$$\tilde{\Omega}^l{}_k = u^l{}_i v^j{}_k \Omega^i{}_j,$$

showing that $\Omega^i{}_j$ are in fact components of a tensorial form.

In terms of a natural basis (dx^i) we have

$$\Omega^i{}_j = L^i{}_{jlm} dx^l \wedge dx^m,$$

and it is easily verified that the coefficients $L^i{}_{jlm}$ defined by this
equation, skew symmetric in l and m, are none other than the
coefficients of the curvature tensor given by (5.17).

By exterior differentiation of the relations

$$d\omega^i + \omega^i{}_j \wedge \omega^j = \Omega^i, \tag{7.11}$$

$$d\omega^i{}_j + \omega^i{}_k \wedge \omega^k{}_j = \Omega^i{}_j, \tag{7.12}$$

we get respectively

$$d\Omega^i - \Omega^i{}_j \wedge \omega^j + \omega^i{}_j \wedge \Omega^j = 0, \tag{7.13}$$

$$d\Omega^i{}_j - \Omega^i{}_k \wedge \omega^k{}_j + \omega^i{}_k \wedge \Omega^k{}_j = 0, \tag{7.14}$$

which can be written as

$$D\Omega^i = \Omega^i{}_j \wedge \omega^j, \tag{7.13'}$$

$$D\Omega^i{}_j = 0. \tag{7.14'}$$

These identities reduce to certain identities discovered by Bianchi
when the connexion has zero torsion, and they will be referred to
as the *generalized Bianchi identities*. In fact, relative to the natural
basis (dx^i), equation (7.13') reduces *in the case of zero torsion* to the
identity

$$B^i{}_{jkl} + B^i{}_{klj} + B^i{}_{ljk} = 0, \tag{7.15}$$

a relation which could have been obtained directly from (5.17).
Similarly, (7.14') reduces to

$$B^i{}_{jkl,m} + B^i{}_{jlm,k} + B^i{}_{jmk,l} = 0, \tag{7.16}$$

which is often called *Bianchi's identity*. It arises naturally from the
condition
$$B^i{}_{jkl,m}\,dx^k \wedge dx^l \wedge dx^m = 0.$$

Equation (7.16) can also be obtained by the methods of classical tensor calculus.

APPENDIX VI. 1

Tangent vectors to manifolds of class ∞

In section 2 a tangent vector $\boldsymbol{\lambda}$ was defined as a real-valued linear mapping of $\mathscr{D}^s(P)$, the space of real-valued differentiable functions of class s at P, with the property that if $g_1, g_2, ..., g_k$ belong to $\mathscr{D}^s(P)$ and if $f(g_1, g_2, ..., g_k)$ is of class s in each g_j, then

$$\lambda\big(f(g_1, g_2, ..., g_k)\big) = \sum_{j=1}^{k} \left(\frac{\partial f}{\partial g_j}\right)_P \lambda(g_j). \qquad (1)$$

In particular, by taking $f(g_1, g_2) = g_1 g_2$ we have

$$\lambda(g_1 g_2) = g_2(P)\lambda(g_1) + g_1(P)\lambda(g_2). \qquad (2)$$

We shall now prove that for manifolds of class ∞ (and *a fortiori* of class ω) equation (2) implies equation (1).

Writing g_1 as the unit function and $g_2 = g$, we obtain from (2)

$$\lambda(g) = g(P)\lambda(1) + 1\lambda(g).$$

Provided that $g(P) \neq 0$, this implies that $\lambda(1) = 0$. Since $\boldsymbol{\lambda}$ is a linear mapping, we have, for any constant function a,

$$\lambda(a) = \lambda(a.1) = a\lambda(1) = 0, \qquad (3)$$

and thus a tangent vector maps a constant function into zero.

Let $(x^1, x^2, ..., x^n)$ be local coordinates valid in a neighbourhood of P and let the coordinates of P be $x_0 = (x_0^1, x_0^2, ..., x_0^n)$. Since any function f of $\mathscr{D}^\infty(P)$ is of class ∞ in some neighbourhood V of P, we can find a star-shaped neighbourhood S of P which is contained in V. By saying that S is *star-shaped* we mean that if the point (x^i) belongs to S then so do the points $\big(x_0^i + t(x^i - x_0^i)\big)$ for $0 \leqslant t \leqslant 1$.

We have as an identity

$$f(x) = f(x_0) + \int_0^1 \frac{d}{dt}\{f(x_0 + t(x - x_0))\}\,dt$$

$$= f(x_0) + \sum_{j=1}^{n} (x - x_0^j)\int_0^1 \frac{\partial}{\partial x^j}\{f(x_0 + t(x - x_0))\}\,dt.$$

Thus, for x belonging to S, we have

$$f(x) = f(x_0) + \sum_{j=1}^{n} (x^j - x_0^j) g_j(x), \tag{4}$$

where each of the functions $g_j(x)$ belongs to $\mathscr{D}^{\infty}(P)$ and

$$g_j(x_0) = \left(\frac{\partial f}{\partial x^j}\right)_P. \tag{5}$$

Since $\boldsymbol{\lambda}$ is a linear operator, we have from (4),

$$\lambda f = \lambda f(x_0) + \lambda \sum_{j=1}^{n} (x^j - x_0^j) g_j(x)$$

$$= 0 + \sum_{j=1}^{n} g_j(x_0) \lambda(x^j), \quad \text{using (2) and (3)},$$

$$= \sum_{j=1}^{n} \left(\frac{\partial f}{\partial x^j}\right)_P \lambda(x^j), \quad \text{using (5)}.$$

Write
$$\lambda^j = \lambda(x^j).$$

We then have, the derivatives being evaluated at P,

$$\lambda(f(g_1, g_2, ..., g_k)) = \sum_{i=1}^{n} \lambda^i \frac{\partial}{\partial x^i} \{f(g_1, g_2, ..., g_k)\}$$

$$= \sum_{i=1}^{n} \lambda^i \sum_{j=1}^{k} \frac{\partial f}{\partial g_j} \frac{\partial g_j}{\partial x^i}$$

$$= \sum_{j=1}^{k} \frac{\partial f}{\partial g_j} \sum_{i=1}^{n} \lambda^i \frac{\partial g_j}{\partial x^i}$$

$$= \sum_{j=1}^{k} \frac{\partial f}{\partial g_j} \lambda(g_j),$$

and equation (1) has been obtained.

We have therefore proved that in the case of manifolds of class ∞ or ω condition (1) can be relaxed to condition (2). This, however, is not the case ıor manifolds of finite class, as has been shown in a paper by A. G. Walker and W. F. Newns (1956).

APPENDIX VI. 2

Tensor-connexions

In this chapter a connexion has been regarded as giving a natural isomorphism of the tangent space at P onto the tangent space at Q. Instead of considering isomorphisms of the tangent spaces (which induce isomorphisms of the corresponding spaces of tensors of arbitrary rank) one can define a *tensor-connexion* as giving a natural isomorphism between the space of tensors of type (m, p) at P and the space of tensors of the same type at Q. This isomorphism between the space of tensors need not be induced by an isomorphism of the tangent spaces at P and Q, so a tensor-connexion is a more general concept than the connexions previously considered.

However, it is easily seen that the analysis containing equations (4.1) to (4.5) readily applies to tensor-connexions provided that λ is interpreted as an element of the space of tensors of type (m, p) instead of as a contravariant vector. For example, for the tensor field of type $(2, 0)$ with components $a^{ik}(x)$, the absolute differential could be defined by

$$Da^{ik} = da^{ik} + L^{ik}_{rst} a^{rs} dx^t, \tag{1}$$

where the n^5 functions L^{ik}_{rst} are the coefficients of a tensor-connexion relative to a system of coordinates (x^i). Relative to a new system of coordinates $(x^{i'})$ the tensor-connexion has coefficients $L^{i'k'}_{r's't'}$ given by

$$L^{i'k'}_{r's't'} = L^{ik}_{rst} p^{i'}{}_i p^{k'}{}_k\, p^r{}_{r'} p^s{}_{s'} p^t{}_{t'} - p^{i'}_{rt} \delta^{k'}{}_{s'}\, p^r{}_{r'} p^t{}_{t'} - p^{k'}_{st} \delta^{i'}{}_{r'}\, p^s{}_{s'} p^t{}_{t'}, \tag{2}$$

where $\qquad (p^{i'}{}_i) = (\partial x^{i'}/\partial x^i), \qquad (p^{i'}_{rt}) = (\partial^2 x^{i'}/\partial x^r\, \partial x^t).$

Similar laws can be given for covariant or mixed tensor fields.

The idea of a tensor-connexion was first introduced by E. Bompiani (1946). A paper by Aldo Cossu (1956) makes a study of those particular tensor-connexions which may be written in the form

$$L^{ik}_{rst} = M^i_{rt} \delta^k{}_s + N^k_{st} \delta^i{}_r,$$

where M^i_{rt}, N^k_{st} are the coefficients of affine connexions. When $M^i_{rt} = N^i_{rt}$, the law of covariant differentiation with respect to the tensor-connexion is the same as the 'ordinary' law for covariant differentiation of second-order contravariant tensors, with respect to the affine connexion with coefficients M^i_{rt}.

It seems possible to define, as in section 4, tensor-connexions which are homogeneous of degree 1 in $\dot{x}$ but not linear in $\dot{x}$. This should give a generalization of Finsler geometry, but as far as the writer knows this possibility has not been investigated.

REFERENCES

BOMPIANI, E., *R. C. Accad. Lincei*, (8) **1** (1946) 478–82.

COSSU, A., *Rend. Mat. Appl.* (5) **15** (1956) 190–210.

EHRESMANN, C., 'Les connexions infinitésimales dans un espace fibré différentiable', *Coll. de topologie Bruxelles* (1950), pp. 29–55.

MILNOR, J., 'On manifolds homeomorphic to the 7-sphere', *Ann. Math.* (2) **64** (1956) 399–405.

WALKER, A. G., and NEWNS, W. F., *J. London Math. Soc.* (1956) 400–7.

MISCELLANEOUS EXERCISES VI

1. Two symmetric connexions Γ and $\bar{\Gamma}$ which are related by

$$\bar{\Gamma}^i_{jk} = \Gamma^i_{jk} + \delta^i_j \psi_k + \delta^i_k \psi_j,$$

where ψ_i are the components of an arbitrary covariant vector, are said to be *projectively related*. Show that the curvature tensors are related by

$$\bar{B}^i_{jkl} = B^i_{jkl} + \delta^i_j(\psi_{lk} - \psi_{kl}) + \delta^i_l \psi_{jk} - \delta^i_k \psi_{jl},$$

where $\psi_{jk} = \psi_{j,k} - \psi_j \psi_k$ and the comma denotes covariant differentiation with respect to the connexion Γ.

2. Show also that

$$\bar{B}_{jk} = B_{jk} + n\psi_{jk} - \psi_{kj};$$

$$\bar{\beta}_{kl} = \beta_{kl} + \frac{(n+1)}{2}\left(\frac{\partial \psi_k}{\partial x^l} - \frac{\partial \psi_l}{\partial x^k}\right);$$

$$\psi_{jk} = \frac{1}{(n-1)}(\bar{B}_{jk} - B_{jk}) - \frac{2}{(n^2-1)}(\bar{\beta}_{jk} - \beta_{jk});$$

$$\psi_{jk} - \psi_{kj} = \frac{2}{(n+1)}(\bar{\beta}_{jk} - \beta_{jk}).$$

3. Use the results of Exercises 1 and 2 to show that the *Weyl curvature tensor* with components

$$W^i_{jkl} = B^i_{jkl} + \frac{2}{(n+1)}\delta^i_j \beta_{kl} + \frac{1}{(n-1)}(\delta^i_k B_{jl} - \delta^i_l B_{jk}) + \frac{2}{(n^2-1)}(\delta^i_l \beta_{jk} - \delta^i_k \beta_{jl})$$

has the property that $\qquad\qquad \bar{W}^i_{jkl} = W^i_{jkl},$

i.e. *all connexions which are projectively related have the same Weyl tensor.*

4. Obtain the following expression for the components of the Weyl tensor:

$$W^i_{jkl} = B^i_{jkl} + \frac{1}{(n+1)}\delta^i_j(B_{kl} - B_{lk}) + \frac{1}{(n^2-1)}[\delta^i_k(nB_{jl} + B_{lj}) - \delta^i_l(nB_{jk} + B_{kj})].$$

Show that

$$W^i_{jkl} + W^i_{jlk} = 0,$$

$$W^i_{jkl} + W^i_{klj} + W^i_{ljk} = 0,$$

$$W^i_{ikl} = W^i_{jki} = 0.$$

5. Prove that
$$W^i_{jkl,i} = \frac{n-2}{n-1}\Big[B_{jk,l} - B_{jl,k} + \frac{2}{(n+1)}(\beta_{jl,k} - \beta_{jk,l})\Big].$$

[Hint: use the formula for the Weyl tensor given in Exercise 3, differentiate covariantly with respect to Γ^i_{jk}, and use the Bianchi identity (7.16).]

6. Given any mixed tensor field h^i_j show that
$$H^i_{jk} = -\tfrac{1}{8}(h^p_j\,\partial_p h^i_k - h^p_k\,\partial_p h^i_j) + \tfrac{1}{8}(h^i_p\,\partial_j h^p_k - h^i_p\,\partial_k h^p_j)$$
is a tensor field, where $\partial_p h^i_k = \dfrac{\partial h^i_k}{\partial x^p}$.

[Hint: express the partial derivatives in terms of covariant derivatives with respect to an arbitrary symmetric affine connexion.]

7. Coefficients h^i_{jrs} are defined by
$$h^i_{jrs} = -\tfrac{1}{2}\partial_j H^i_{rs} + \tfrac{1}{2}h^i_p(h^q_j\,\partial_q H^p_{rs} - H^q_{rs}\,\partial_q h^p_j + H^p_{qs}\,\partial_r h^q_j - H^p_{qr}\,\partial_s h^q_j),$$
where $h^i_j h^j_k = -\delta^i_k$, and H^i_{jk} is defined in terms of the tensor h^i_j by the formula given in Exercise 6.

A. G. Walker has defined† the *torsional derivative* $T^{ih\cdots}_{jk\cdots\|rs}$ of any tensor field with components $T^{ih\cdots}_{jk\cdots}$ by the formula
$$T^{i\cdots}_{j\cdots\|rs} = H^p_{rs}\,\partial_p T^{ih\cdots}_{jk\cdots} + T^{ph\cdots}_{jk\cdots}\,h^i_{prs} + T^{ip\cdots}_{jk\cdots}\,h^h_{prs} + \ldots - T^{ih\cdots}_{pk\cdots}\,h^p_{jrs} - T^{ih\cdots}_{jp\cdots}\,h^p_{krs} - \ldots,$$
where the dots indicate that there is a term for each suffix in $T^{ih\cdots}_{jk\cdots}$, as in the formula for covariant differentiation. Prove that $T^{ih\cdots}_{jk\cdots\|rs}$ are the components of a tensor of the type indicated by the suffixes.

8. Prove that $h^i_{j\|rs} = 0$ and also that $\delta^i_{j\|rs} = 0$.

9. Show that the tensors with components $H^i_{jk\|rs}$, $H^i_{jk\|ri}$ are not zero.

† A. G. Walker, *C.R. Acad. Sci. Paris* (1957) 1213–5.

VII

RIEMANNIAN GEOMETRY

1. Riemannian manifolds

DIFFERENTIAL geometry is concerned with the study of geometric objects defined on differential manifolds. One of the simplest geometric objects is a field of non-singular, symmetric, second-order covariant tensors, and the branch of differential geometry which studies the structures associated with this object is called Riemannian geometry. Alternatively, this object is given by associating with the tangent space at each point x an inner product ϕ_x having properties given in Chapter V, section 7, with the additional property that ϕ_x varies differentiably with x. A differentiable manifold equipped with such an inner product is called a *Riemannian* manifold.

The literature of Riemannian geometry is very vast, and in this chapter it will be possible to summarize only the essential features of the subject. For a more detailed treatment the reader is referred to Eisenhart (1950) and Cartan (1946).

Except for section 20, the whole of this chapter is concerned with *local* Riemannian geometry, i.e. only with the differential geometric properties of a part of a differentiable manifold which can be covered by one system of coordinates.

2. Metric

The second-order tensor determined by ϕ is non-singular but not necessarily definite, so that corresponding to a non-zero tangent vector λ the number $\phi(\lambda, \lambda)$ may be positive, negative, or zero. If $\phi(\lambda, \lambda) = 0$ then λ is said to be a *null vector*. The *length* of λ (*magnitude* of λ), denoted by $|\lambda|$, is defined by

$$|\lambda|^2 = e\phi(\lambda, \lambda) \qquad (2.1)$$

where
$$
\begin{aligned}
e &= +1, \quad \text{if } \phi(\lambda, \lambda) > 0, \\
&= -1, \quad \text{if } \phi(\lambda, \lambda) < 0, \\
&= 0, \quad\;\; \text{if } \phi(\lambda, \lambda) = 0.
\end{aligned}
$$

The symbol e is called the *indicator* of the vector λ.

In terms of coordinates, if ϕ has components g_{ij} and λ has components λ^i, then

$$|\lambda|^2 = eg_{ij}\lambda^i\lambda^j. \tag{2.2}$$

Similarly if μ has components μ^i, the scalar product of λ and μ is

$$\phi(\lambda, \mu) = g_{ij}\lambda^i\mu^j. \tag{2.3}$$

Since the tensor with components g_{ij} is non-singular, as in Chapter V, section 8, there exists a reciprocal tensor with components g^{ij} defined by

$$g^{ij}g_{jk} = \delta^i{}_k, \tag{2.4}$$

and these two tensors can be used for defining associated tensors. Thus, if $\lambda_i = g_{ij}\lambda^j$, $\mu_i = g_{ij}\mu^j$, then

$$\phi(\lambda, \mu) = g_{ij}\lambda^i\mu^j = \lambda^i\mu_i = \lambda_i\mu^i = g^{ij}\lambda_i\mu_j. \tag{2.5}$$

Let C be a differentiable curve on a Riemannian manifold defined in a coordinate neighbourhood U by equations $x^i = f^i(t)$. Let t_0 be the parameter of P_0 on the curve and let t_1 be the parameter of some point P_1. Then the length s of the arc of the curve from P_0 to P_1 is *defined* by the equation

$$s = \int_{t_0}^{t_1} \left[eg_{ij}\frac{dx^i}{dt}\frac{dx^j}{dt}\right]^{\frac{1}{2}}dt. \tag{2.6}$$

Symbolically, in terms of differentials, this equation may be written

$$ds^2 = eg_{ij}dx^idx^j, \tag{2.7}$$

and ds may be regarded as the infinitesimal distance from the points of coordinates (x^i), (x^i+dx^i). It should be remembered, however, that ds regarded as a function of (x^i) is not a perfect differential. If the tangent vector to C at each point is a null vector, C is called a *null curve*.

Many writers on Riemannian geometry confine their attention to the particular case when the metric is positive definite, as this certainly simplifies many results. However, quite apart from the fact that in applications to relativity theory the metric is necessarily indefinite, it will be seen later in this chapter that indefinite metrics do in fact arise quite naturally in the development of the subject. On the other hand, certain results obtained with an indefinite metric seem unnatural, as they contradict one's geometric intuition which is based on the positive definite metric of Euclidean space. For example, the angle θ between two tangent vectors $\boldsymbol{\lambda}, \boldsymbol{\mu}$ is given by

$$\cos \theta = \phi(\boldsymbol{\lambda}, \boldsymbol{\mu})/[e\phi(\boldsymbol{\lambda}, \boldsymbol{\lambda})]^{\frac{1}{2}}[e\phi(\boldsymbol{\mu}, \boldsymbol{\mu})]^{\frac{1}{2}} \qquad (2.8)$$

but if the metric is indefinite $\cos \theta$ may be numerically greater than 1.

EXERCISE 2.1. Show that three-dimensional space with the usual Euclidean metric forms a Riemannian space.

EXERCISE 2.2. Prove that if the metric is positive definite, then the value of $\cos \theta$ given by (2.8) is such that $-1 \leqslant \cos \theta \leqslant 1$.

It will be remembered (Chapter IV, section 5) that a set of points S carries the structure of a metric space when there is a real-valued function $\rho : S \times S \to R_1$ with the properties

(i) $\rho(A, B) = 0$ if and only if $A = B$,

(ii) $\rho(A, B) = \rho(B, A)$,

(iii) $\rho(A, C) \leqslant \rho(A, B) + \rho(B, C)$,

for all points A, B, C of S. A Riemannian space with a *definite* metric may be made into a metric space by defining $\rho(P, Q)$ to be the least bound of the distances from P to Q measured along all differentiable curves joining P to Q. If the metric is indefinite this construction is no longer possible, because two different points P, Q on the same null curve will have zero distance in contradiction to (i).

3. The fundamental theorem of local Riemannian geometry

The fundamental theorem of local Riemannian geometry states that with a given Riemannian metric there is uniquely associated a symmetric affine connexion with the property that parallel transport preserves scalar products. More precisely, let P be any point on the manifold M and let C be any curve through P of class $\geqslant 1$. Let $\boldsymbol{\lambda}_0$, $\boldsymbol{\mu}_0$ be two arbitrary tangent vectors to M at P. Relative to a suitable coordinate neighbourhood U containing P and part of C, the equations of C will be of the form $x^i = x^i(t)$, $a \leqslant t \leqslant b$, with P corresponding to $t = t_0$, Let $\boldsymbol{\lambda}_0$, $\boldsymbol{\mu}_0$ have components λ_0^i, μ_0^i respectively. Let Γ_{jk}^i be the components of a symmetric affine connexion defined over U.

The differential equation

$$\frac{D\lambda^i}{dt} = \frac{d\lambda^i}{dt} + \Gamma_{jk}^i \lambda^j \frac{dx^k}{dt} = 0 \qquad (3.1)$$

admits a unique solution $\boldsymbol{\lambda} = \boldsymbol{\lambda}(t)$ satisfying the initial condition

$\lambda(t_0) = \lambda_0$, and this defines a parallel vector field along C. Similarly, the equation $D\mu^i/dt = 0$ gives rise to a unique vector field $\mu(t)$ satisfying $\mu(t_0) = \mu_0$. Then the *fundamental theorem asserts that there exists one and only one symmetric connexion* Γ^i_{jk} *having the property*

$$\frac{D}{dt}(g_{ij}\lambda^i\mu^j) = 0 \qquad (3.2)$$

along C. Moreover, this condition must be satisfied along all C^1-curves through P, for arbitrary vectors λ, μ at P, and at every point P in some neighbourhood.

Since $D\lambda^i/dt = D\mu^i/dt = 0$ along C, equations (3.2) give

$$\frac{Dg_{ij}}{dt}\lambda^i\mu^j = 0,$$

i.e.

$$g_{ij,k}\frac{dx^k}{dt}\lambda^i\mu^j = 0.$$

Since these conditions must hold at P for the direction (dx^k/dt) of every curve C, and for arbitrary vectors λ, μ, it follows that at P we must have

$$g_{ij,k} = 0. \qquad (3.3)$$

Hence a necessary condition for parallel transport to preserve scalar product is that the connexion coefficients must satisfy

$$\frac{\partial g_{ij}}{\partial x^k} - g_{hj}\,\Gamma^h_{ik} - g_{ih}\,\Gamma^h_{jk} = 0, \qquad (3.4)$$

at all points P in some neighbourhood.

Conversely, if the connexion coefficients satisfy equations (3.4), equivalent to (3.3), then equation (3.2) shows that the scalar product of vectors will be invariant under parallel transport. Consequently we solve equations (3.4) for the connexion coefficients.

Let new symbols $[k, ij]$ be defined by

$$[k, ij] = g_{hk}\,\Gamma^h_{ij}. \qquad (3.5)$$

Then

$$\Gamma^k_{ij} = g^{hk}[h, ij], \qquad (3.6)$$

and equation (3.4) may be written in the form

$$\frac{\partial g_{ij}}{\partial x^k} = [j, ik] + [i, jk]. \qquad (3.7)$$

Cyclic permutation of i, j, k gives

$$\frac{\partial g_{jk}}{\partial x^i} = [k, ji] + [j, ki], \qquad (3.7')$$

and a further cyclic permutation gives

$$\frac{\partial g_{ki}}{\partial x^j} = [i, kj] + [k, ij]. \tag{3.7''}$$

Add (3.7′) to (3.7″) and subtract (3.7) to obtain

$$\frac{\partial g_{jk}}{\partial x^i} + \frac{\partial g_{ki}}{\partial x^j} - \frac{\partial g_{ij}}{\partial x^k} = 2[k, ij],$$

where use has been made of (3.5) to imply the symmetry of $[k, ij]$ in suffixes i and j. From (3.6), the connexion coefficients are given by

$$\Gamma_{ji}^k = \tfrac{1}{2} g^{hk} \left(\frac{\partial g_{ih}}{\partial x^j} + \frac{\partial g_{jh}}{\partial x^i} - \frac{\partial g_{ij}}{\partial x^h} \right). \tag{3.8}$$

The connexion defined by (3.8) has the required properties. This completes the proof of the fundamental theorem.

In particular, the length of a vector is invariant under parallel transport, and the angle between two unit vectors is similarly preserved. In future, by covariant differentiation in a Riemannian space we shall mean differentiation with respect to this *Riemannian connexion*. In order to distinguish the Riemannian connexion from other symmetric connexions we shall use the notation $\begin{Bmatrix} i \\ j\,k \end{Bmatrix}$ instead of Γ_{jk}^i. These symbols, known as Christoffel symbols, have already been introduced in Part 1.

Since $D(\delta^k{}_i) = 0$ it follows that

$$g_{ij} Dg^{jk} = 0,$$

which on multiplication by g^{hi} gives $Dg^{hk} = 0$, i.e.

$$g^{hk}{}_{,j} = 0. \tag{3.9}$$

Equations (3.3), (3.9) are important because they show that the two operations of covariant differentiation and association with respect to the metric tensor are commutative. For example, if $a^i{}_{jk}$ are the components of a tensor field, then

$$g_{ih}(a^i{}_{jk})_{,m} = (g_{ih} a^i{}_{jk})_{,m}$$

and

$$g^{hj}(a^i{}_{jk})_{,m} = (g^{hj} a^i{}_{jk})_{,m}.$$

Using the well-known rule for differentiating a determinant we have

$$\partial g/\partial x^k = g g^{ij} \partial g_{ij}/\partial x^k = g g^{ij}([j, ik] + [i, jk]) \text{ using } (3.7),$$

$$= g \left(\begin{Bmatrix} i \\ i\,k \end{Bmatrix} + \begin{Bmatrix} j \\ j\,k \end{Bmatrix} \right) \text{ using } (3.6),$$

$$= 2g \begin{Bmatrix} i \\ i\,k \end{Bmatrix}$$

This leads to the useful formula

$$\begin{Bmatrix} i \\ i\,k \end{Bmatrix} = \frac{1}{2g}\frac{\partial g}{\partial x^k} = \frac{\partial}{\partial x^k}(\log \sqrt{g}). \qquad (3.10)$$

4. Differential parameters

In this section it is shown how the operators *grad*, *div*, *curl*, and *nabla* of vector analysis can be readily generalized to a Riemannian manifold. If f is a real-valued function defined over a coordinate neighbourhood U, i.e. a 0-form, then the coefficients of the exterior derivative df relative to the natural basis (dx^i) are the components of a covariant vector called the *gradient of f*. Thus

$$(\operatorname{grad} f)_i = (\partial f/\partial x^i). \qquad (4.1)$$

The set of points whose coordinates satisfy an equation $f = 0$ form a submanifold such that $(\partial f/\partial x^i)\,dx^i = 0$. Thus the gradient vector at a point on the hypersurface $f = 0$ is parallel to the normal.

If $\boldsymbol{\lambda}$ is a covariant vector field, the coefficients of the exterior derivative of the associated 1-form are components of a skew-symmetric tensor field called the *curl* of $\boldsymbol{\lambda}$. Thus for a covariant field with components λ_i, the associated 1-form is given by

$$\Phi = \lambda_i\,dx^i$$

and the exterior derivative

$$d\Phi = d\lambda_i \wedge dx^i = \frac{\partial \lambda_i}{\partial x^j}dx^j \wedge dx^i.$$

Thus

$$(\operatorname{curl} \lambda)_{ij} = \left(\frac{\partial \lambda_i}{\partial x^j} - \frac{\partial \lambda_j}{\partial x^i}\right). \qquad (4.2)$$

Evidently the curl of a gradient field is zero since $d(df) = 0$. Conversely, if $\operatorname{curl}\boldsymbol{\lambda} = 0$, then $\partial\lambda_i/\partial x^j = \partial\lambda_j/\partial x^i$, from which it follows that $\lambda_i\,dx^i$ is an exact differential, say df, and hence (λ_i) is a gradient vector.

The *divergence* of a contravariant vector field is defined by

$$\operatorname{div}\boldsymbol{\lambda} = \lambda^i{}_{,i}, \qquad (4.3)$$

which may be written

$$\operatorname{div}\boldsymbol{\lambda} = \frac{\partial \lambda^i}{\partial x^i} + \begin{Bmatrix} i \\ j\,i \end{Bmatrix}\lambda^j$$

and, from (3.10), $\qquad \operatorname{div}\boldsymbol{\lambda} = \dfrac{1}{\sqrt{g}}\dfrac{\partial}{\partial x^i}(\sqrt{g}\,\lambda^i). \qquad (4.3')$

The *Laplacian operator* Δ_2 is defined by

$$\Delta_2 f = g^{ij} f_{,ij}, \tag{4.4}$$

which is equivalent to

$$\Delta_2 f = \frac{1}{\sqrt{g}} \frac{\partial}{\partial x^i} (\sqrt{g}\, g^{ij} f_{,j}). \tag{4.4'}$$

Sometimes Δ_2 is called a *differential parameter of the second order*. Two *first-order differential parameters* which often occur in calculations are defined by

$$\Delta_1 f = g^{ij}\, \partial_i f\, \partial_j f = g^{ij} f_{,i} f_{,j}, \tag{4.5}$$

and

$$\Delta_1(f, \phi) = g^{ij} \partial_i f\, \partial_j \phi = g^{ij} f_{,i} \phi_{,j}. \tag{4.6}$$

Evidently $\Delta_1 f$ measures the length of the normal vectors to the hypersurface $f = 0$, while the vanishing of $\Delta_1(f, \phi)$ expresses the fact that the hypersurfaces $f = 0$, $\phi = 0$ cut orthogonally.

EXERCISE 4.1. Show that the generalized operators defined in this section reduce to the well-known operators in vector calculus when the metric is that of Euclidean space of three dimensions.

5. Curvature tensors

In Chapter VI, section 5, it was shown how various tensors are associated with an affine connexion; in particular, these tensors are associated with the Riemannian connexion given by (3.6). Since this connexion is symmetric the curvature tensor $L^i{}_{jkl}$ reduces to $B^i{}_{jkl}$. However, owing to the special form of (3.6) there are several other simplifications so it is convenient to use the symbol $R^i{}_{jkl}$ instead of $B^i{}_{jkl}$ in this case. Equations (3.10) and VI (5.29) show that in the Riemannian case $\beta_{jk} = 0$, so from VI (5.31) $S_{kl} = 0$.

Thus, if $R_{jk} = R^i{}_{jki}$, we have

$$R^i{}_{jkl} = -R^i{}_{jlk}, \tag{5.1}$$

$$R^i{}_{jkl} + R^i{}_{klj} + R^i{}_{ljk} = 0, \tag{5.2}$$

$$R^i{}_{jkl,m} + R^i{}_{jlm,k} + R^i{}_{jmk,l} = 0, \tag{5.3}$$

$$R^i{}_{ikl} = 0, \tag{5.4}$$

$$R_{jk} = R_{kj}. \tag{5.5}$$

Equation (5.1) is satisfied by all affine connexions, (5.2) and (5.3) by all symmetric connexions, (5.4) and (5.5) by Riemannian connexions.

It is convenient to introduce an associated covariant curvature tensor defined by
$$R_{ijkl} = g_{ih} R^h{}_{jkl}. \tag{5.6}$$
This tensor has the following symmetry properties which are easily deduced from those above:

$$R_{ijkl} = -R_{ijlk}, \tag{5.7}$$

$$R_{ijkl} + R_{iklj} + R_{iljk} = 0, \tag{5.8}$$

$$R_{ijkl} = -R_{jikl}, \tag{5.9}$$

$$R_{ijkl} = R_{klij}, \tag{5.10}$$

$$R_{hijk,l} + R_{hikl,j} + R_{hilj,k} = 0. \tag{5.11}$$

The tensor with components R_{ijkl} is called the *Riemann-Christoffel curvature tensor*, while the tensor with components R_{ij} is called the *Ricci curvature tensor*. Two further curvature tensors can be obtained from the Ricci tensor, namely the Ricci tensor of type $(1, 1)$ with components
$$R^i{}_j = g^{ik} R_{kj}, \tag{5.12}$$
and the *scalar curvature* R given by
$$R = R^i{}_i = g^{ij} R_{ij}. \tag{5.13}$$

From equations (5.1), (5.3) it follows that
$$R^h{}_{ijk,l} - R^h{}_{ilk,j} + g^{hm} R_{milj,k} = 0.$$
Contracting for h and k we obtain
$$R_{ij,l} - R_{il,j} + g^{hm} R_{milj,h} = 0.$$
Multiplying by g^{il} we get
$$R^l{}_{j,l} - R_{,j} + R^l{}_{j,l} = 0,$$
i.e.
$$R^l{}_{j,l} = \tfrac{1}{2} \partial R / \partial x^j. \tag{5.14}$$
If we define a new tensor by
$$T^l{}_j = R^l{}_j - \tfrac{1}{2}(R - 2\lambda)\delta^l{}_j, \tag{5.15}$$
where λ is a constant, then (5.14) implies that
$$T^l{}_{j,l} = 0. \tag{5.16}$$
This equation plays an important role in relativity theory.

6. Geodesics

As in the special case of the surfaces considered in Chapter II, in any n-dimensional Riemannian space there are special intrinsic curves called *geodesics* which are curves of shortest distance. As in that chapter we shall show that the problem of finding a curve of shortest distance leads to definite differential equations for functions $x^i = x^i(t)$ which define a curve, but the question of the existence of an arc which actually has a minimum length is left unanswered.

Appeal is made to the calculus of variations as in Chapter II —in fact, the treatment is so analogous and the extension to n dimensions so straightforward that it is sufficient to give just a brief outline here.

By considering the variation of the integral

$$I = \int_{t_0}^{t_1} f(x^1, x^2, ..., x^n; \dot{x}^1, \dot{x}^2, ..., \dot{x}^n) \, dt$$

we arrive at Euler's equations

$$\frac{\partial f}{\partial x^i} - \frac{d}{dt}\left(\frac{\partial f}{\partial \dot{x}^i}\right) = 0 \quad (i = 1, 2, ..., n). \tag{6.1}$$

Writing $T = \frac{1}{2}g_{ij}\dot{x}^i\dot{x}^j$, and substituting $f = \sqrt{(2eT)}$ in (6.1), we get

$$U_i \equiv \frac{d}{dt}\left(\frac{\partial T}{\partial \dot{x}^i}\right) - \frac{\partial T}{\partial x^i} = \frac{1}{2T}\frac{dT}{dt}\left(\frac{\partial T}{\partial \dot{x}^i}\right), \tag{6.2}$$

which is a generalization of equation (10.4) of Chapter II.

Also

$$\dot{x}^i U_i = \frac{d}{dt}\left(\dot{x}^i\frac{\partial T}{\partial \dot{x}^i}\right) - \ddot{x}^i\frac{\partial T}{\partial \dot{x}^i} - \dot{x}^i\frac{\partial T}{\partial x^i} = \frac{d}{dt}(2T) - \frac{dT}{dt},$$

so we have a generalization of II (10.5) to

$$\dot{x}^i U_i = \frac{dT}{dt}. \tag{6.3}$$

Since the right-hand expressions of (6.2) satisfy the same identity it follows that the n equations in (6.2) are not independent.

In the particular case when t is the arc length s, equations (6.2) reduce to
$$U_i = 0. \tag{6.4}$$

We now obtain an alternative expression for equations (6.2). We have

$$\frac{d}{dt}(g_{ij}\dot{x}^j) - \frac{1}{2}\frac{\partial g_{jk}}{\partial x^i}\dot{x}^j\dot{x}^k = \frac{1}{2T}\frac{dT}{dt}g_{ij}\dot{x}^j,$$

i.e.
$$g_{ij}\ddot{x}^j + [i, jk]\dot{x}^j\dot{x}^k = \frac{1}{2T}\frac{dT}{dt}g_{ij}\dot{x}^j,$$

which on multiplying by g^{il} gives

$$\ddot{x}^l + \begin{Bmatrix} l \\ j\,k \end{Bmatrix} \dot{x}^j \dot{x}^k = \frac{1}{2T} \frac{dT}{dt} \dot{x}^l. \tag{6.5}$$

In particular when the parameter is equal to the arc length s, $T = \frac{1}{2}$ and so (6.5) reduces to

$$\frac{d^2 x^l}{ds^2} + \begin{Bmatrix} l \\ j\,k \end{Bmatrix} \frac{dx^j}{ds} \frac{dx^k}{ds} = 0. \tag{6.6}$$

Moreover, it follows from (6.2) that a first integral of these equations is $T = $ constant. Equation (6.6) also expresses the fact that a tangent vector to a geodesic at a point remains tangent to the geodesic when displaced by parallel transport along the curve. Hence the tangent vectors at the various points along a geodesic all have the same indicator, and we can thus talk about *the indicator of the geodesic*. If the indicator is zero the curve is called a *null geodesic*. In the general theory of relativity null geodesics play an important role as they represent light paths.

Many questions involving geodesics mentioned in Chapters II and IV can be generalized immediately to n-dimensional Riemannian manifolds. For example, at what stage does the distance measured along a geodesic from a point P_0 to a variable point P cease to be the 'shortest distance' between P_0 and P? How many geodesics pass through two specified points? Under what circumstances can any geodesic be prolonged to an infinite length? We shall not deal with these questions here; however, the methods used in Chapter II enable us to prove that through any point P there is a geodesic in a prescribed direction, and, moreover, that any two points P and Q can be joined by a geodesic provided that they are 'not too far apart'.

EXERCISE 6.1. In a Riemannian space with definite metric, s is the geodesic distance from a point x_0 to a point x sufficiently near x_0. If s is regarded as a function of x, prove that:

(i) $\Delta_1(s) = 1$,

(ii) $\Delta_2 \phi(s) = \phi'(s)\Delta_2 s + \phi''(s)$,

where ϕ is any function of s of class 2, and Δ_1, Δ_2 are the differential operators defined in section 4.

7. Geodesic curvature

On a Riemannian manifold with positive definite metric, let a curve C be given by equations $x^i = x^i(s)$, the parameter being the arc length. Then the unit tangent vector at a point P on C is

$$\mathbf{t} = (dx^i/ds). \tag{7.1}$$

The intrinsic derivative of $\mathbf{t}$ along the curve is

$$\frac{Dt^i}{ds} = \left(\frac{dx^i}{ds}\right)_{,k}\frac{dx^k}{ds} = \frac{d^2x^i}{ds^2} + \left\{\begin{matrix} i \\ j\,k \end{matrix}\right\}\frac{dx^j}{ds}\frac{dx^k}{ds}. \tag{7.2}$$

If the curve C is a geodesic, it follows from (6.6) that Dt^i/ds is zero. If C is not a geodesic, then the magnitude of the vector (7.2) serves as a measure of the deviation of C from a geodesic. If we write

$$\frac{Dt^i}{ds} = \frac{d^2x^i}{ds^2} + \left\{\begin{matrix} i \\ j\,k \end{matrix}\right\}\frac{dx^j}{ds}\frac{dx^k}{ds} = \kappa_g n^i{}_g, \tag{7.3}$$

where $(n^i{}_g)$ is a unit vector, then $\kappa_g n^i{}_g$ are the components of the *geodesic curvature vector*. The length κ_g of this vector is called the *geodesic curvature*, while $n^i{}_g$ are the components of the *geodesic normal* $\mathbf{n}_g$. That $\mathbf{n}_g$ is normal to $\mathbf{t}$ follows at once from multiplying (7.3) by $(g_{ij}t^j)$ and using the result that $Dt^2/ds = g_{ij}t^j Dt^i/ds = 0$ (since $\mathbf{t}$ is a unit vector).

8. Geometrical interpretation of the curvature tensor

Let $\boldsymbol{\lambda}, \boldsymbol{\mu}$ be two unit tangent vectors at a point P on an n-dimensional Riemannian manifold, and consider the pencil of geodesics through P whose directions lie in the two-dimensional subspace of the tangent space at P determined by the directions $\boldsymbol{\lambda}, \boldsymbol{\mu}$. The Riemannian curvature at P in the directions $\boldsymbol{\lambda}, \boldsymbol{\mu}$ was defined by Riemann to be the Gaussian curvature K at P of this two-dimensional subspace. After some calculation[†] it can be shown that

$$K = \frac{R_{hijk}\lambda^h\mu^i\lambda^j\mu^k}{(g_{hj}g_{ik} - g_{hk}g_{ij})\lambda^h\mu^i\lambda^j\mu^k} \tag{8.1}$$

and this relation justifies the use of the word *curvature* for the tensor with components R_{hijk}.

If the curvature K at P is independent of the particular directions $\boldsymbol{\lambda}, \boldsymbol{\mu}$, then

$$R_{hijk} = K(g_{hj}g_{ik} - g_{hk}g_{ij}). \tag{8.2}$$

A result originally due to Schur states that if the curvature at P is independent of the directions $\boldsymbol{\lambda}, \boldsymbol{\mu}$ and if this is true for all points

† For details of the calculation the reader can consult, for example, L. P. Eisenhart (1950), p. 79.

P of a region, then K is constant over that region provided that $n > 2$. To prove this take the covariant derivative to get

$$R_{hijk,l} = K_{,l}(g_{hj}g_{ik} - g_{hk}g_{ij}).\qquad(8.3)$$

Use (8.3) and (5.11) to obtain

$$K_{,l}(g_{hj}g_{ik} - g_{hk}g_{ij}) + K_{,j}(g_{hk}g_{il} - g_{hl}g_{ik}) + K_{,k}(g_{hl}g_{ij} - g_{hj}g_{il}) = 0.$$

Multiply through by g^{hj} to get

$$(n-2)[K_{,l}g_{ik} - K_{,k}g_{il}] = 0.\qquad(8.4)$$

Hence, since $n > 2$, $\qquad K_{,l}g_{ik} = K_{,k}g_{il},\qquad(8.5)$

and $\qquad\qquad\qquad K_{,l}\delta^h{}_k = K_{,k}\delta^h{}_l,\qquad(8.6)$

from which $K_{,l} = 0$. It follows that K is constant, giving the required result. A Riemannian space for which the curvature tensor can be written in the form (8.2) is called a *space of constant curvature*.

9. Special Riemannian spaces

Riemannian spaces whose metrics satisfy some additional restriction have been subjected to very considerable investigation. These restrictions may have a simple geometrical interpretation or they may be formally analytic in nature. Perhaps the simplest special case is obtained by requiring the curvature tensor to be zero at all points, i.e.

$$R_{ijkl} = 0.\qquad(9.1)$$

Such a space is locally Euclidean, that is it admits a metric of the form

$$ds^2 = \sum_{\alpha=1}^{n} (dx^\alpha)^2.\qquad(9.2)$$

Another example is a space of constant curvature for which the curvature tensor is expressible in the form (8.2). Here the geometric interpretation of the restriction is obvious.

Another example is a space whose Ricci tensor satisfies an equation

$$R_{ij} = \lambda g_{ij}\qquad(9.3)$$

for some constant λ. Such a space is called an *Einstein* space.

On multiplying (8.2) by g^{hk} we get

$$R_{ij} = K(1-n)g_{ij},\qquad(9.4)$$

so that a space of constant curvature is necessarily an Einstein space.

EXERCISE 9.1. Prove that the space with metric

$$ds^2 = A\,dr^2 + r^2 d\theta^2 + r^2 \sin^2\theta\,d\phi^2 - A^{-1}\,dt^2 \qquad (9.5)$$

where
$$A = \left[1 + \frac{ar^2}{3} + \frac{c}{r}\right]^{-1}, \quad a \neq 0,$$

is an Einstein space which is not of constant curvature unless $c = 0$.

Exercise 9.1 shows that a space of constant curvature is more special than an Einstein space. It is easily verified that *every two-dimensional space is an Einstein space* and that *every three-dimensional Einstein space is necessarily of constant curvature*. There are many problems involving Einstein spaces of four dimensions which remain unsolved—for example, the most general form of metric is unknown.

Another special type of space which has received considerable attention is a *harmonic* space, defined as follows. Let s be the geodesic distance from a point x_0 to a point x sufficiently near x_0 and let e $(= \pm 1$ or $0)$ be the indicator of the geodesic $x_0 x$. Then if $\Omega = \frac{1}{2}es^2$ is regarded as a function of x, the space is *harmonic at* x_0 if $\Delta_2\Omega = f(\Omega, x_0)$ for some function f. When the space is harmonic at x_0 for each point x_0 it is called a *harmonic space*. It can be proved that *every harmonic space is an Einstein space*, but *not every Einstein space is harmonic*. Also it can be proved that *every space of constant curvature is harmonic*, but *not every harmonic space is of constant curvature*. Thus harmonic spaces form a new type of spaces more special than Einstein spaces but less special than spaces of constant curvature.

Another type of space whose properties have been extensively studied are the *symmetric spaces* characterized by the equation

$$R_{hijk,l} = 0. \qquad (9.6)$$

Harmonic spaces of four dimensions with a positive definite metric are known to be symmetric, but whether this is true for dimensions exceeding four is unknown.

A further type which we mention are the *recurrent spaces* defined by the condition

$$R_{hijk,l} = \kappa_l R_{hijk} \qquad (9.7)$$

for some covariant vector (κ_l). These spaces have many interesting properties, but one of the unsolved problems is to obtain a natural geometrical interpretation of the so-called recurrence vector (κ_l).

The bibliography at the end of this chapter will assist readers who wish to obtain detailed knowledge of the properties of harmonic spaces and recurrent spaces.

10. Parallel vectors

At a point P on a C^∞ n-dimensional Riemannian manifold M consider a C^∞-curve γ lying in some coordinate neighbourhood U of P. Let $\boldsymbol{\lambda}$ be any tangent vector to M at P, and consider the parallel vector field along γ obtained by parallel transport of $\boldsymbol{\lambda}$. If Q is a point on γ then the vector of this parallel vector field at Q depends on the initial vector $\boldsymbol{\lambda}$ and in general also on the curve γ. However, when the vector at Q is independent of the curve γ and when this is the case for every point Q belonging to U, then we say that U admits a *parallel vector field*. Evidently the whole field over U can be generated by parallel transport from the vector of the field at an arbitrary point in U.

Suppose that a parallel vector field is given over a coordinate neighbourhood U by the functions $\lambda^i(x)$. Then, since $D\boldsymbol{\lambda} = \mathbf{0}$, we must have

$$\lambda^i{}_{,k} = \frac{\partial \lambda^i}{\partial x^k} + \begin{Bmatrix} i \\ j\,k \end{Bmatrix}\lambda^j = 0. \tag{10.1}$$

In general it will be impossible to find non-trivial solutions of (10.1), and in order that solutions exist it is necessary that certain conditions of integrability are satisfied.

From equation VI (5.18) we have the *Ricci identity*

$$\lambda^h{}_{,jk} - \lambda^h{}_{,kj} = -R^h{}_{ljk}\lambda^l, \tag{10.2}$$

and it follows that any solution of (10.1) must also satisfy

$$R^h{}_{ljk}\lambda^l = 0. \tag{10.3}$$

When the space has zero curvature, (10.3) is satisfied identically and the system of equations (10.1) is always integrable. Thus in a Euclidean space a parallel vector field is determined over U by prescribing arbitrary initial values of (λ^i). However, in a non-Euclidean space it may still be possible to satisfy equations (10.1), (10.3) simultaneously. By differentiating (10.3) covariantly and using (10.1) we find

$$R^h{}_{ljk,m_1}\lambda^l = 0. \tag{10.4}$$

Similarly, by covariant differentiation of (10.4) and use of (10.1), (10.3), (10.4) we find

$$R^h{}_{ljk,m_1 m_2}\lambda^l = 0. \tag{10.5}$$

In this way we arrive at the sequence of conditions which are necessary and sufficient for the existence of a parallel vector field:

$$\left.\begin{array}{c} R^i{}_{ljk}\lambda^l = 0 \\ R^i{}_{ljk,m_1}\lambda^l = 0 \\ R^i{}_{ljk,m_1m_2}\lambda^l = 0 \\ \cdot \quad \cdot \quad \cdot \quad \cdot \\ R^i{}_{ljk,m_1m_2\ldots m_r}\lambda^l = 0 \\ \cdot \quad \cdot \quad \cdot \quad \cdot \quad \cdot \end{array}\right\}. \tag{10.6}$$

When equations (10.6) are inconsistent there is no solution of (10.1) and hence the space admits no parallel vector field. However, when the equations are consistent there will exist one or possibly many independent parallel vector fields. For example, in a symmetric space ($R^i{}_{ljk,m_1} = 0$) equations (10.6) reduce to the two equations (10.1) and (10.2), and any consistent solution of these two equations will give rise to a parallel vector field.

EXERCISE 10.1. Show that a necessary and sufficient condition that the space with metric

$$ds^2 = (dx^1)^2 + g_{rs}dx^r dx^s \quad (r, s = 2, 3, \ldots, n)$$

shall admit a parallel vector field with components $\lambda^i = \delta^i_1$ is that g_{rs} shall be independent of the coordinate x^1.

11. Vector subspaces

We shall now be concerned with generalizing the concept of a parallel field of vectors—instead of a vector at each point P we shall consider an r-dimensional vector subspace of the tangent space at P. This theory has been developed in particular by A. G. Walker to apply to Riemannian spaces with a metric which may be indefinite. It might be thought that properties of parallelism with definite metrics also hold, by some process of continuity, for spaces with indefinite metrics. This, however, is not the case as will be seen. In the following pages great use has been made of papers by A. G. Walker on the subject, and his treatment has been very closely followed.

An *r-plane* p at P is defined to be an r-dimensional subspace of the tangent space T_P. Usually we shall be concerned only with contravariant planes, but occasionally it is useful to refer to a subspace of T_P^* as an r-plane.

If p, p' are two planes at P, their *intersection* $p \cap p'$ is the set of vectors belonging to both p and p'. Similarly, their *sum* $p+p'$ is the set of vectors obtained by adding a vector of p to a vector of p'. Evidently $p \cap p'$ and $p+p'$ are both planes, and if the dimensionalities of $p, p', p \cap p'$ are r, s, t respectively, then the dimensionality of $p+p'$ is $r+s-t$.

The Riemannian metric provides a means of defining orthogonality of vectors in T_P. Two planes at P are *orthogonal* when every vector of one is orthogonal to every vector of the other. The $(n-r)$-plane of all vectors orthogonal to an r-plane p is called the *conjugate* plane of p. Evidently if p' is conjugate to p, then p is conjugate to p' and the two planes p, p' are orthogonal.

A plane is *null* if it is self-orthogonal. This implies that every vector (λ^i) of a null plane is null, i.e. $g_{ij}\lambda^i\lambda^j = 0$, and also that if (μ^j) is any other vector of the null plane then $g_{ij}\lambda^i\mu^j = 0$. The *null part* of a plane p is the intersection of p and its conjugate. It follows that the null part of a plane is a null plane. The 0-plane consisting of the zero vector is null and it belongs to the null part of every plane. A plane other than the 0-plane whose null part is the 0-plane is called *non-null*. A plane which is not non-null is said to be *partially null*.

We are now in a position to prove

THEOREM 11.1. *The null part of an r-plane in M cannot have dimensionality exceeding* $(n-r)$. *In particular the dimensionality of a null plane cannot exceed* $\frac{1}{2}n$.

To prove the theorem, let the null part p^* of an r-plane p have dimension s $(s \leqslant r)$. It follows that the plane p' conjugate to p has dimension $n-r$, and the sum $p+p'$ has dimension $n-s$. Also $p+p'$ is conjugate to p^*. Since the $(n-r)$-plane p' contains the s-plane p^*, we have $(n-r) \geqslant s$, i.e. $s \leqslant (n-r)$, and the theorem is proved.

The r-plane at P is spanned by a basis of r linearly independent vectors $\boldsymbol{\lambda}_a$ $(a = 1, 2, ..., r)$. When the basis has the property that the r vectors are mutually orthogonal and each vector is either unit or null, it is said to be *normal*. It follows that when the null part of an r-plane is an s-plane p^*, every normal basis of p consists of s null vectors forming a basis of p^* and $(n-s)$ non-null vectors.

Let the components of an arbitrary tensor $\mathbf{T}$ at P relative to the basis of T_P be $T^{abc..}_{ijk..}$. Then we may associate with any particular suffix of $\mathbf{T}$, say i, the planes whose vectors $\boldsymbol{\lambda}$ have components λ^i satisfying

$$T^{abc..}_{ijk..}\lambda^i = 0. \tag{11.1}$$

Consider now the matrix $(T^{abc..}_{ijk..})$, where i is the row suffix and all other suffixes denote columns, and suppose that its rank is r. Then equation (11.1) determines uniquely a plane of dimensions $(n-r)$ which contains all vectors $\boldsymbol{\lambda}$ satisfying (11.1)—this plane will be called *associated* with the tensor $\mathbf{T}$ and the index i. However, it is found to be more convenient to consider the conjugate plane of the associated plane, and this conjugate plane is said to be the plane *generated* by $\mathbf{T}$ and the index i. With this definition, the plane generated by $\mathbf{T}$ has the same dimension as the rank of the matrix (T).

As an example, consider the case when $\mathbf{T}$ is a covariant vector with components T_i. Then the plane generated by $\mathbf{T}$ is the 1-plane consisting of all vectors $\alpha\mathbf{T}$—in this case the 1-plane consists of *covariant* vectors.

Evidently a tensor of order exceeding 1 will generate a plane corresponding to each suffix. However, if the tensor is symmetric or skew symmetric in every pair of suffixes the planes generated by the different suffixes will be identical. (This is also the case for tensors like the Riemann curvature tensor which have symmetry properties which enable every suffix to be brought to the first position.) Hence a multivector $(p^{i_1...i_r})$ determines an r-plane. Conversely a basis of an r-plane determines a multivector of order r, for if λ^i_a $(a = 1,...,r)$ are components of a basis $(\boldsymbol{\lambda}_a)$, the multivector is defined by

$$p^{i_1...i_r} = e^{a_1a_2...a_r}\lambda^{i_1}_{a_1}\lambda^{i_2}_{a_2}...\lambda^{i_r}_{a_r}, \tag{11.2}$$

where the symbol $e^{a_1a_2...a_r}$ takes the values $+1$ or -1 if $(a_1, a_2,..., a_r)$ is respectively an even or odd permutation of $(1, 2,..., r)$, and is otherwise 0.

The measure of a multivector is defined as

$$\left|\frac{1}{r!}\, p^{i_1...i_r} p_{i_1...i_r}\right|^{\frac{1}{2}}. \tag{11.3}$$

EXERCISE 11.1. Prove that the multivectors corresponding to different bases will differ from each other by multiplicative scalars.

EXERCISE 11.2. Show that relative to a normal basis of an r-plane p, the measure of the corresponding multivector is 1 when p is non-null and 0 when p is partially null.

12. Parallel fields of planes

A *field of r-planes of class* ∞, defined over some coordinate neighbourhood U of M, is a set of r-planes, one at each point P in U, such that the vectors λ_a of any basis are functions of class ∞ of the coordinates valid in U. A field of r-planes is called an r-*dimensional distribution* in the terminology of C. Chevalley (1946).

A field of r-planes is *parallel* with respect to the Riemannian metric over U if an arbitrary vector in the plane at P is displaced into a vector in the plane at Q by parallel transport along *any* curve γ joining P to Q. For a fixed vector at P the final vector at Q may well vary with the choice of the curve joining P to Q. However, if the vector at Q is independent of the choice of γ, and if this is the case for every vector at P and for every pair of points P, Q, the field of planes is said to be *strictly parallel*. Evidently a strictly parallel field of r-planes implies the existence of r parallel vector fields of the type discussed in section 10, and thus does not introduce an essentially new idea.

It will be observed that if p and p' denote two parallel fields of planes, then the fields $p \cap p'$, $p+p'$ (obtained by taking the intersection and sum of the planes at each point) will also be parallel. Moreover, since parallel transport preserves orthogonality of vectors, it follows that if two parallel fields are orthogonal at any one point then they are orthogonal at every point. Moreover, since a normal basis remains normal when displaced by parallel transport, it follows that the planes of a parallel field are either all null, non-null, or all partially null to the same degree. If the planes are partially null, then their null parts form a null parallel field. These results are summarized in the following:

THEOREM 12.1. *Associated with a given parallel field of planes are*

  (i) *the field of all conjugate planes,*
  (ii) *the field of all null parts,*

both (i) *and* (ii) *forming parallel fields.*

We now prove the following decomposition theorem:

THEOREM 12.2. *If a parallel field of r-planes contains a parallel field of* NON-NULL *s-planes ($s < r$), then there exists a parallel field of $(r-s)$ planes orthogonal to the s-planes such that at each point the r-plane is decomposed into the sum of the s-plane and the $(r-s)$-plane at that point.*

To prove the theorem suppose that an r-plane p contains a non-null s-plane p' ($s < r$), and let p'' be the $(n-s)$-plane conjugate to p'. Then $p \cap p''$ has dimension $t \geqslant r-s$. Since p' is given non-null, it follows that $p \cap p''$ and p' are disjoint and hence $(p \cap p'') + p'$ has dimension $t + s$. But since this plane is contained in p, evidently $t + s \leqslant r$, which with the previous inequality gives $t = r - s$, and $(p \cap p'') + p' = p$. This gives the required decomposition. Moreover, if p and p' are planes of two parallel fields, then p'' and hence $p \cap p''$ will be parallel. This completes the proof of the theorem.

It should be noted that the proof makes full use of the hypothesis that the field of s-planes is non-null. If this hypothesis is relaxed, the theorem is no longer valid.

We now obtain an analytical expression for the condition that a field of planes shall be parallel.

THEOREM 12.3. *The field of planes with basis* $\left(\underset{a}{\lambda^i}\right)$ *is parallel if and only if*

$$\underset{a}{\lambda^i}_{,k} = A^b_{ak} \underset{b}{\lambda^i} \tag{12.1}$$

for some functions A^b_{ak}, *where* a, b *run over the range* $1, 2, ..., r$.

To prove the theorem, consider a curve γ passing through P and let t^k be components of a vector along the tangent to γ at P. Let (μ_i) be a covariant vector orthogonal to the r-plane at P so that $\left(\underset{a}{\mu_i \lambda^i}\right)_P = 0$, and displace (μ_i) along γ by parallel transport. Then, if the r-planes are parallel, $\underset{a}{\mu_i \lambda^i} = 0$ at all points of γ, and so $\underset{a}{\mu_i D\lambda^i}/ds = 0$ along γ, i.e.

$$\underset{a}{\mu_i \lambda^i}_{,k} t^k = 0$$

at P. Since this holds for all directions (t^k) and for all vectors (μ_i) orthogonal to the r-plane at P, relations (12.1) follow.

Conversely, if (12.1) holds, it follows that

$$\frac{D}{ds}\left(\underset{a}{\mu_i \lambda^i}\right) = \underset{a}{\mu_i \lambda^i}_{,k}\frac{dx^k}{ds} = \mu_i A^b_{ak} \underset{b}{\lambda^i}\frac{dx^k}{ds}.$$

Since $\underset{a}{\mu_i \lambda^i} = 0$ at P, it follows that $\underset{a}{\mu_i \lambda^i} = 0$ for parallel transport along any curve, and the sufficiency is proved.

[It may be noted that although in this proof we have used the terminology of Riemannian geometry, a slight adaptation shows that the theorem remains valid when parallelism is with respect to an arbitrary affine connexion.]

The reader is warned against confusing a parallel field of 1-*planes with a parallel vector field.* Let $\boldsymbol{\lambda}$ be a basis of a parallel field of 1-planes so that

$$\lambda^i_{,k} = \lambda^i a_k \tag{12.2}$$

for some vector (a_k). If we write $\lambda^i = \phi\mu^i$, then (12.2) becomes

$$\phi_{,k}\mu^i + \phi\mu^i_{,k} = \phi\mu^i a_k,$$

i.e. $$\mu^i_{,k} = \mu^i[a_k - \partial_k(\log\phi)].$$

Now if the vector with components a_k is a gradient—say $a_k = \partial_k(\log\phi)$ for some function ϕ—the effect of writing $\mu^i = \lambda^i/\phi$ is to obtain $\mu^i_{,k} = 0$. Thus when a_k are components of a gradient, a parallel field of 1-planes implies the existence of a parallel vector field.

We now prove that when the basis vector $\boldsymbol{\lambda}$ is non-null, then the vector (a_k) is necessarily a gradient. This follows at once from the relation

$$g_{ij}\lambda^i\lambda^j a_k = g_{ij}\lambda^i_{,k}\lambda^j = \tfrac{1}{2}\partial_k(g_{ij}\lambda^i\lambda^j),$$

so $$a_k = \partial_k\log|g_{ij}\lambda^i\lambda^j|^{\frac{1}{2}}.$$

Thus *a parallel field of non-null 1-planes implies the existence of a parallel field of vectors.*

However, when $\boldsymbol{\lambda}$ is null, (a_k) is not necessarily a gradient as is shown by the following exercise.

EXERCISE 12.1. Show that the 1-planes $(\alpha\delta^i_3)$ are parallel with respect to the metric

$$ds^2 = \psi(x,y,z)\,dx^2 + dy^2 + 2\,dx\,dz,$$

but that (a_k) is a gradient only when ψ has the special form

$$zf(x) + g(x,y).$$

13. Recurrent tensors

A tensor of any order will be called *recurrent* if its components satisfy

$$T^{\cdots}_{i\cdots,k} = T^{\cdots}_{i\cdots}a_k, \qquad (13.1)$$

for some vector (a_k). An example of a recurrent tensor has already been seen in section 9 where a recurrent space had the property

$$R_{ijkl,m} = R_{ijkl}\kappa_m. \qquad (13.2)$$

A tensor whose covariant derivative is zero is called a *constant* tensor. Evidently a constant tensor is a particular case of a recurrent tensor. Recurrent tensors are closely related to fields of parallel planes as can be seen from the following:

THEOREM 13.1. *The field of planes generated by a recurrent tensor is parallel.*

To prove the theorem, let $T_{i..}$ be components of a recurrent tensor **T**. The field of planes associated with $(T_{i..})$ and the suffix i is given by
$$T_{i..}\lambda^i = 0. \tag{13.3}$$
Differentiate this equation covariantly and use (13.3) to obtain
$$T_{i..}\lambda^i_{,k} = 0. \tag{13.4}$$
The vectors which satisfy (13.3) lie in the conjugate plane to the plane generated by **T**, and so from (13.4) it follows that when we regard the suffix k as fixed, the set $(\lambda^i_{,k})$ may be considered as components of a vector which lies in the conjugate plane. It follows that the basis vectors of the conjugate plane satisfy a relation of the type (12.1). This implies that the conjugate planes are parallel and hence the planes generated by **T** are parallel. This completes the proof of the theorem.

The next theorem shows that, conversely, every parallel field of planes is generated by some recurrent tensor.

THEOREM 13.2. *The multivector corresponding to a parallel field of planes is recurrent.*

This result follows immediately by differentiating (11.2) covariantly and using (12.1), when we obtain
$$p^{i_1 \cdots i_r}_{,k} = p^{i_1 \cdots i_r} a_k,$$
where
$$a_k = A^a_{ak}.$$

We now show that any parallel field of *non-null* r-planes can be generated by a symmetric second-order covariant tensor with zero covariant derivative.

THEOREM 13.3. *Let λ^i_a be components of vectors $\boldsymbol{\lambda}_a$ which form a normal basis of a parallel field of* NON-NULL *r-planes. Then the tensor* **T** *with components T_{ij} defined by*
$$T_{ij} = T_{ji} = \sum e_a \lambda^i_a \lambda^j_a \quad \text{where} \quad e_a = g_{ij}\lambda^i_a \lambda^j_a = \pm 1,$$
generates the given field of r-planes and is constant.

To prove the theorem, consider the components μ^j of vectors $\boldsymbol{\mu}$ which lie in the conjugate plane, i.e. solutions of
$$T_{ij}\mu^j = 0.$$
Substitute for T_{ij} to obtain
$$\sum_{a=1}^{r} e_a \lambda^i_a \left(\lambda^j_a \mu^j \right) = 0.$$

But since the basis vectors are linearly independent, this implies

$$\lambda_j \underset{a}{\mu^j} = 0,$$

so μ is orthogonal to each of the basis vectors. Thus μ must belong to the r-plane generated by $\mathbf{T}$, and so $\mathbf{T}$ generates the given field of r-planes.

Covariant differentiation of T_{ij} gives

$$T_{ij,k} = \sum_{a=1}^{r} e_a \Big(\underset{a}{\lambda_{i,k}} \underset{a}{\lambda_j} + \underset{a}{\lambda_i} \underset{a}{\lambda_{j,k}} \Big)$$

$$= \sum_{a=1}^{r} e_a A_{ak}^b \Big(\underset{b}{\lambda_i} \underset{a}{\lambda_j} + \underset{b}{\lambda_j} \underset{a}{\lambda_i} \Big). \qquad (13.5)$$

Covariant differentiation of the relation

$$\underset{a}{\lambda_i} \underset{b}{\lambda^i} = e_a \delta_{ab}$$

gives

$$\underset{a}{\lambda_{i,k}} \underset{b}{\lambda^i} + \underset{a}{\lambda_i} \underset{b}{\lambda^i}_{,k} = 0,$$

which with (12.1) gives

$$A_{ak}^c \underset{c}{\lambda_i} \underset{b}{\lambda^i} + A_{bk}^c \underset{c}{\lambda^i} \underset{a}{\lambda_i} = 0,$$

i.e.

$$A_{ak}^c e_b \delta_{bc} + A_{bk}^c e_a \delta_{ac} = 0.$$

This equation with (13.5) gives the required relation $T_{ij,k} = 0$ and the proof of the theorem is now complete.

However, the corresponding theorem for partially null planes is false, even when the requirement is weakened from a constant tensor to a recurrent tensor. This can be seen as follows.

Consider the space with metric

$$ds^2 = y^2\,dx^2 + dy^2 + 2\,dx\,dz. \qquad (13.6)$$

This admits a parallel field of null 1-planes generated by (δ_3^i), and hence a conjugate field of partially null 2-planes. If the tensor $\mathbf{T}$ with components $T_{ij} = T_{ji}$ generates this field of 2-planes, then its rank must be 2 and, moreover, $T_{ij}\delta_3^j = 0$, i.e. $T_{i3} = 0$. The components of the tensor $\mathbf{T}$ must satisfy the conditions

$$T_{ij} = T_{ji}, \quad T_{i3} = 0, \quad T_{11}T_{22} - (T_{12})^2 \neq 0,$$

$$T_{ij,k} = T_{ij} a_k$$

for some functions a_k. Suppose that (a_k) is a gradient, say

$$a_k = -\partial_k(\log \phi).$$

Then if we write $T'_{ij} = \phi T_{ij}$, T'_{ij} are components of a constant tensor. Hence when (a_k) is a gradient, there is no essential loss of generality in taking $(a_k) = 0$. In this case the components of $\mathbf{T}$ satisfy the equations

$$T_{i3} = 0, \quad \partial_1 T_{11} = -2y T_{12}, \quad \partial_1 T_{12} = -y T_{22},$$
$$\partial_2 T_{11} = \partial_3 T_{11} = \partial_2 T_{12} = \partial_3 T_{12} = 0,$$

and evidently the only solution is $T_{ij} = 0$.

Suppose now that the functions a_k are not the components of a gradient. Then $\partial_k T_{22} = T_{22} a_k$, which implies $T_{22} = 0$, otherwise (a_k) would be a gradient. Also $\partial_k T_{12} = T_{12} a_k$, which similarly implies $T_{12} = 0$. Hence $T_{11} T_{22} - (T_{12})^2 = 0$, which contradicts the hypothesis on the rank of (T_{ij}). Thus, in either case, there is no non-trivial solution.

From Theorem 13.1 it follows that a symmetric recurrent tensor $\mathbf{T}$ with components T_{ij} generates a parallel field of r-planes, but this field is trivial unless the rank of (T_{ij}) differs from n, i.e. unless (T_{ij}) is a singular matrix. The following theorem shows that a non-trivial field of parallel planes exists when $|T_{ij}| \neq 0$, although of course the field is not generated by $\mathbf{T}$.

THEOREM 13.4. *If M admits a recurrent tensor (T_{ij}) other than a multiple of the metric tensor (g_{ij}), then it admits a non-trivial parallel field of planes.*

If (T_{ij}) is singular the result follows from Theorem 13.1. Suppose that (T_{ij}) is non-singular, i.e. $|T_{ij}| \neq 0$, and that $T_{ij,k} = T_{ij} a_k$. Write $\phi = |T_{ij}|/|g_{ij}|$ so that ϕ is homogeneous in the T's of degree n. Hence, $\phi_{,k} = n\phi a_k$, so that (a_k) is a gradient. If $S_{ij} = \phi^{-1/n} T_{ij}$, then $S_{ij,k} = 0$. It follows that we have constructed a constant tensor with components S_{ij} with the properties that it is non-singular, and not proportional to the metric tensor.

Consider now the polynomial

$$\tau(\rho) = |\rho g_{ij} - S_{ij}|/|g_{ij}| = \rho^n + \tau_1 \rho^{n-1} + \ldots + \tau_n.$$

Each coefficient τ_r is a sum of scalar products of the S's and the g's and is constant because $S_{ij,k} = 0$. It follows that every root of the equation $\tau(\rho) = 0$ is a constant, and there cannot be zero roots since $|S_{ij}| \neq 0$. Let ρ be any one root, and write

$$U_{ij} = \rho g_{ij} - S_{ij}.$$

Then $U_{ij,k} = 0$, and $|U_{ij}| = 0$ by construction. It follows that the planes associated with (U_{ij}) are parallel and non-trivial, and so Theorem 13.4 is proved.

An obvious problem in Riemannian geometry is to investigate whether there are constant symmetric second-order covariant tensors apart from multiples of the metric tensor. Theorem 13.4 shows that in attempting to solve both this problem and the more general problem involving recurrent tensors, attention should be focused on the parallel fields of planes. If these planes do not exist, then there are no recurrent second order symmetric tensors other than multiples of the metric tensor.

14. Integrable distributions

When a differentiable manifold M admits a distribution of r-planes, it may happen that these r-planes (or contact elements) fit together in such a way that they determine an r-dimensional submanifold of M, the r-plane at P being the tangent space at P to this submanifold. In this case the distribution is said to be *integrable* or *involutive*. The condition of integrability may be expressed as follows:

The distribution D is integrable when for any two vector fields $\boldsymbol{\lambda}$, $\boldsymbol{\mu}$ belonging to D, the vector field $[\boldsymbol{\lambda}, \boldsymbol{\mu}]$ also belongs to D.

Instead of requiring this condition to be satisfied by any two vector fields belonging to D, it is sufficient for the condition to be satisfied by vector fields forming a basis. To prove this, let $\boldsymbol{\lambda}_\alpha$ ($\alpha = 1, 2, ..., r$) be a basis for the vector fields of D, and let $\boldsymbol{\lambda}$, $\boldsymbol{\mu}$ be any two vector fields of D so that

$$\boldsymbol{\lambda} = u^\alpha \boldsymbol{\lambda}_\alpha, \qquad \boldsymbol{\mu} = v^\alpha \boldsymbol{\lambda}_\alpha.$$

Then $\quad [\boldsymbol{\lambda}, \boldsymbol{\mu}] = \left(u^\alpha \underset{\alpha}{\lambda^i} \partial_i v^\beta\right)\boldsymbol{\lambda}_\beta - \left(v^\beta \underset{\beta}{\lambda^i} \partial_i u^\alpha\right)\boldsymbol{\lambda}_\alpha + [\boldsymbol{\lambda}_\alpha, \boldsymbol{\lambda}_\beta]u^\alpha v^\beta. \quad (14.1)$

If the condition of integrability is satisfied by the basis vectors, i.e. if
$$[\boldsymbol{\lambda}_\alpha, \boldsymbol{\lambda}_\beta] = c^\gamma_{\alpha\beta} \boldsymbol{\lambda}_\gamma \qquad (14.2)$$

for some functions $c^\gamma_{\alpha\beta}$, then the right-hand member of (14.1) is a linear combination of basis vectors. Hence $[\boldsymbol{\lambda}, \boldsymbol{\mu}]$ belongs to the distribution.

We now prove

THEOREM 14.1. *A distribution of r-planes which is parallel with respect to a Riemannian connexion is necessarily integrable.*

Let $\underset{\alpha}{\lambda^i}$ be components of vectors $\boldsymbol{\lambda}_\alpha$ which form a basis for the parallel field of r-planes. Then

$$[\boldsymbol{\lambda}_\alpha, \boldsymbol{\lambda}_\beta]^j = \underset{\alpha}{\lambda^i} \partial_i \underset{\beta}{\lambda^j} - \underset{\beta}{\lambda^i} \partial_i \underset{\alpha}{\lambda^j}. \qquad (14.3)$$

We know that
$$\partial_i \lambda^j_{\,\beta} = \lambda^j_{\,\beta},_i - \left\{ {j \atop k\ i} \right\} \lambda^k_{\,\beta}, \tag{14.4}$$

so by substituting in (14.3) we get

$$[\boldsymbol{\lambda}_\alpha, \boldsymbol{\lambda}_\beta]^j = \lambda^i_{\,\alpha}\left(\lambda^j_{\,\beta},_i - \left\{ {j \atop k\ i} \right\}\lambda^k_{\,\beta}\right) - \lambda^i_{\,\beta}\left(\lambda^j_{\,\alpha},_i - \left\{ {j \atop k\ i} \right\}\lambda^k_{\,\alpha}\right)$$

$$= \lambda^i_{\,\alpha}\lambda^j_{\,\beta},_i - \lambda^i_{\,\beta}\lambda^j_{\,\alpha},_i, \tag{14.5}$$

the connexion coefficients cancelling owing to their symmetry. Thus the partial derivatives in (14.3) are replaced by covariant derivatives in (14.5). By substituting for $\lambda^j_{\,\alpha},_i$ from (12.1) we obtain

$$[\boldsymbol{\lambda}_\alpha, \boldsymbol{\lambda}_\beta] = c^\gamma_{\alpha\beta}\boldsymbol{\lambda}_\gamma$$

where
$$c^\gamma_{\alpha\beta} = \lambda^i_{\,\alpha}A^\gamma_{\beta i} - \lambda^i_{\,\beta}A^\gamma_{\alpha i},$$

and the integrability follows immediately.

It may be noticed that in this proof use has been made only of the symmetry of the Riemannian connexion, and the same proof shows that a *distribution of r-planes parallel with respect to a symmetric connexion is necessarily integrable*.

Instead of determining an r-dimensional distribution D by a basis of r linearly independent vector fields $\boldsymbol{\lambda}_\alpha$, D is equally determined by $(n-r)$ linearly independent 1-forms $\omega^{\alpha''}$ ($\alpha'' = r+1$, $r+2,...,n$). A vector $\boldsymbol{\lambda}$ belongs to D if and only if $\omega^{\alpha''}(\boldsymbol{\lambda}) = 0$. In the terminology of the theory of partial differential equations, L is given by the system of $(n-r)$ total differential equations (Pfaffian forms)
$$\omega^{\alpha''} = 0. \tag{14.6}$$

Now it is known from the theory of such equations that the condition that this system of equations shall be completely integrable is that
$$d\omega^{\alpha''} = 0 \tag{14.6'}$$

shall be satisfied identically as a consequence of (14.6).

Applying the identity VI (3.19)
$$\lambda\{\omega(\boldsymbol{\mu})\} - \mu\{\omega(\boldsymbol{\lambda})\} = \omega[\boldsymbol{\lambda}, \boldsymbol{\mu}] + 2\,d\omega(\boldsymbol{\lambda}, \boldsymbol{\mu}) \tag{14.7}$$

to the $(n-r)$ forms $\omega^{\alpha''}$, we see that if $\boldsymbol{\lambda}, \boldsymbol{\mu}$ are vector fields belonging to D, then
$$\omega^{\alpha''}[\boldsymbol{\lambda}, \boldsymbol{\mu}] = -2\,d\omega^{\alpha''}(\boldsymbol{\lambda}, \boldsymbol{\mu}). \tag{14.8}$$

Hence $d\omega^{\alpha''} = 0$ if and only if $[\boldsymbol{\lambda}, \boldsymbol{\mu}]$ belongs to D. It follows that *the distribution is integrable if and only if the corresponding system of total partial differential equations is completely integrable.* This justifies our use of the word 'integrable'.

THEOREM 14.2. *If D_1, D_2 are two integrable distributions, then the intersection $D_1 \cap D_2$ is also integrable.*

The proof is trivial, for if $\lambda \in D_1 \cap D_2$ we have $\lambda \in D_1$ and $\lambda \in D_2$; similarly $\mu \in D_1 \cap D_2$ implies $\mu \in D_1$ and $\mu \in D_2$. Hence $[\lambda, \mu] \in D_1$ and $[\lambda, \mu] \in D_2$, so $[\lambda, \mu] \in D_1 \cap D_2$.

It should be noted that Theorem 14.2 is false if we replace intersection by sum, and this fact has important consequences in the theory of parallel planes. Suppose, for example, that we are given two integrable distributions D_1 and D_2 and we seek a Riemannian (or more generally a symmetric) connexion with respect to which they are parallel. If such a connexion exists, it follows that the two distributions $D_1 \cap D_2$ and $D_1 + D_2$ are parallel, and from Theorem 14.1 it is necessary that $D_1 \cap D_2$ and $D_1 + D_2$ should be integrable. The integrability of $D_1 \cap D_2$ is ensured by Theorem 14.2, but in general nothing can be said about $D_1 + D_2$. So, unless $D_1 + D_2$ is also integrable (when for example $D_1 + D_2$ spans the complete tangent space), no Riemannian (symmetric) connexion will exist which makes D_1 and D_2 simultaneously parallel.

More generally, suppose we have a system $\{D_\rho\}$ $(\rho = 1, 2, ..., p)$, of disjoint distributions defined over M. Then, as we have seen, in order that the system $\{D_\rho\}$ shall be parallel with respect to a Riemannian (symmetric) connexion, it is necessary that each distribution D_ρ shall be integrable. However, this condition is not sufficient because if D_ρ, D_σ are two disjoint distributions, each being parallel, then the sum $D_\rho + D_\sigma$ is necessarily parallel, but the sum of two integrable distributions is not necessarily integrable. A. G. Walker has proposed the following definition of integrability of the system $\{D_\rho\}$. *A system of distributions is integrable if every distribution formed by taking sums of the distributions is integrable.*

If we write

$$D = \sum_{\rho=1}^{p} D_\rho, \qquad D^\sigma = \sum_{\rho \neq \sigma} D_\rho, \qquad (14.9)$$

then it follows that the system $\{D_\rho\}$ is integrable if and only if each distribution $D, D^1, D^2, ..., D^p$ is separately integrable. This follows from the fact that every sum of distributions of the system $\{D_\rho\}$ is the intersection of distributions of the system $\{D, D^\sigma\}$, and the result that the intersection of integrable distributions is integrable.

Connexions and distributions

Suppose an n-dimensional manifold M admits globally two disjoint distributions D, $\bar{D}$ which span the tangent space to M at each point. Let base vectors for D_x, $\bar{D}_x$ at any point x be given by

$$(\lambda_{\alpha'}^i(x)), \quad (\lambda_{\alpha''}^i(x)) \qquad (\alpha' = 1,...,r; \; \alpha'' = r+1,...,n).$$

Denote by Λ the $n \times n$ matrix (λ_α^i), and write $\Lambda^{-1} = (\mu_i^\alpha)$. Define tensors $\mathbf{a}$, $\bar{\mathbf{a}}$ at x by

$$a^i_j = \sum_{\alpha'=1}^{r} \lambda_{\alpha'}^i \mu_j^{\alpha'}, \qquad \bar{a}^i_j = \sum_{\alpha''=r+1}^{n} \lambda_{\alpha''}^i \mu_j^{\alpha''}.$$

EXERCISE 14.1. Prove that

  (i) $\mathbf{a}$, $\bar{\mathbf{a}}$ are independent of the particular bases chosen for the two planes at x;

  (ii) $\mathbf{a}^2 = \mathbf{a}$, $\quad \bar{\mathbf{a}}^2 = \bar{\mathbf{a}}$, $\quad \mathbf{a}\bar{\mathbf{a}} = \bar{\mathbf{a}}\mathbf{a} = 0$, $\quad \mathbf{a} + \bar{\mathbf{a}} = \mathbf{I}$;

  (iii) $\operatorname{rank} \mathbf{a} = r$, $\quad \operatorname{rank} \bar{\mathbf{a}} = (n-r)$.

The tensors $\mathbf{a}$, $\bar{\mathbf{a}}$ are called projection tensors associated with the distributions D and $\bar{D}$. By means of these tensors any vector $\boldsymbol{\lambda}$ lying in the tangent space T_x can be resolved uniquely in the form

$$\boldsymbol{\lambda} = \boldsymbol{\lambda}' + \boldsymbol{\lambda}'', \tag{14.10}$$

where $\boldsymbol{\lambda}' \in D$ and $\boldsymbol{\lambda}'' \in \bar{D}$. We have

$$\mathbf{a}\boldsymbol{\lambda} = \boldsymbol{\lambda}', \qquad \bar{\mathbf{a}}\boldsymbol{\lambda} = \boldsymbol{\lambda}'', \tag{14.11}$$

and equations (14.10), (14.11) could have been used to *define* the tensors $\mathbf{a}$ and $\bar{\mathbf{a}}$.

It is convenient to use the following notation used by A. G. Walker (1955). If (λ^i), (μ_i) are vectors at the point x, we write

$$\lambda^{i'} = a^i_{\,p}\lambda^p, \quad \lambda^{i''} = \bar{a}^i_{\,p}\lambda^p, \quad \mu_{i'} = a^p_{\,i}\mu_p, \quad \mu_{i''} = \bar{a}^p_{\,i}\mu_p. \tag{14.12}$$

We have $\qquad \lambda^i = \lambda^{i'} + \lambda^{i''}, \qquad \mu_i = \mu_{i'} + \mu_{i''}, \tag{14.13}$

and the vectors $(\lambda^{i'})$, $(\lambda^{i''})$ are respectively the resolved parts of (λ^i) along the planes of D and $\bar{D}$ at the point x. *This process of resolution into components* in D and $\bar{D}$ can be applied to an arbitrary tensor with components $T^{ij\cdots}_{k\cdots}$. There are two projections for each suffix, so that the tensor may be partly or wholly projected in a number of ways. For example,

$$T^{i'j\cdots}_{k\cdots} = a^i_{\,p}T^{pj\cdots}_{k\cdots}, \qquad T^{i'j'\cdots}_{k'\cdots} = a^i_{\,p}\bar{a}^j_{\,q}\bar{a}^r_{\,k}T^{pq\cdots}_{r\cdots}.$$

Owing to the identity $\mathbf{a} + \bar{\mathbf{a}} = \mathbf{I}$, i.e. $a^i_j + \bar{a}^i_j = \delta^i_j$, there is a relation between projections which may be written symbolically $i' + i'' = i$. Thus

$$\lambda^i = \lambda^{i'} + \lambda^{i''}, \qquad \mu_i = \mu_{i'} + \mu_{i''},$$
$$T_{k'}^{ij} = T_{k'}^{ij} + T_{k'}^{ij} = T_{k'}^{ij} + T_{k'}^{ij'} + T_{k'}^{ij} + T_{k'}^{ij'}.$$

One advantage of this notation is that the summation convention still applies regardless of primes—thus

$$\mu_i \lambda^{i'} = \mu_i a^i_p \lambda^p = \mu_{i'} \lambda^i = \mu_{i'} \lambda^{i'}$$

since $\mathbf{a}^2 = \mathbf{a}$. It follows that

$$\mu_{i'} \lambda^{i''} = 0 \quad \text{and} \quad \mu_{i''} \lambda^{i'} = 0.$$

The notation may also be used with covariant differentiation with respect to an affine connexion with components L_{jk}^i provided that

$$T_{j'|k}^i \text{ denotes } (T_{p|k}^i) a^p_j \text{ and NOT } (T_p^i a^p_j)_{|k},$$

where the solidus denotes covariant differentiation.

We now obtain a number of identities satisfied by the covariant derivative of the tensor $\mathbf{a}$ with respect to an arbitrary connexion. Since $\delta^i_{j|k} = 0$ and $a^i_j + \bar{a}^i_j = \delta^i_j$, we have

$$a^i_{j|k} + \bar{a}^i_{j|k} = 0. \tag{14.14}$$

Also, since $\mathbf{a}\bar{\mathbf{a}} = 0$, we have

$$a^i_{p|k} \bar{a}^p_j = -a^i_p \bar{a}^p_{j|k} = a^i_p a^p_{j|k},$$

so that

$$a^i_{j''|k} = a^i_{j|k}. \tag{14.15}$$

Similarly, from $\bar{\mathbf{a}}\mathbf{a} = 0$ we obtain

$$a^i_{j'|k} = a^i_{j|k}. \tag{14.16}$$

In particular, relations (14.14), (14.15), (14.16) still hold when the arbitrary connexion (L_{jk}^i) is a symmetric connexion (Γ_{jk}^i). In this case we denote covariant differentiation by a comma, and for the tensor components $a^i_{j,k}$ we shall write a^i_{jk}. Then (14.14) becomes

$$\bar{a}^i_{j,k} = -a^i_{jk}, \tag{14.17}$$

and (14.15), (14.16) give

$$a^{i'}_{jk} = a^i_{j''k}, \qquad a^{i''}_{jk} = a^i_{j'k},$$

from which

$$a^i_{j'k} = a^i_{j''k} = 0. \tag{14.18}$$

These relations hold in particular when the connexion coefficients are Christoffel symbols of a Riemannian metric.

Some of the equations concerning integrability and parallelism can be written more concisely in terms of this notation. For example, the condition that the distribution D is integrable is that for any vectors $\boldsymbol{\lambda}$, $\boldsymbol{\mu}$ the vector $[a\boldsymbol{\lambda}, a\boldsymbol{\mu}]$ must lie in D, i.e. its projection in $\bar{D}$ must be zero. By replacing partial derivatives by covariant derivatives with respect to some arbitrary symmetric connexion (Γ^i_{jk}), as in (14.5), this condition is seen to be

$$\bar{a}^h{}_j(a\lambda)^i(a\mu)^j{}_{,i} - \bar{a}^h{}_j(a\mu)^i(a\lambda)^j{}_{,i} = 0,$$

i.e. $\quad \bar{a}^h{}_j a^i{}_m \lambda^m \{a^j{}_n \mu^n{}_{,i} + \mu^n a^j{}_{ni}\} - \bar{a}^h{}_j a^i{}_n \mu^n \{a^j{}_m \lambda^m{}_{,i} + a^j{}_{mi} \lambda^m\} = 0.$

Since $\bar{\mathbf{a}}\mathbf{a} = 0$, this simplifies to

$$(\bar{a}^h{}_j a^i{}_m a^j{}_{ni} - \bar{a}^h{}_j a^i{}_n a^j{}_{mi})\lambda^m \mu^n = 0.$$

As this condition must be satisfied by arbitrary vectors $\boldsymbol{\lambda}$, $\boldsymbol{\mu}$, we have

$$\bar{a}^h{}_j a^i{}_m a^j{}_{ni} = \bar{a}^h{}_j a^i{}_n a^j{}_{mi},$$

i.e. $\qquad\qquad a^{h''}{}_{nm'} = a^{h''}{}_{mn'}.$

On using (14.18) this simplifies to $a^{h''}{}_{n'm'} = a^{h''}{}_{m'n'}$. *Thus the condition for D to be integrable is*

$$a^i{}_{j'k'} = a^i{}_{k'j'}. \tag{14.19}$$

Similarly, the condition for $\bar{D}$ to be integrable is

$$a^i{}_{j''k''} = a^i{}_{k''j''}. \tag{14.20}$$

Equations (14.19) and (14.20) are easily deduced from

$$a^i{}_{jk} = a^i{}_{kj}, \tag{14.21}$$

a *sufficient* condition that D and $\bar{D}$ are both integrable.

We now find the condition that D is parallel with respect to an arbitrary connexion with components L^i_{jk}. If λ^p are components of an arbitrary vector $\boldsymbol{\lambda}$, $\bar{a}^i{}_p \lambda^p$ is its projection in $\bar{D}$. If D is parallel, then† the absolute differential satisfies the condition $D(\bar{a}^i{}_p \lambda^p) = 0$ when $D\lambda^i = 0$ for all dx^i and all λ^i satisfying $\bar{a}^i{}_p \lambda^p = 0$. Hence $\bar{a}^i{}_{p|k} \lambda^p = 0$ for all (λ^p) belonging to D. Thus the condition that D is parallel with respect to L is $\bar{a}^i{}_{j'|k} = 0$. Using (14.14), this condition is equivalent to

$$a^i{}_{j'|k} = 0. \tag{14.22}$$

Similarly, the condition that $\bar{D}$ is parallel with respect to L is

$$a^i{}_{j''|k} = 0. \tag{14.23}$$

The condition for *both* D and $\bar{D}$ to be parallel with respect to L is

$$a^i{}_{j|k} = 0. \tag{14.24}$$

---

† The use of the same symbol D both for the distribution and the absolute differential will cause no confusion.

When L is a symmetric connexion Γ, conditions (14.22), (14.23), (14.24) reduce to

$$a^i_{j'k} = 0, \tag{14.25}$$

$$a^i_{j''k} = 0, \tag{14.26}$$

$$a^i_{jk} = 0. \tag{14.27}$$

Since condition (14.25) implies (14.19), we have an immediate generalization of Theorem 14.1 where the Riemannian connexion is replaced by an arbitrary symmetric connexion.

EXERCISE 14.2. Show that if Γ is an arbitrary symmetric connexion, then the connexion $L = \Gamma + S$, where

$$S^i_{jk} = -a^i_{j'k} - a^i_{k'j''} + a^i_{j''k} + a^i_{k''j'},$$

has the property that D and $\bar{D}$ are both parallel with respect to L, and, moreover, L is symmetric when D and $\bar{D}$ are both integrable.

Relative parallelism

A differentiable curve on the manifold M with the property that the direction of the tangent vector at any point x on the curve lies in D_x, is called an *integral curve of D*. (If D is integrable these curves lie in the integral submanifolds determined by D.)

A vector field $\boldsymbol{\lambda}$ is *parallel relative to D* with respect to a connexion L if, for any points x, y on an integral curve of D, the vectors $\boldsymbol{\lambda}_x$, $\boldsymbol{\lambda}_y$ are parallel relative to this curve.

In terms of local coordinates and projection tensors $\mathbf{a}$, $\bar{\mathbf{a}}$, the vector field $\boldsymbol{\lambda}$ is parallel relative to D if $D\lambda^i = 0$ for all dx^i satisfying $\bar{a}^i_j dx^j = 0$. Since $D\lambda^i = \lambda^i_{|j} dx^j$ the condition for relative parallelism is

$$\lambda^i_{|j'} = 0. \tag{14.28}$$

Similarly, the vector field $\boldsymbol{\lambda}$ is parallel relative to $\bar{D}$ if

$$\lambda^i_{|j''} = 0. \tag{14.29}$$

The conditions (14.28), (14.29) are each weaker than the condition for ordinary parallelism

$$\lambda^i_{|j} = 0, \tag{14.30}$$

which is equivalent to *both* these conditions.

The idea of relative parallelism can evidently be extended from a vector field to a distribution D^*. We shall say that D^* *is parallel relative to D with respect to a connexion L if the planes of D^* at points of any integral curve of D are parallel relative to this curve.* Suppose that D^* determines the projection tensor $\mathbf{b}$ such that for all vectors $\boldsymbol{\lambda}$ belonging to D^* we have $b^i_j \lambda^j = \lambda^i$, then we require $D(b^i_j \lambda^j) = 0$

when $D\lambda^j = 0$ for all λ belonging to D^* and all dx^i such that $\bar{a}^i{}_j dx^j = 0$. This implies $b^i{}_{j|k}\lambda^j dx^k = 0$ for all allowable λ^j and dx^k, i.e.
$$b^i{}_{p|q} b^p{}_j a^q{}_k = 0. \tag{14.31}$$
This is the condition to be satisfied by $\mathbf{b}$ for the associated distribution to be parallel relative to D.

The distribution D is *self-parallel*† with respect to L if it is parallel relative to itself. The condition for this may be obtained from (14.31) by replacing $\mathbf{b}$ by $\mathbf{a}$, and is therefore
$$a^i{}_{j'|k'} = 0. \tag{14.32}$$
Clearly condition (14.22) implies (14.32), so if D is parallel it is also self-parallel. Moreover, if D is self-parallel with respect to a symmetric connexion, then (14.32) may be written $a^i{}_{j'k'} = 0$, and, from (14.19), D is necessarily integrable.

The condition that $\bar{D}$ is parallel relative to D is obtained from (14.31) by replacing $\mathbf{b}$ by $\bar{\mathbf{a}}$, and is therefore equivalent to
$$a^i{}_{j''|k'} = 0. \tag{14.33}$$
This does not imply that $\bar{D}$ is integrable when the connexion is symmetric.

The condition that D is parallel relative to $\bar{D}$ may be obtained by interchanging the roles of $\mathbf{a}$ and $\bar{\mathbf{a}}$ in (14.33). This gives $\bar{a}^i{}_{j'|k''} = 0$ which is equivalent to
$$a^i{}_{j'|k''} = 0. \tag{14.34}$$
When D is parallel relative to $\bar{D}$ and $\bar{D}$ is parallel relative to D, both (14.33) and (14.34) hold, but the distribution need not be integrable when the connexion is symmetric.

If D is parallel and at the same time $\bar{D}$ is parallel relative to D, then $a^i{}_{j'|k} = 0$ and $a^i{}_{j''|k'} = 0$, which may be written
$$a^i{}_{j'|k} = 0, \qquad a^i{}_{j|k'} = 0. \tag{14.35}$$
When the connexion is symmetric these imply that D is integrable but $\bar{D}$ need not be integrable.

We have seen that self-parallelism is weaker than parallelism. We now introduce the idea of *path-parallelism* which is weaker than self-parallelism. We have already seen that associated with the Christoffel symbols $\begin{Bmatrix} i \\ j\ k \end{Bmatrix}$ of a Riemannian metric is a system of auto-parallels or geodesics given by the equations
$$\frac{d^2 x^i}{ds^2} + \begin{Bmatrix} i \\ j\ k \end{Bmatrix} \frac{dx^j}{ds}\frac{dx^k}{ds} = 0.$$

† This is called *semi-parallel* in Walker's papers.

In a similar way we may associate with any connexion L a system of auto-parallels or *paths* given by the equations

$$\frac{d^2x^i}{dt^2} + L^i_{jk}\frac{dx^j}{dt}\frac{dx^k}{dt} = 0.$$

As with geodesics, for any point x and direction $\boldsymbol{\lambda}$ at x there is just one path through x tangent to $\boldsymbol{\lambda}$.

We say that D is *path-parallel* with respect to L if for every point x of M and every vector $\boldsymbol{\lambda}$ belonging to D_x, the path determined by x and $\boldsymbol{\lambda}$ is an integral curve of D.

Writing $\lambda^i = dx^i/dt$ for the components of a tangent vector to a path, then $D\lambda^i = 0$, and we require $D(a^i{}_j\lambda^j) = 0$ when $\bar{a}^i{}_j\lambda^j = 0$, i.e. we require $a^i{}_{j|k}\lambda^j\lambda^k = 0$ for all vectors (λ^i) satisfying $\bar{a}^i{}_j\lambda^j = 0$. Hence $a^i{}_{j'|k'}\lambda^j\lambda^k = 0$ for all (λ^i), i.e.

$$a^i{}_{j'|k'} + a^i{}_{k'|j'} = 0. \tag{14.36}$$

This is the condition for D to be path-parallel with respect to L. Condition (14.32) implies (14.36), so that if D is self-parallel it is certainly path-parallel.

We now prove the following theorem.

THEOREM 14.3. *If L is symmetric and if D is integrable, path-parallel, and parallel relative to $\bar{D}$ with respect to L, then D is parallel.*

Since L is symmetric and D is integrable, we have, from (14.19), $a^i{}_{j'k'} = a^i{}_{k'j'}$. Moreover, since D is path-parallel we have from (14.36), $a^i{}_{j'k'} + a^i{}_{k'j'} = 0$. Hence $a^i{}_{j'k'} = 0$. Since D is parallel relative to $\bar{D}$ we have from (14.34), $a^i{}_{j'k''} = 0$. Thus we have $a^i{}_{j'k} = 0$, which from (14.22) is precisely the condition that D is parallel. This completes the proof of the theorem.

The distributions D, $\bar{D}$ are orthogonal with respect to a Riemannian metric $\mathbf{g}$ if at every point x, $g_{ij}\lambda^i\mu^j = 0$ for all $\boldsymbol{\lambda} \in D_x$ and $\boldsymbol{\mu} \in \bar{D}_x$, the components g_{ij}, λ^i, μ^j being relative to some coordinate system in a neighbourhood of x. The condition of orthogonality is $g_{ij}a^i{}_k\lambda^k\bar{a}^j{}_h\mu^h = 0$ for all vectors $\boldsymbol{\lambda}$, $\boldsymbol{\mu}$, and is therefore

$$g_{i'j''} = 0. \tag{14.37}$$

We now observe *that for any complementary distributions D, $\bar{D}$ there is a positive definite metric $\mathbf{g}$ with respect to which D, $\bar{D}$ are orthogonal.* If $\mathbf{h}$ is any positive definite metric, a suitable form for $\mathbf{g}$ is given by

$$g_{ij} = h_{i'j'} + h_{i''j''}. \tag{14.38}$$

The ideas of relative parallelism and path-parallelism are due to A. G. Walker.

EXERCISE 14.3. If D, $\bar{D}$ are complementary distributions orthogonal with respect to a Riemannian metric $\mathbf{g}$, show that the connexion L defined by

$$L^i_{jk} = \begin{Bmatrix} i \\ j\,k \end{Bmatrix} + a^i_{j''k} - a^i_{j'k},$$

where covariant differentiation is with respect to $\begin{Bmatrix} i \\ j\,k \end{Bmatrix}$ calculated from the metric, has the property that D and $\bar{D}$ are parallel with respect to L, and $\mathbf{g}$ is constant, i.e. $g_{ij|k} = 0$. Show that L is symmetric if and only if $a^i_{jk} = 0$, in which case $L^i_{jk} = \begin{Bmatrix} i \\ j\,k \end{Bmatrix}$.

15. Riemann extensions

In developing the theory of parallel fields of null planes, A. G. Walker introduced the notion of a Riemann extension of a differentiable manifold which carries a certain differential-geometric structure. A Riemann extension provides a solution of the general problem of embedding a manifold M carrying a given structure in a manifold M' carrying another structure, the embedding being carried out in such a way that the structure on M' induces in a natural way the given structure on M.

Let us take an example to illustrate this. Suppose M is a Riemannian manifold and M' a second Riemannian manifold. The problem is one of embedding M in M' in such a way that the metric of M is induced from that of M' by the inclusion mapping. In particular, given M, a significant problem is to find a Euclidean M' satisfying the conditions.

As a second example, suppose M is a differentiable manifold carrying some given structure not as rich as a Riemannian structure, as happens when M carries a symmetric affine connexion not derived from a Riemannian metric. The problem is to embed M in a Riemannian manifold M' in such a way that the structure on M is described in a natural manner by the metric of M'. In this way the non-Riemannian properties of M are related to Riemannian properties of M', and this relation may possibly be used to extend to non-Riemannian spaces concepts which are primarily Riemannian.

Although the problem of Riemann extension can be treated

from a more general viewpoint, we shall describe the subject as it arose historically as a by-product of the theory of fields of parallel null planes. Suppose M' is a $2n$-dimensional Riemannian manifold of class ∞ which admits a parallel field of null planes of maximum dimension, viz. n. Then it was shown by Walker (1950) that locally a coordinate system can be chosen so that the metric tensor of M' takes the form given by the matrix†

$$(g_{ij}) = \begin{pmatrix} g_{\alpha\beta} & I \\ I & 0 \end{pmatrix}, \tag{15.1}$$

where I is the unit $n \times n$ matrix, and $(g_{\alpha\beta})$ is any symmetric matrix whose elements are C^∞-functions of $2n$ coordinates (x^i). The parallel field of null n-planes is determined by the basis vectors $(\delta^i_{\alpha'})$.

It is found that the canonical form (15.1) remains unchanged by certain restricted transformations of coordinates. If we use the notation ξ_α for $x^{\alpha'}$, then the most general transformation with this property which takes (x^α, ξ_α) into $(x^{*\alpha}, \xi^*_\alpha)$ is

$$x^{*\alpha} = f^\alpha(x), \qquad \xi^*_\alpha = \frac{\partial x^\beta}{\partial x^{*\alpha}}(\xi_\beta + \Phi_\beta(x)), \tag{15.2}$$

where $f^\alpha(x)$, $\Phi_\beta(x)$ are arbitrary functions of class ∞ in the (x^α) coordinates such that the Jacobian $|\partial x^{*\alpha}/\partial x^\beta|$ is non-zero. It will be noticed that when the functions Φ_α are taken to be zero, equation (15.2) gives the ordinary transformation law for the components ξ_α of a covariant vector.

The effect of such a transformation on the metric (15.1) is to replace $g_{\alpha\beta}$ by $g^*_{\alpha\beta}$ where

$$g^*_{\alpha\beta} = \frac{\partial x^\gamma}{\partial x^{*\alpha}}\frac{\partial x^\delta}{\partial x^{*\beta}}\left(g_{\gamma\delta} - \frac{\partial \Phi_\gamma}{\partial x^\delta} - \frac{\partial \Phi_\delta}{\partial x^\gamma}\right) - \frac{2\partial^2 x^\gamma}{\partial x^{*\alpha}\,\partial x^{*\beta}}(\xi_\gamma + \Phi_\gamma). \tag{15.3}$$

Locally the coordinates (x^α) can be taken to define a point of an n-dimensional manifold M_0. It follows from the nature of (15.3) that $g_{\alpha\beta}$ are not the components of a tensor on M_0. However, the form of $g_{\alpha\beta}$ as functions of ξ_α may be chosen so that (15.3) gives some desired transformation of the structure of M_0. For example, if $g_{\alpha\beta}$ are polynomials in ξ_α of degree $r \geqslant 1$, then from (15.3) it follows that $g^*_{\alpha\beta}$ are also polynomials of degree r in ξ^*_α. Writing

$$g_{\alpha\beta} = c_{\alpha\beta}(x) - 2\Gamma^\gamma_{\alpha\beta}\xi_\gamma + c^{\gamma\delta}_{\alpha\beta}(x)\xi_\gamma\xi_\delta + ..., \tag{15.4}$$

it follows in the special case when $\Phi_\alpha = 0$ that the coefficients

† In the remainder of this section it will be convenient to use Roman suffixes i, j, k for the range $1,... 2n$; Greek suffixes α, β, γ for the range $1,..., n$; and primed Greek suffixes for the range $n+1,..., 2n$, such that $\alpha' = \alpha + n$.

$\Gamma^{\gamma}_{\alpha\beta}$ transform as connexion coefficients while all other coefficients in (15.4) transform as components of tensors. The situation becomes more complicated in the case $\Phi_{\alpha} \neq 0$, $r > 1$, but for our purpose it will be sufficient to confine attention to the special case when $\Phi_{\alpha} = 0$ and hence ξ_{α} transform as components of a covariant vector.

Let M be a given manifold and let M' be the manifold formed by the set of all covariant tangent vectors at all points of M. A point z of M' is described locally by the coordinates $(x^{\alpha}, \xi_{\alpha})$, where (x^{α}) are the coordinates of some point x of M relative to some suitable coordinate neighbourhood and ξ_{α} are the components of a covariant tangent vector at x relative to this coordinate system. With the manifold M' we associate the Riemannian metric given by (15.1), with $(g_{\alpha\beta})$ suitably chosen to incorporate the appropriate structure given on M. This metric is independent of any particular coordinate system in the coordinate neighbourhoods of M, and is defined over the whole of M'. *The space M' is called a Riemann extension of M.*

In the particular case when (15.4) reduces to

$$g_{\alpha\beta} = -2\Gamma^{\gamma}_{\alpha\beta}(x)\xi_{\gamma} \tag{15.5}$$

the corresponding extension is called *restricted*. For the remainder of this section we shall be concerned only with such restricted extensions, and by 'extension' we shall mean 'restricted extension'. By means of (15.1) and (15.5), a symmetric affine connexion with coefficients $\Gamma^{\gamma}_{\alpha\beta}$ defined over M gives rise to a definite Riemannian extension over M'. Moreover, starting with this Riemannian structure over M' we can recover the affine connexion over M.

Several interesting questions now arise concerning the relation between special types of affine connexions on M and the corresponding special metrics on M'. We state without proof a number of theorems which illustrate such results.

For convenience we denote the n-dimensional space M which carries a symmetric affine connexion by A^n, and denote the $2n$-dimensional Riemann extension by R^{2n}.

THEOREM 15.1. *The curvature tensor of A^n is zero if and only if the curvature tensor of R^{2n} is zero.*

THEOREM 15.2 (first proved by Z. Afifi). *R^{2n} is symmetric if and only if A^n is symmetric.*

In other words, $R^h_{ijk,l} = 0$ if and only if $B^{\alpha}_{\beta\gamma\delta,\epsilon} = 0$.

A harmonic Riemannian space has already been defined in section 9. Such a space is said to be *simply harmonic* if, for every point x_0, $\Delta_2\Omega = n$. It has been shown by E. M. Patterson that certain simply harmonic spaces can be constructed as Riemann extensions. In particular Patterson proved

THEOREM 15.3. *If A^n has the connexion of a simply harmonic Riemannian space, then the Riemann extension of A^n is simply harmonic.*

He also proved conversely that if the Riemann extension of A^n is simply harmonic, then the curvature tensor of A^n satisfies the same sequence of conditions as the curvature tensor of a simply harmonic Riemannian space. Evidently an affine space whose curvature tensor satisfied this sequence of conditions could be *defined* as a simply harmonic affine space, but a more pleasing alternative is to define a simply harmonic affine space as one whose Riemann extension is simply harmonic. In this way we have extended to the affine space A^n concepts which are primarily Riemannian.

Most of the results of sections 10–15 have been obtained during the last ten years, and many unsolved problems remain. It will be noticed that the ideas of harmonic space, parallel planes, recurrent tensors, and Riemann extensions are closely related, and that theorems obtained in investigating one of these topics often find application or interpretation in the others. It also illustrates the fact that by confining attention in Riemannian geometry to positive definite metrics much interesting mathematics is overlooked.

16. É. Cartan's approach to Riemannian geometry

In this and the following sections we give a brief account of É. Cartan's approach to Riemannian geometry. The essential idea is to reduce, as far as possible, problems of Riemannian geometry to problems of Euclidean geometry. In this way Cartan was able to give a simple geometrical interpretation of a number of properties which had previously been hidden under what he called 'les débauches d'indices'. Although he himself never refrained from detailed calculations, he disliked much of the computational work on differential geometry which was very fashionable during the period 1920–35, especially as much of this work was of little geometrical interest. This was a natural attitude to be adopted by a powerful mathematician who possessed a remarkable geometrical intuition which enabled him to see the geometrical content of very complicated calculations. In fact Cartan often used geometrical arguments to replace some of the calculations, and a reader who does not possess this remarkable gift is often baffled by his arguments. It is hoped that as a result of reading this brief introduction the reader will be encouraged to make a serious study of Cartan's book (1928; 2nd edition 1946).

Consider a certain region of Euclidean space E_n referred to a system of curvilinear coordinates (x^i). At each point P in this region there is a natural basis of tangent vectors $(\mathbf{e}_i)$ which span the tangent space to E_n at P. Of course the tangent space to E_n at P can be identified with E_n itself, but it will be convenient for certain purposes to treat these spaces as distinct. The vector $\mathbf{e}_i$ is directed along the curve at P whose equation is $x^j = \text{constant}, j \neq i$.

It is convenient to use the symbol $\mathbf{P}$ to denote the position vector of the point P referred to some fixed point O of E_n. Then the vectors of the natural frame at P are given by

$$\mathbf{e}_i = \frac{\partial}{\partial x^i}\mathbf{P}. \tag{16.1}$$

It should be remembered that (16.1) gives n equations, one for each index i; in particular, $(\partial\mathbf{P}/\partial x^i)$ are *not* the components of a covariant vector which is the gradient of a scalar function.

If the point P moves along a curve γ of parameter t we have

$$\frac{d\mathbf{P}}{dt} = \frac{\partial\mathbf{P}}{\partial x^i}\frac{dx^i}{dt} = \mathbf{e}_i\frac{dx^i}{dt},$$

or, in terms of differentials,

$$d\mathbf{P} = \mathbf{e}_i\,dx^i. \tag{16.2}$$

This equation may be interpreted as expressing the displacement vector $\overrightarrow{PP'}$ determined by points P, P' with coordinates (x^i), $(x^i + dx^i)$ respectively in terms of the natural frame at P. However, in the terminology of Chapter VI we consider $d\mathbf{P}$ as a vectorial 1-form. The vectors $\mathbf{e}_i$ form a basis for the tangent space T_P at P, while the forms dx^i form the dual basis for the space T_P^* of covariant vectors or 1-forms at P. The elements $\mathbf{e}_i \otimes dx^j$ form a basis for the tensor product $T_P \otimes T_P^*$, and in terms of this basis the components of $d\mathbf{P}$ are seen from (16.2) to be $\delta^i{}_j$. Alternatively, we can regard $d\mathbf{P}$ as that vectorial 1-form which gives the identity mapping of T_P onto itself.

It appears that É. Cartan regarded such concepts as infinitesimal elements, small quantities of the second order, etc., as quite precise entities, and his geometrical insight was amply powerful to prevent his falling into the numerous pitfalls which normally accompany such concepts. For example, he would regard (16.2) as *established* by treating the geometrical interpretation as fundamental, and interpreting this geometrical fact symbolically.

If we differentiate (16.1) partially with respect to x^j we get

$$\frac{\partial \mathbf{e}_i}{\partial x^j} = \frac{\partial^2 \mathbf{P}}{\partial x^j \partial x^i}. \tag{16.3}$$

As P moves along a curve γ of parameter t, there will be a natural frame at each point on the curve and in particular the vector $\mathbf{e}_i$ will vary along γ, i.e. $\mathbf{e}_i$ will be a function of t. By differentiating (16.1) with respect to t we get

$$\frac{d\mathbf{e}_i}{dt} = \frac{\partial^2 \mathbf{P}}{\partial x^j \partial x^i}\frac{dx^j}{dt} = \frac{\partial \mathbf{e}_i}{\partial x^j}\frac{dx^j}{dt},$$

or, in terms of differentials,

$$d\mathbf{e}_i = \frac{\partial \mathbf{e}_i}{\partial x^j}dx^j. \tag{16.4}$$

Now the element $d\mathbf{e}_j$, for a fixed j, is a vectorial 1-form and must be expressible in terms of the basis $\mathbf{e}_i \otimes dx^k$. If the components of $d\mathbf{e}_j$ relative to this basis are L^i_{jk}, then from (16.4) we have

$$\frac{\partial \mathbf{e}_j}{\partial x^k} = L^i_{jk}\,\mathbf{e}_i. \tag{16.5}$$

If we define a matrix $(\omega^i{}_j)$ by the equation

$$(\omega^i{}_j) = (L^i_{jk}\,dx^k), \tag{16.6}$$

then (16.4) can be written in the form

$$d\mathbf{e}_j = \omega^i{}_j\,\mathbf{e}_i. \tag{16.7}$$

Geometrically this expresses the difference between the jth basis vector of the natural frame at P with coordinates (x^i) and the jth basis vector of the natural frame at a neighbouring point P' with coordinates $(x^i + dx^i)$, this difference being referred to the natural frame at P. Alternatively, the matrix $(\omega^i{}_j)$ gives a mapping of the tangent space at P' onto the tangent space at P. However, in the terminology of Chapter VI, such a mapping gives an affine connexion and the functions L^i_{jk} of (16.5) are none other than the connexion coefficients referred to the natural basis.

The coefficients g_{ij} of the metric tensor at each point P satisfy the equation

$$\mathbf{e}_i \cdot \mathbf{e}_j = g_{ij}. \tag{16.8}$$

If we consider the motion of P along the curve γ of parameter t we have

$$\frac{d\mathbf{e}_i}{dt} \cdot \mathbf{e}_j + \mathbf{e}_i \cdot \frac{d\mathbf{e}_j}{dt} = \frac{dg_{ij}}{dt},$$

or, in terms of differentials,

$$d\mathbf{e}_i \cdot \mathbf{e}_j + \mathbf{e}_i \cdot d\mathbf{e}_j = dg_{ij}.$$

Using (16.7) and (16.8) this equation takes the form

$$g_{jk}\omega^k{}_i + g_{ik}\omega^k{}_j = dg_{ij}. \tag{16.9}$$

In terms of the matrix of 1-forms (ω_{ij}) defined by

$$\omega_{ij} = g_{ik}\omega^k{}_j, \tag{16.10}$$

equation (16.9) becomes

$$\omega_{ij} + \omega_{ji} = dg_{ij}. \tag{16.11}$$

Incidentally we note from the equation preceding (16.2) that

$$\frac{d\mathbf{P}}{dt} \cdot \frac{d\mathbf{P}}{dt} = \mathbf{e}_i \frac{dx^i}{dt} \cdot \mathbf{e}_j \frac{dx^j}{dt} = g_{ij}\frac{dx^i}{dt}\frac{dx^j}{dt} \quad \text{(from (16.8))},$$

or, in terms of differentials,

$$(d\mathbf{P})^2 = g_{ij}dx^i dx^j = ds^2. \tag{16.12}$$

From (16.3) we have, after changing suffixes,

$$\frac{\partial \mathbf{e}_j}{\partial x^k} = \frac{\partial \mathbf{e}_k}{\partial x^j},$$

which with (16.5) gives

$$L^i_{jk} = L^i_{kj}, \tag{16.13}$$

showing that the coefficients are symmetric.

Write

$$[i, jh] = g_{ik}L^k_{jh}, \tag{16.14}$$

so that $[i, jh]$ is symmetric in j and h.

Then

$$\omega_{ij} = g_{ih}L^h_{jk}dx^k = [i, jk]dx^k,$$

and (16.9) becomes

$$[i, jk] + [j, ik] = \partial g_{ij}/\partial x^k. \tag{16.15}$$

Cyclic permutation of the suffixes i, j, k gives

$$[j, ki] + [k, ji] = \partial g_{jk}/\partial x^i, \tag{16.15'}$$

and again permutation gives

$$[k, ij] + [i, kj] = \partial g_{ki}/\partial x^j. \tag{16.15''}$$

Add (16.15'), (16.15''), and subtract (16.15) to get

$$[k, ij] = \frac{1}{2}\left(\frac{\partial g_{ik}}{\partial x^j} + \frac{\partial g_{jk}}{\partial x^i} - \frac{\partial g_{ij}}{\partial x^k}\right), \tag{16.16}$$

where we have also used the symmetry of $[k, ij]$. Thus, as has been anticipated by the notation, the symbols $[k, ij]$ are none other than the Christoffel symbols previously defined. Thus we have

$$L^i_{jk} = g^{ih}[h, jk] = \begin{Bmatrix} i \\ j\,k \end{Bmatrix}. \tag{16.17}$$

By differentiating (16.5) we get

$$\frac{\partial^2 \mathbf{e}_j}{\partial x^h \partial x^k} = \frac{\partial L^i_{jk}}{\partial x^h}\mathbf{e}_i + L^i_{jk}\frac{\partial \mathbf{e}_i}{\partial x^h}$$

$$= \frac{\partial L^i_{jk}}{\partial x^h}\mathbf{e}_i + L^i_{jk}\,L^t_{ih}\,\mathbf{e}_t$$

$$= \left(\frac{\partial L^i_{jk}}{\partial x^h} + L^t_{jk}\,L^i_{th}\right)\mathbf{e}_i.$$

Using the symmetry of the left-hand member in h and k we get

$$\frac{\partial L^i_{jk}}{\partial x^h} - \frac{\partial L^i_{jh}}{\partial x^k} + L^t_{jk}\,L^i_{th} - L^t_{jh}\,L^i_{tk} = 0, \tag{16.18}$$

which expresses the fact that in a Euclidean space the curvature tensor is identically zero. Cartan proved that this condition is both *necessary* and *sufficient* that a region of a space shall be Euclidean. We have proved the necessity but we shall not prove the sufficiency here.

Consider now a field of vectors $\mathbf{x}$ defined over a region of E_n, and let the vector of the field at P have components X^i referred to the natural frame at P, so that

$$\mathbf{x} = X^i \mathbf{e}_i. \tag{16.19}$$

As P moves along some curve γ in the region with parameter t we have

$$\frac{d\mathbf{x}}{dt} = \frac{dX^i}{dt}\mathbf{e}_i + X^i\frac{d\mathbf{e}_i}{dt} = \left(\frac{dX^i}{dt} + L^i_{jk}\,X^j\frac{dx^k}{dt}\right)\mathbf{e}_i.$$

Writing

$$\frac{DX^i}{dt} = \frac{dX^i}{dt} + L^i_{jk}\,X^j\frac{dx^k}{dt}, \tag{16.20}$$

we say DX^i/dt are the components of the *intrinsic derivative* of **x** along γ. In terms of differentials we have

$$DX^i = dX^i + L^i_{jk} X^j dx^k, \tag{16.21}$$

or $$DX^i = dX^i + \omega^i{}_j X^j. \tag{16.22}$$

The vectorial form (DX^i) is called the *absolute differential* or *covariant differential* of the vector field **x**. In particular we note that (16.7), which expresses the differentials of the basis vectors, could be written as $$D\mathbf{e}_j = \omega^i{}_j \mathbf{e}_i. \tag{16.23}$$

The vector field **x** along γ will be parallel if $DX^i/dt = 0$ along γ. From (16.20), $$\frac{DX^i}{dt} = \left(\frac{\partial X^i}{\partial x^k} + L^i_{jk} X^j\right)\frac{dx^k}{dt}.$$

Write $$D_k X^i = \frac{\partial X^i}{\partial x^k} + L^i_{jk} X^j; \tag{16.24}$$

then $D_k X^i$ is called the covariant derivative of the vector field **x** with respect to x^k.

Suppose Y_i are the components of a field of covariant vectors defined over some region of E_n, and let **x** be an arbitrary parallel vector field defined over the same region. As in the earlier sections of this chapter, we write

$$D(X^i Y_i) = d(X^i Y_i) = dX^i Y_i + X^i dY_i,$$

from which $$(DY_i - dY_i + \omega^j{}_i Y_j)X^i = 0.$$

Since this must be satisfied by arbitrary (X^i), we get

$$DY_i = dY_i - \omega^j{}_i Y_j, \tag{16.25}$$

which is the analogue of (16.22) for a covariant vector field. In a similar way we obtain

$$Da^i{}_j = da^i{}_j + \omega^i{}_k a^k{}_j - \omega^k{}_j a^i{}_k, \tag{16.26}$$

with analogous formulae for a tensor field of type (p, q).

17. Euclidean tangent metrics

Let A be a point with coordinates (x^i_0) belonging to some coordinate neighbourhood N of a given n-dimensional Riemannian manifold M, whose metric in N is given by

$$ds^2 = g_{ij} dx^i dx^j. \tag{17.1}$$

A metric tensor (γ_{ij}) defined over N will be called a *Euclidean metric* if equation (16.18) is satisfied by the Christoffel symbols associated with (γ_{ij}).

A metric tensor (γ_{ij}) defined over N is said to be a *tangent metric* to the *Riemannian metric tensor* (g_{ij}) at the point A when

$$\gamma_{ij}(x_0) = g_{ij}(x_0). \qquad (17.2)$$

We wish to consider Euclidean tangent metrics to (g_{ij}) at A. These tangent metrics certainly exist, for if we define $\gamma_{ij}(x)$ for all x belonging to N by

$$\gamma_{ij}(x) = g_{ij}(x_0),$$

then the components γ_{ij} are constant over N and (16.18) is certainly satisfied. On the other hand there is evidently an infinity of tangent metrics to (g_{ij}) at A, since the restriction (17.2) is imposed only at the single point. However, from the form of (17.2) it follows that the set of all Euclidean tangent metrics to (g_{ij}) at A is independent of the system of coordinates (x^i). A Euclidean space when metrized by a Euclidean tangent metric at A will be called a *Euclidean tangent space at A*. There is an infinite number of Euclidean tangent spaces at A, but as we shall consider only properties common to the whole set of tangent spaces at A there will be no confusion if we refer to 'the Euclidean tangent space at A'.

As an example, consider the Riemannian metric

$$ds^2 = (1+x)\,dx^2 + (1+y)\,dy^2 + (1+z)\,dz^2 \qquad (17.3)$$

defined in some neighbourhood of the origin of coordinates A. Then one Euclidean tangent metric at A is

$$ds^2 = dx^2 + dy^2 + dz^2, \qquad (17.4)$$

for the metric coefficients of (17.3) and (17.4) agree at the point A. We note, incidentally, that the first derivatives of the two metric coefficients do not agree at A.

Suppose now we have at A two tangent vectors $\boldsymbol{\lambda}$, $\boldsymbol{\mu}$ to M. These vectors can be considered to be in the Euclidean tangent space at A, and it is known that in terms of n-dimensional Euclidean geometry the angle θ between these vectors is given by

$$\cos\theta = \frac{\gamma_{ij}(x_0)\lambda^i\mu^j}{\{\gamma_{ij}(x_0)\lambda^i\lambda^j\}^{\frac{1}{2}}\{\gamma_{ij}(x_0)\mu^i\mu^j\}^{\frac{1}{2}}}. \qquad (17.5)$$

In view of (17.2) this could equally well be written

$$\cos\theta = \frac{g_{ij}\lambda^i\mu^j}{\{g_{ij}\lambda^i\lambda^j\}^{\frac{1}{2}}\{g_{ij}\mu^i\mu^j\}^{\frac{1}{2}}}. \qquad (17.6)$$

Cartan now *defines* the angle between the tangent vectors $\boldsymbol{\lambda}$, $\boldsymbol{\mu}$ in the Riemannian space to be given by (17.6). In a similar way, any result of n-dimensional Euclidean geometry concerning angles made at the same point A by lines, surfaces, etc., can be generalized immediately to any Riemannian space. The only restriction is that the geometric configuration must involve only the metric tensor (g_{ij}) (and not its derivatives) at the point A.

We leave it to the reader to verify that at a point A

   (i) the volume element $\sqrt{g}\, dx^1\, dx^2 \ldots dx^n$,

   (ii) the divergence of a vector field (X^i),

may be defined first in the Euclidean tangent space at A, and then naturally extended to the Riemannian space. However, the idea of parallelism of vectors $\boldsymbol{\lambda}$, $\boldsymbol{\mu}$ localized at *different* points A, A' cannot be defined from the concept of Euclidean tangent space, and for this reason we now consider tangent spaces which satisfy further conditions.

18. Euclidean osculating metrics

A metric tensor (γ_{ij}) defined over N is said to be an *osculating metric to the Riemannian metric tensor (g_{ij}) at the point A* when

$$\text{(i)} \qquad (\gamma_{ij})_0 = (g_{ij})_0, \tag{18.1}$$

and
$$\text{(ii)} \quad (\partial_k \gamma_{ij})_0 = (\partial_k g_{ij})_0. \tag{18.2}$$

We shall be concerned only with Euclidean osculating metrics. A Euclidean osculating metric is a Euclidean tangent metric which satisfies the additional restriction (ii). In view of equation (16.16) an alternative condition to (ii) is

$$\text{(ii)}' \quad \begin{Bmatrix} k \\ i\,j \end{Bmatrix}_\gamma = \begin{Bmatrix} k \\ i\,j \end{Bmatrix}_g, \tag{18.3}$$

i.e. the Christoffel symbols must have the same values at A.

If we have a metric tangent to (g_{ij}) at A which is not an osculating metric, then the change of coordinates from (x^i) to (u^i) given by

$$u^i = (x^i - x_0^i) + \tfrac{1}{2} \begin{Bmatrix} i \\ j\,k \end{Bmatrix}_0 (x^j - x_0^j)(x^k - x_0^k), \tag{18.4}$$

leads to a tangent metric whose coefficients *and their first order derivatives* agree with the corresponding terms of the Riemannian metric.

Thus osculating metrics certainly exist, and it is easy to see that they form an infinite set. Cartan considered only properties which are common to the whole set of osculating metrics since these are properties of the Riemannian space involving only the values of (g_{ij}) and $\begin{Bmatrix} k \\ i\,j \end{Bmatrix}$ at a point A. A Euclidean space which carries a Euclidean osculating metric is called a *Euclidean osculating space*, and since we shall be concerned only with properties common to the whole set of Euclidean osculating spaces at A, there will be no confusion in referring to 'the Euclidean osculating space at A'.

Before dealing with the general case, consider the two-dimensional surface in E_3 whose equation is

$$z = f(x, y)$$

referred to rectangular Cartesian coordinates whose origin is the point A on the surface, such that Oz is along the surface normal. Writing

$$p = \partial z / \partial x, \qquad q = \partial z / \partial y,$$

the metric (g_{ij}) induced from E_3 is

$$ds^2 = (1+p^2)\,dx^2 + 2pq\,dx\,dy + (1+q^2)\,dy^2 .$$

Compare this with the Euclidean metric (γ_{ij})

$$ds^2 = dx^2 + dy^2.$$

Since $p = q = 0$ at A, (γ_{ij}) is a tangent metric to (g_{ij}) at A. Moreover, it is easily verified that the first derivatives of (g_{ij}) and (γ_{ij}) are all zero at A, so (γ_{ij}) is an osculating metric, and in this case the Euclidean osculating space is just the tangent plane to the surface at A. Corresponding to a point $P\ (x, y, z)$ on the surface near A, there is uniquely determined a point $\bar{P}$ in the osculating space with coordinates $(x, y, 0)$. In this way we get a representation of a neighbourhood of the surface near A on a neighbourhood of the Euclidean osculating space. Roughly speaking, we regard this Euclidean neighbourhood as an approximation to the neighbourhood of the surface near A. Two curves on the surface intersecting at some angle at A will have as their images in the osculating space $\bar{S}$ at A two curves which intersect at the same angle. Moreover, a curve γ in the surface passing through A which carries a field of tangent vectors to S will have as its image in the osculating space a curve $\bar{\gamma}$ carrying a field of tangent vectors—i.e. a curve $\bar{\gamma}$ in the tangent plane which carries a field of vectors lying in the plane. It is evidently not true that when the vector field in the plane is parallel, then the corresponding vector field in S will be parallel, for otherwise every surface would be isometric with the Euclidean plane. However, it is true in the following restricted sense for points of S belonging to a neighbourhood of A. Let A' be a point on the surface S near A in the sense that the coordinates of A and A' differ only by terms of the order of ϵ. The points A, A' will have representatives $\bar{A}$ $(=A)$ and $\bar{A}'$ in the Euclidean osculating space at A. The straight line joining $\bar{A}$ to $\bar{A}'$ in the Euclidean osculating space will correspond to some curve γ in S. Suppose now we have a tangent vector $\boldsymbol{\lambda}_A$ to the surface S at A and we carry this along γ to A' by parallel transport to obtain a vector $\boldsymbol{\lambda}_{A'}$ tangent to S at A'.

Let $\boldsymbol{\lambda}^*_{\bar{A}'}$ be the tangent vector to $\bar{S}$ at $\bar{A}'$ obtained from the tangent vector $\boldsymbol{\lambda}_A$ at A (regarded now as tangent to $\bar{S}$) by translation in the Euclidean space, and let $\boldsymbol{\lambda}^*_{A'}$ be the image of this vector in S at A'. Then the components of the two vectors $\boldsymbol{\lambda}_{A'}$, $\boldsymbol{\lambda}^*_{A'}$ will differ only by terms involving ϵ^2 and higher powers. In this sense the Riemannian geometry of an ϵ-neighbourhood of S is represented by the Euclidean geometry of a neighbourhood of $\bar{S}$. Any property of S at A which depends only on (g_{ij}) and $(\partial_k g_{ij})$ at A can be described equally well by the corresponding property of $\bar{S}$ at A. The same property at A' near A can be described in terms of $\bar{S}$ only as an approximation to within first-order terms.

We now return to consider the general case of a Riemannian manifold M with a Riemannian metric (g_{ij}) defined in a neighbourhood N containing the point A. As in the two-dimensional example just considered, any point A' whose coordinates (x^i) differ from those of A (x_0^i) by terms of order ϵ will be represented in the Euclidean osculating space at A by a point $\bar{A}'$, whose coordinates with respect to the natural frame at A are $(x^i - x_0^i)$. Evidently the vectors $\mathbf{e}_i$ at A, considered as tangent vectors to M, provide a frame of reference for the Euclidean osculating space at A.

Again, the arguments used in the particular example of the two-dimensional surface apply in the general case. In fact we can apply the theory of the structure of a Euclidean space E_n, developed in section 16, to the Euclidean osculating space at each point belonging to the coordinate neighbourhood N. For example, corresponding to a vector field $\boldsymbol{\lambda}$ defined over N in the Riemannian space, the absolute differential of the field at a general point A is given by

$$D\lambda^i = d\lambda^i + \begin{Bmatrix} i \\ j\,k \end{Bmatrix} \lambda^j \, dx^k, \tag{18.5}$$

where the Christoffel symbols are considered in the first place to be calculated from the Euclidean osculating metric at A, and then identified with the corresponding symbols calculated from the Riemannian metric at A.

Equation (16.2), $$d\mathbf{P} = \mathbf{e}_i \, dx^i, \qquad (18.6)$$

is still valid in a Riemannian space, but we must be careful to interpret it correctly. We prefer to regard $d\mathbf{P}$ as that vectorial 1-form which gives the identity mapping of T_P onto itself. The geometrical interpretation is that given immediately after (16.2) in section 16 where the Euclidean space is now identified with the Euclidean osculating space at P.

From (18.6) we have

$$ds^2 = \mathbf{e}_i . \mathbf{e}_j \, dx^i dx^j = g_{ij} \, dx^i dx^j. \qquad (18.7)$$

Equation (16.23), $$D\mathbf{e}_j = \omega^i{}_j \mathbf{e}_i, \qquad (18.8)$$

can be given a similar geometrical interpretation in terms of the Euclidean osculating space. However, in the spirit of our treatment in Chapter VI, we prefer to regard (18.8) as defining an affine connexion, i.e. a mapping of the tangent space at P' onto the tangent space at P. It will be noted that we have considered equation (16.23) and not equation (16.7) which is equivalent to it in Euclidean space. We do this because, *in a Riemannian space*, equations (16.18) no longer hold and $d\mathbf{e}_j$ is no longer a perfect differential.

In concluding this section we remark that by using Cartan's approach to Riemannian geometry many of the properties which we obtained previously by purely analytical methods receive a natural geometrical interpretation. However, unless the reader possesses a geometrical insight comparable with that of Cartan, he is advised to use Cartan's geometrical arguments with care and to regard results as tentative until they have been established by the calculus of tensors and exterior forms developed in Chapter VI.

19. The equations of structure

We now collect together the so-called equations of structure of Riemannian geometry, which may be written

$$d\mathbf{P} = \mathbf{e}_i \, dx^i, \qquad (19.1)$$

$$\mathbf{e}_i . \mathbf{e}_j = g_{ij}, \qquad (19.2)$$

$$ds^2 = d\mathbf{P} . d\mathbf{P} = g_{ij} \, dx^i dx^j, \qquad (19.3)$$

$$D\mathbf{e}_i = \omega^j{}_i \mathbf{e}_j, \qquad (19.4)$$

$$\omega_{ij} = g_{ik}\omega^k{}_j, \tag{19.5}$$

$$\omega_{ij} + \omega_{ji} = dg_{ij}, \tag{19.6}$$

$$\omega^i{}_j \wedge dx^j = 0, \tag{19.7}$$

$$\Omega^i{}_j = d\omega^i{}_j + \omega^i{}_k \wedge \omega^k{}_j, \tag{19.8}$$

$$\Omega^i{}_j = R^i{}_{jkl}\, dx^k \wedge dx^l, \tag{19.9}$$

$$\Omega_{ij} = g_{ik}\Omega^k{}_j = R_{ijkl}\, dx^k \wedge dx^l. \tag{19.10}$$

So far throughout this chapter the frames $(\mathbf{e}_i)$ have been naturally related to the coordinate system, i.e. at each point P the vector $\mathbf{e}_i$ has been directed along the parametric curve $x^j = \text{constant}$, $j \neq i$. However, the theory of connexions described in Chapter VI dealt with general frames $(\mathbf{e}_i)$, and in Riemannian geometry it is often convenient to use frames which are not related to any coordinate system. Accordingly we rewrite the equations of structure in the following form suitable for use with a general frame $(\mathbf{e}_i)$ with its co-frame of 1-forms (ω^i).

$$d\mathbf{P} = \mathbf{e}_i\,\omega^i, \tag{19.1'}$$

$$\mathbf{e}_i.\mathbf{e}_j = g_{ij}, \tag{19.2'}$$

$$ds^2 = d\mathbf{P}.d\mathbf{P} = g_{ij}\,\omega^i\omega^j, \tag{19.3'}$$

$$D\mathbf{e}_i = \omega^j{}_i\,\mathbf{e}_j, \tag{19.4'}$$

$$\omega_{ij} = g_{ik}\omega^k{}_j, \tag{19.5'}$$

$$\omega_{ij} + \omega_{ji} = dg_{ij}, \tag{19.6'}$$

$$d\omega^i + \omega^i{}_j \wedge \omega^j = 0, \tag{19.7'}$$

$$\Omega^i{}_j = d\omega^i{}_j + \omega^i{}_k \wedge \omega^k{}_j, \tag{19.8'}$$

$$\Omega^i{}_j = \bar{R}^i{}_{jkl}\,\omega^k \wedge \omega^l, \tag{19.9'}$$

$$\Omega_{ij} = \bar{R}_{ijkl}\,\omega^k \wedge \omega^l. \tag{19.10'}$$

These equations reduce to the previous set when $(\omega^i) = (dx^i)$.

Sometimes it is convenient to use an *orthonormal* set of frames, i.e. frames such that at each point P the basis vectors are of unit length and mutually orthogonal. In this case, we see from (19.2') that the components of the metric tensor assume the simple form $g_{ij} = \delta_{ij}$, and the metric is given by

$$ds^2 = \sum_{i=1}^{n} (\omega^i)^2. \tag{19.11}$$

Moreover, all tensors associated with the metric tensor now have identical components, and all suffixes can be written as covariant suffixes. From (19.6') we have

$$\omega_{ij} + \omega_{ji} = 0, \tag{19.12}$$

so that when the frames are orthogonal the connexion matrix of
1-forms is skew symmetric.

When frames other than natural frames are used, the connexion
coefficients are no longer given in terms of the metric by the
Christoffel symbols. We now obtain formulae for these coefficients
when the frames are orthonormal. Since $(\omega_k \wedge \omega_h)$ is a basis for
2-forms, we may write

$$d\omega_i = \tfrac{1}{2}c_{khi}\,\omega_k \wedge \omega_h, \tag{19.13}$$

where

$$c_{khi}+c_{hki} = 0. \tag{19.14}$$

Let γ_{ijk} be the connexion coefficients referred to the co-frame (ω^i)
so that

$$\omega_{ij} = \gamma_{ijk}\,\omega_k. \tag{19.15}$$

From equation (19.12) it follows that

$$\gamma_{ijk}+\gamma_{jik} = 0. \tag{19.16}$$

From equation (19.7′) we have

$$\tfrac{1}{2}c_{kji}\,\omega_k \wedge \omega_j + \gamma_{ijk}\,\omega_k \wedge \omega_j = 0.$$

Since the forms $\omega_k \wedge \omega_j$ are linearly independent, we must have

$$\tfrac{1}{2}c_{kji}+\gamma_{ijk} = \tfrac{1}{2}c_{jki}+\gamma_{ikj},$$

from which

$$c_{kji} = \gamma_{ikj}-\gamma_{ijk}. \tag{19.17}$$

Using (19.17) we see that

$$\tfrac{1}{2}(c_{ijk}-c_{jki}-c_{kij}) = \tfrac{1}{2}(\gamma_{kij}-\gamma_{kji}-\gamma_{ijk}+\gamma_{ikj}-\gamma_{jki}+\gamma_{jik})$$
$$= \gamma_{jik}.$$

Making use of (19.16) we obtain the required formula

$$\gamma_{ijk} = \tfrac{1}{2}(c_{jki}+c_{kij}-c_{ijk}). \tag{19.18}$$

É. Cartan (1927) proved by means of orthonormal frames that
a given n-dimensional Riemannian space can be locally embedded
in a Euclidean space E_m where $m = \tfrac{1}{2}n(n+1)$. When $n = 2$, this
gives the first fundamental existence theorem mentioned on page
52.

20. Global Riemannian geometry

Nearly all the theorems discussed in this chapter have been
concerned with properties of a Riemannian metric defined over a
coordinate neighbourhood. Although the theory of Riemann
extensions is really a global theory, we have not stressed this point.

Moreover, although it is trivial that fields of r-planes exist locally over one coordinate neighbourhood, questions of the *global* existence of such fields constitute an important branch of the study of the topology of differentiable manifolds. In order to deal with such problems in differential geometry it was necessary to create new mathematical methods which gave rise to the modern *theory of fibre-bundles*, the introduction of which has had repercussions in regions of mathematics well outside the realms of differential geometry.

In an introductory text of this nature it is impossible to deal adequately with global Riemannian geometry. This is unfortunate because the subject is full of extremely interesting unsolved problems, and present research interest in differential geometry is focused on global rather than on local problems. There seems little doubt that the theory of fibre-bundles together with the calculus of exterior forms will remain powerful tools for many investigations in global theory. Indeed the present global theory of connexions makes great use of both these tools. The theory of Lie Groups has also proved to be a powerful weapon in attacking global problems of differential geometry, while the recently developed theory of *stacks* or *faisceaux* may also prove extremely fruitful.

In this very brief reference to the subject, attention will be restricted to three theorems—the first being an existence theorem, while the other two give definite relations between differential-geometric invariants of a Riemannian manifold and certain topological invariants of the manifold.

THEOREM 20.1. *A differentiable manifold of class r always admits a positive definite Riemannian metric of class $r-1$.*

An elegant proof of this theorem was given by N. E. Steenrod by means of the theory of fibre bundles. His proof also holds when $r = \infty$ but not when $r = \omega$. However, a paper by C. B. Morrey, *Annals of Mathematics*, **68** (1958) 159–201, shows that the theorem is also true for compact analytic manifolds.

Some results are available on conditions for the global existence of indefinite metrics—in particular it is known that a necessary and sufficient condition for the global existence of a hyperbolic metric (signature $(n-2)$) over a compact orientable manifold is that the Euler characteristic of M shall be zero.

The second theorem concerns a generalization to $2n$-dimensional

Riemannian manifolds of the Gauss–Bonnet formula for surfaces,

$$\frac{1}{2\pi} \int_S K\sqrt{g}\,du^1 du^2 = \chi, \tag{20.1}$$

obtained in Chapter IV.

THEOREM 20.2. *If M is a compact, orientable, $2n$-dimensional Riemannian manifold, then*

$$\frac{2}{\omega^{2n}} \int_M K_T \sqrt{g}\,dx^1 dx^2...dx^{2n} = \chi, \tag{20.2}$$

where χ is the Euler characteristic of M, ω^{2n} is the area of a $2n$-dimensional sphere of unit radius given by

$$\omega^{2n} = \frac{\pi^n 2^{2n+1} n!}{2n!}, \tag{20.3}$$

and K_T is the 'total curvature' given by

$$K_T = \frac{R_{i_1 i_2 j_1 j_2} R_{i_3 i_4 j_3 j_4}...R_{i_{2n-1} i_{2n} j_{2n-1} j_{2n}} \epsilon^{i_1...i_{2n}} \epsilon^{j_1...j_{2n}}}{2n!\,2^n}. \tag{20.4}$$

The ϵ-tensors in (20.4) are defined by

$$\epsilon^{i_1 i_2...i_{2n}} = e^{i_1 i_2...i_{2n}}/\sqrt{g},$$

where $e^{i_1 i_2...i_{2n}}$ is $+1$ if $i_1,...,i_{2n}$ is an even permutation of $1, 2,...,2n$;
        is -1 if $i_1,...,i_{2n}$ is an odd permutation of $1, 2,..., 2n$;
        and is 0 otherwise.

This generalized Gauss–Bonnet theorem was first obtained under the assumption that M was a hypersurface of E_{2n+1}. Later it was established under the more general assumption that M was embedded in E_{2n+p}. The first intrinsic proof which made no assumptions about M being embedded in a Euclidean space was given by Allendoerfer and Weil in 1943. A much simpler proof was given in 1945 by S. S. Chern which made use of the theory of fibre-bundles and exterior differential forms.

The third theorem is due to W. V. D. Hodge, who defined for fields of skew symmetric covariant tensors **T** two derived tensor fields, d**T** and δ**T**, which generalize the curl and the divergence of vector fields. For the skew symmetric tensor **T** with components $T_{i_1...i_r}$ these derived tensors have components

$$T_{i_1...i_r,i_{r+1}} \epsilon^{i_1...i_{r+1} i_{r+2}...i_n}$$

and

$$T_{i_1...i_r,i_{r+1}} g^{i_r i_{r+1}}$$

respectively.

The tensor **T** is said to be *harmonic* if $d\mathbf{T} = 0$ and $\delta\mathbf{T} = 0$. Then Hodge's main result is the following:

THEOREM 20.3. *The r-th Betti number of a compact orientable manifold M with a positive definite Riemannian metric is equal to the number of linearly independent harmonic tensors of order r defined globally over M.*

This theorem leads to a method of computing the Betti numbers of M† (topological invariants) from considerations of the local differential geometric structure supported by M. The proof originally given by Hodge about 1940 was later simplified by Bidal and de Rham. It should be noted that the harmonic tensors of Theorem 20.3 were defined relative to a positive definite metric. Whether the theorem can be adapted to deal with indefinite metrics remains an unsolved problem.

REFERENCES

CARTAN, É., *Ann. Soc. Pol. Math.* **6** (1927) 1–7; *Œuvres Complètes*, vol. iii, pp. 1091–7.

—— *Leçons sur la géométrie des espaces de Riemann*, Gauthier–Villars (1928; 2nd edition 1946).

CHEVALLEY, C., *Theory of Lie Groups*, vol. i, Princeton University Press (1946).

EISENHART, L. P., *Riemannian Geometry*, Princeton University Press (1926; 2nd edition 1950).

WALKER, A. G., *Quart. J. Math.* (2) **1** (1950) 69–79.

WYLIE, S., and HILTON, P., *Introduction to Algebraic Topology*, Cambridge University Press (1960).

BIBLIOGRAPHY FOR CHAPTER VII
Harmonic Spaces

ALLAMIGEON, A. C., 'Espaces harmoniques décomposables', *Comptes Rendus*, **245** (1957) 1498–1500.

COPSON, E. T., and RUSE, H. S., 'Harmonic Riemannian spaces', *Proc. Roy. Soc. Edin.* **60** (1939–40) 117–33.

LEDGER, A. J., 'Harmonic homogeneous spaces of Lie groups', *J. Lond. Math. Soc.* **29** (1954) 345–7.

—— 'Symmetric harmonic spaces' ibid. **32** (1957) 53–56.

LICHNEROWICZ, A., 'Sur les espaces riemanniens complètement harmoniques', *Bull. Soc. math. Fr.* **72** (1944) 146–68.

—— 'Equations de Laplace et espaces harmoniques', *Premier colloque sur les équations aux dérivées partielles, Louvain 1953*, pp. 9–23; Georges Thone, Liège; Masson & Cie, Paris (1954).

† See, for example, S. Wylie and P. Hilton (1959).

LICHNEROWICZ, A., and WALKER, A. G., 'Sur les espaces riemanniens harmoniques de type hyperbolique normal', *Comptes Rendus*, **221** (1945) 394–6.

PATTERSON, E. M., 'An existence theorem on simply harmonic spaces', *J. Lond. Math. Soc.* **26** (1951) 238–40.

RUSE, H. S., 'On the "elementary" solution of Laplace's equation', *Proc. Edin. Math. Soc.* **2** (1930–1) 135–9.

—— 'The Riemann tensor in a completely harmonic V_4', *Proc. Roy. Soc. Edin.* A, **62** (1944–5) 156–63.

—— 'On simply harmonic spaces', *J. Lond. Math. Soc.* **21** (1946) 243–7.

—— 'On simply harmonic "kappa spaces" of four dimensions', *Proc. Lond. Math. Soc.* **50** (1949) 317–29.

—— 'Simply harmonic affine spaces of symmetric connection', *Publ. Math. Debrecen*, **2** (1952) 169–74.

WALKER, A. G., 'Note on a distance invariant and the calculation of Ruse's invariant', *Proc. Edin. Math. Soc.* **7** (1942) 16–26.

—— 'On completely harmonic spaces, *J. Lond. Math. Soc.* **20** (1945) 159–63.

—— 'A particular harmonic Riemannian space', ibid. 93–99.

—— 'Symmetric harmonic spaces', ibid. **21** (1946) 47–57.

—— 'On Lichnerowicz's conjecture for harmonic 4-spaces', ibid. **24** (1948–9) 21–28.

WILLMORE, T. J., 'Mean-value theorems in harmonic Riemannian spaces', ibid. **25** (1950) 54–57.

—— 'Quelques propriétés locales et globales des espaces riemanniens harmoniques', *Coll. internat. du C.N.R.S., Strasbourg* (1953), pp. 89–95.

—— 'Some properties of harmonic Riemannian manifolds', *Convegno di Geometria Differenziale, Venezia* (1953), pp. 141–7.

Recurrent Spaces

MATSUMOTO, M., 'Riemann spaces of recurrent and separated curvature and their imbedding', *Mem. Coll. Sci. Univ. Kyoto. Ser. A. Math.* **27** (1952) 175–88.

MOGI, I., 'A remark on recurrent curvature spaces', *Kodai Math. Seminar Rep.* (1950) 73–74.

PATTERSON, E. M., 'On symmetric recurrent tensors of the second order', *Quart. J. Math.* (2) **2** (1951) 151–8.

—— 'Some theorems on Ricci-recurrent spaces', *J. Lond. Math. Soc.* **27** (1952) 287–95.

RUSE, H. S., 'The Riemann complex in a four-dimensional space of recurrent curvature', *Proc. Lond. Math. Soc.* (2) **53** (1951) 13–31.

—— 'A classification of K^*-spaces', ibid., pp. 212–29.

WALKER, A. G., 'On Ruse's spaces of recurrent curvature', ibid. (2) **52** (1950) 36–54.

WONG, Y. C., 'A class of non-Riemannian K^*-spaces', ibid. (3) **3** (1953) 118–28.

Parallel Distributions

LEVINE, J., 'Fields of parallel vectors in conformally flat spaces', *Duke Math. J.* **17** (1950) 15–20.

DE RHAM, G., 'Sur la réducibilité d'un espace de Riemann', *Comment. Math. Helvet.* **26** (1952), 328–44.

RUSE, H. S., 'On parallel fields of planes in a Riemannian space', *Quart. J. Math.* **20** (1949), 215–34.

—— 'Parallel planes in a Riemannian V_n', *Proc. Roy. Soc. Edin.* **63** (1950) 78–92.

WALKER, A. G., 'On parallel fields of partially null vector spaces', *Quart. J. Math.* **20** (1949) 135–45.

—— 'A canonical form for a Riemannian space with a parallel field of null planes', ibid. (2) **1** (1950) 69–79.

—— 'Canonical forms II: Parallel partially null planes', ibid., pp. 147–52.

—— 'Sur la fibration des variétés riemanniennes', *C. R. Acad. Sci. Paris*, **232** (1951) 1465–7.

—— 'The fibring of Riemannian manifolds', *Proc. Lond. Math. Soc.* (3) **3** (1953) 1–19.

—— 'Connexions for parallel distributions in the large', I, *Quart. J. Math.* (2) **6** (1955) 301–8: II, ibid. (2) **9** (1958) 221–31.

WILLMORE, T. J., 'Les plans parallèles dans les espaces riemanniens globaux', *C. R. Acad. Sci. Paris*, **232** (1951) 298–9.

—— 'Parallel distributions on manifolds', *Proc. Lond. Math. Soc.* (3) **6** (1956) 191–204.

—— 'Connexions for systems of parallel distributions', *Quart. J. Math.* (2) **7** (1956) 269–76.

—— 'Systems of parallel distributions', *J. Lond. Math. Soc.* **32** (1957) 153–6.

Riemann Extensions

PATTERSON, E. M., and WALKER, A. G., 'Riemann extensions', *Quart. J. Math.* (2) **2** (1952) 19–28.

—— 'Simply harmonic Riemann extensions', *J. Lond. Math. Soc.* **27** (1952) 102–7.

WALKER, A. G., 'Riemann extensions of non-Riemannian spaces', *Convegno di Geometria Differenziale, Venezia* (1953), pp. 64–70.

MISCELLANEOUS EXERCISES VII

1. A Riemannian space is *recurrent* if $R_{hijk,l} = \kappa_l R_{hijk}$ where (κ_l) is a non-zero vector.

Prove that $R_{hijk}\kappa_l + R_{hikl}\kappa_j + R_{hilj}\kappa_k = 0$.

[HINT: Use Bianchi's identity.]

2. A Riemannian space V_n is *decomposable* if it can be expressed as a product $V_r \times V_{n-r}$ for some r, i.e. if coordinates can be found so that its metric takes the form

$$ds^2 = \sum_{\alpha,\beta=1}^{r} g_{\alpha\beta} dx^\alpha dx^\beta + \sum_{\lambda,\mu=r+1}^{n} g_{\lambda\mu} dx^\lambda dx^\mu$$

where the components $g_{\alpha\beta}$ are functions of $x^1,\ x^2,...,x^r$ only, and the components $g_{\lambda\mu}$ are functions of $x^{r+1},\ x^{r+2},...,x^n$ only.

Prove that the Christoffel symbols and the components of the curvature tensor and its covariant derivatives in V_n are zero unless all suffixes belong to the same range $1,...,r$ or $r+1,...,n$. Prove also that when all suffixes belong to the range $1,...,r$, then the symbols and tensor components are the same for V_r as for V_n, and covariant differentiation with respect to $x^1,...,x^r$ is the same in V_r as in V_n.

3. If a recurrent Riemannian space is decomposable show that one of the decomposition spaces is flat and the other is a recurrent space.

4. Prove that every V_n which admits $(n-1)$ independent parallel vector fields is flat.

5. Prove that every V_2 of non-constant Gaussian curvature is recurrent.

6. Prove that the V_3 with metric

$$ds^2 = \psi\,dx^2 + 2dxdy + dz^2,$$

where ψ is any function of x and z, is recurrent.

7. Prove that the V_4 with metric

$$ds^2 = \psi\,dx^2 + 2dxdy + 2dzdt,$$

where ψ is any function of x and z, is recurrent.

8. Prove that the components of the curvature tensor of a Riemannian space satisfy the identity

$$R_{hijk,lm} - R_{hijk,ml} + R_{jklm,hi} - R_{jklm,ih} + R_{lmhi,jk} - R_{lmhi,kj} = 0.$$

[HINT: Use the Ricci identity.]

9. Show that the recurrence vector (κ_l) of a recurrent space is a gradient.

10. If a non-singular symmetric covariant tensor with components a_{ij} has the property that $a_{ij,k} = 0$, where the comma denotes covariant differentiation with respect to the Christoffel symbols $\begin{Bmatrix} i \\ j\,k \end{Bmatrix}$ of a Riemannian metric g_{ij}, prove that the Christoffel symbols calculated from the tensors a_{ij} and g_{ij} are the same.

11. An n-dimensional Riemannian space with coordinates (x^i) admits a parallel vector field (λ^i). Show that the curves which satisfy the equations

$$\frac{dx^1}{\lambda^1} = \frac{dx^2}{\lambda^2} = ... = \frac{dx^n}{\lambda^n}$$

are geodesics, and show that these geodesics are orthogonal to a system of hypersurfaces.

12. In an n-dimensional Riemannian space there is a coordinate system (x^i) such that every geodesic satisfies the equations $d^2x^i/dt^2 = 0$ for some parameter t. Show that the Christoffel symbols are given by

$$\begin{Bmatrix} i \\ j\,k \end{Bmatrix} = \delta^i{}_j\phi_k + \delta^i{}_k\phi_j$$

for some functions ϕ_i, and hence or otherwise show that the space has constant curvature.

13. An n-dimensional Riemannian space $(n > 2)$ has its curvature tensor expressible in the form

$$R_{hijk} = g_{hk}T_{ij} + g_{ij}T_{hk} - g_{hj}T_{ik} - g_{ik}T_{hj}.$$

Show that $T_{ij} = T_{ji}$ and express T_{ij} in terms of g_{ij}, R_{ij}, and R. Prove also that, if $n > 3$,

$$2(n-1)(R_{ij,k} - R_{ik,j}) = g_{ij}R_{,k} - g_{ik}R_{,j}.$$

14. Two disjoint complementary distributions D, $\bar{D}$ are defined over a neighbourhood U of a Riemannian manifold. A system of orthonormal frames is chosen over U such that, at each point P, the basis vectors $\mathbf{e}_\alpha$ span the plane of the distribution D at P while the basis vectors $\mathbf{e}_{\alpha''}$ span the plane of $\bar{D}$ ($\alpha' = 1, 2, ..., r$; $\alpha'' = r+1, r+2, ..., n$). If $(\boldsymbol{\omega}_{\alpha'}, \boldsymbol{\omega}_{\alpha''})$ is the corresponding system of co-frames, show that the planes of D are given by the system of equations $\boldsymbol{\omega}_{\alpha''} = 0$ while the planes of $\bar{D}$ are given by $\boldsymbol{\omega}_{\alpha'} = 0$.

If the Riemannian connexion is given relative to these frames by the matrix of 1-forms (ω_{ij}), prove that a necessary and sufficient condition for D to be parallel is $\omega_{\alpha''\beta'} = 0$. Deduce that when D is parallel so is $\bar{D}$, and that both D and $\bar{D}$ are then necessarily integrable.

APPLICATIONS OF TENSOR METHODS TO SURFACE THEORY

1. The Serret-Frenet formulae

IN this chapter the calculus of tensors and exterior forms will be applied to obtain a number of properties of curves and surfaces which have already been obtained in Part 1 by vector calculus. The new methods are much more powerful than vector methods, but their power can only really be appreciated when they are applied to more general problems not treated in this book, such as the study of an n-dimensional manifold embedded in an m-dimensional Riemannian manifold. Similarly, the theory of harmonic integrals on Riemannian manifolds makes extensive use of the theory of exterior forms, but again this theory is outside our scope. However, it is hoped that by applying tensor methods to familiar problems of surface theory the reader will gain proficiency in their use and hence be able to use them in more general problems which would be intractable by methods used in Part 1.

It was possible to use classical vector calculus in Part 1 because curves and surfaces were defined as embedded in three-dimensional Euclidean space E_3, and a system of rectangular coordinates was available for the whole space. However, it is often advantageous to use other systems of coordinates, e.g. spherical polars, ellipsoidal coordinates, or general curvilinear coordinates, and in these circumstances tensor methods are more suitable.

Let (y^i) be a set of rectangular coordinates in E_3 in terms of which the metric assumes the form

$$ds^2 = \sum_{i=1}^{3} (dy^i)^2. \tag{1.1}$$

Let (x^i) be a set of general curvilinear coordinates, valid over some region R of E_3, which are related to (y^i) by equations

$$y^i = f^i(x^1, x^2, x^3). \tag{1.2}$$

These equations may be interpreted as giving a mapping ϕ of a region R of E_3 onto itself—in fact the identity map. The dual map $\delta\phi$ will map the covariant metric tensor given by (1.1) onto the covariant metric whose components

$$g_{jk} = \sum_{i=1}^{3} \left(\frac{\partial y^i}{\partial x^j}\right)\left(\frac{\partial y^i}{\partial x^k}\right) \tag{1.3}$$

are no longer constants.

In terms of *general curvilinear coordinates* (x^i), a curve γ of class r is given by parametric equations

$$x^i = f^i(u), \tag{1.4}$$

where $a \leqslant u \leqslant b$, the functions $f^i(u)$ are of class r in this interval, and the derivatives $\dot{f}^i(u)$ do not vanish simultaneously at any point in the interval. The arc length of γ between points of parameters a, b is

$$s = \int_a^b (g_{ij}\dot{x}^i\dot{x}^j)^{\frac{1}{2}} \, du, \tag{1.5}$$

while the *unit* tangent vector **t** to γ has components

$$t^i = dx^i/ds. \tag{1.6}$$

Serret–Frenet formulae in curvilinear coordinates

At all points of γ, since **t** is a unit vector,

$$g_{ij}t^it^j = 1, \tag{1.7}$$

and intrinsic differentiation along γ gives

$$g_{ij}\frac{Dt^i}{ds}t^j = 0.$$

It follows that Dt^i/ds are components of a vector normal to the curve. Writing n^i for the components of the unit vector **n** in this direction we get

$$Dt^i/ds = \kappa n^i, \tag{1.8}$$

where κ is the curvature and **n** the principal normal.

Let **b** be the unit binormal, directed so that the set **t**, **n**, **b** is right-handed. Then, since $g_{ij}b^ib^j = 1$, it follows that

$$g_{ij}b^iDb^j/ds = 0,$$

and hence

$$Db^j/ds = \alpha t^j + \beta n^j \tag{1.9}$$

for some functions α and β. Differentiation of the relation

$$g_{ij}t^ib^j = 0$$

leads to

$$g_{ij}\frac{Dt^i}{ds}b^j + g_{ij}t^i\frac{Db^j}{ds} = 0,$$

which from (1.8) reduces to

$$g_{ij} t^i \frac{Db^j}{ds} = 0.$$

Hence $\alpha = 0$, and writing $\beta = -\tau$ equation (1.9) becomes

$$\frac{Db^i}{ds} = -\tau n^i \qquad (1.10)$$

when τ is the torsion.

Finally, since **n** is a unit vector, Dn/ds lies in the plane of **t** and **b**, and hence

$$\frac{Dn^i}{ds} = \gamma t^i + \delta b^i \qquad (1.11)$$

for some functions γ and δ.

Differentiate the relation $g_{ij} b^i n^j = 0$ and use (1.10) to get

$$-\tau + g_{ij} b^i \frac{Dn^j}{ds} = 0;$$

this with (1.11) gives $\delta = \tau$. Differentiate the relation $g_{ij} t^i n^j = 0$ and use (1.8) to get

$$\kappa + g_{ij} t^i \frac{Dn^j}{ds} = 0;$$

this with (1.11) gives $\gamma = -\kappa$. Thus equation (1.11) reduces to

$$\frac{Dn^i}{ds} = \tau b^i - \kappa t^i. \qquad (1.12)$$

Equations (1.8), (1.10), (1.12) can be written in the form

$$\left. \begin{aligned} \frac{D\mathbf{t}}{ds} &= \kappa \mathbf{n} \\ \frac{D\mathbf{n}}{ds} &= \tau \mathbf{b} - \kappa \mathbf{t} \\ \frac{D\mathbf{b}}{ds} &= -\tau \mathbf{n} \end{aligned} \right\}. \qquad (1.13)$$

It will be seen that these are precisely the same as the Frenet formulae derived in Chapter I, except that ordinary differentiation has been replaced by intrinsic differentiation. When the metric is Euclidean the connexion coefficients are all zero, and intrinsic differentiation reduces to ordinary differentiation. However, in our derivation of the Frenet formulae no use has been made of the assumption that the metric (1.3) is induced from a Euclidean metric. It follows that the Frenet formulae for a curve embedded in a three-dimensional Riemannian space are also given by (1.13), where intrinsic differentiation is taken with respect to the given metric.

2. The induced metric

In this chapter a surface S will be regarded as a two-dimensional submanifold of three-dimensional Euclidean space E_3. The inclusion map $\phi : S \to E_3$

$$y^i = \phi^i(u^1, u^2) \quad (i = 1, 2, 3), \tag{2.1}$$

which relates the parameters (u^1, u^2) of a point on S to the rectangular Cartesian coordinates (y^i) of P referred to a suitable set of axes in E_3, is precisely the parametric form of the equation of S. The differential map $d\phi$ maps a tangent vector to S at P into the same tangent vector at P but this time considered as a tangent vector to E_3 at P. It will be convenient to call tangent vectors to S *surface vectors* to distinguish them from general tangent vectors which will be called *space vectors*. This terminology is justified because the space of tangent vectors to E_3 at P can evidently be identified with E_3 itself.

It will be convenient to use Roman suffixes $i, j, k, \dots$ for the range 1, 2, 3, and Greek suffixes $\alpha, \beta, \gamma, \dots$ for the range 1, 2. Using this notation, a surface vector (λ^α) is mapped by $d\phi$ into the tangent vector with components λ^i, where

$$\lambda^i = \frac{\partial y^i}{\partial u^\alpha} \lambda^\alpha. \tag{2.2}$$

It is convenient to write

$$y^i{}_\alpha = \frac{\partial y^i}{\partial u^\alpha}, \tag{2.3}$$

so that (2.2) becomes $\quad \lambda^i = y^i{}_\alpha \lambda^\alpha. \tag{2.4}$

The dual map $\delta\phi$ maps the covariant (Euclidean) metric tensor of E_3 at P into the induced metric tensor at P which is now regarded as a point of S. In terms of coordinates, the components $a_{\alpha\beta}$ of the induced metric tensor **a** are given by

$$a_{\alpha\beta} = \sum_{i=1}^{3} y^i{}_\alpha y^i{}_\beta. \tag{2.5}$$

If E_3 is referred to general curvilinear coordinates (x^i) with corresponding metric tensor (g_{ij}), then the metric induced on S by $\delta\phi$ is given by

$$a_{\alpha\beta} = g_{ij}x^i{}_\alpha x^j{}_\beta. \tag{2.6}$$

The surface S is thus a two-dimensional Riemannian manifold carrying an induced metric given by (2.5) or, more generally, by (2.6).

The surface vector (λ^α) has space components λ^i given by

$$\lambda^i = x^i{}_\alpha \lambda^\alpha. \tag{2.7}$$

Let (μ^β) be a second surface vector having space components μ^j so that

$$\mu^j = x^j{}_\beta \mu^\beta.$$

Then

$$g_{ij}\lambda^i\mu^j = g_{ij}x^i{}_\alpha x^j{}_\beta \lambda^\alpha\mu^\beta \qquad \text{(from (2.7))}$$

$$= a_{\alpha\beta}\lambda^\alpha\mu^\beta \qquad \text{(from (2.6))}.$$

It follows that the length of a surface vector (λ^α) with respect to the surface metric is the same as the length of the vector with spatial components (λ^i) with respect to the spatial metric. Also, if θ is the angle between the unit surface vectors (λ^α), (μ^β); then

$$\cos\theta = g_{ij}\lambda^i\mu^j = a_{\alpha\beta}\lambda^\alpha\mu^\beta. \tag{2.8}$$

Let (t^α) be a unit vector, and let (t_α) be the associated vector with respect to the metric tensor $(a_{\alpha\beta})$. Consider the vector (n^α) defined by

$$n^\alpha = \epsilon^{\beta\alpha}t_\beta.$$

Then

$$a_{\alpha\beta}n^\alpha n^\beta = a_{\alpha\beta}\epsilon^{\gamma\alpha}t_\gamma \epsilon^{\delta\beta}t_\delta$$

$$= \frac{a_{\alpha\beta}}{a}t_\gamma t_\delta e^{\gamma\alpha}e^{\delta\beta}$$

$$= a^{\gamma\delta}t_\gamma t_\delta = 1.$$

Thus (n^α) is a unit vector. Also

$$a_{\alpha\beta}n^\alpha t^\beta = a_{\alpha\beta}\epsilon^{\gamma\alpha}t_\gamma t^\beta = \epsilon^{\gamma\alpha}t_\gamma t_\alpha = 0.$$

Hence $(\epsilon^{\beta\alpha}t_\beta)$ is a *unit vector orthogonal to* (t^α).

The ϵ-tensor may be used to define an orientation associated with two vectors (λ^α), (μ^β) in the following manner. The rotation from (λ^α) to (μ^β) is defined to be *positive* if $\epsilon_{\alpha\beta}\lambda^\alpha\mu^\beta$ is positive.

The intrinsic properties of S may be obtained immediately from the previous chapter by regarding S as a two-dimensional Riemannian manifold with metric given by (2.6). In particular it follows that the differential equations of the geodesics are

$$\frac{d^2u^\alpha}{ds^2} + \left\{ \begin{matrix} \alpha \\ \beta\,\gamma \end{matrix} \right\} \frac{du^\beta}{ds}\frac{du^\gamma}{ds} = 0. \tag{2.9}$$

Let γ be a curve on S and denote by (t^α) the unit tangent vector to γ at P. Since

$$a_{\alpha\beta}\, t^\alpha t^\beta = 1,$$

it follows by taking the intrinsic derivative along γ that

$$a_{\alpha\beta}\frac{Dt^\alpha}{ds}\, t^\beta = 0.$$

Thus (Dt^α/ds) is orthogonal to (t^α) and we may write

$$\frac{Dt^\alpha}{ds} = \kappa_g\, n_g^\alpha, \tag{2.10}$$

where (n_g^α) is a unit vector orthogonal to the tangent chosen in a direction so that the rotation (t^α, n_g^α) is positive. It follows from Chapter VII, section 7, that the geodesic curvature vector has components $\kappa_g\, n_g^\alpha$ and κ_g is the geodesic curvature of γ at P.

Since $a_{\alpha\beta}\, t^\alpha n_g^\beta = 0$, it follows by intrinsic differentiation that

$$a_{\alpha\beta}\frac{Dt^\alpha}{ds}\, n_g^\beta + a_{\alpha\beta}\, t^\alpha\frac{Dn_g^\beta}{ds} = 0.$$

Using (2.10) we have

$$a_{\alpha\beta}\, t^\alpha\frac{Dn_g^\beta}{ds} = -\kappa_g. \tag{2.11}$$

Also, since (n_g^β) is a unit vector, we may write

$$\frac{Dn_g^\beta}{ds} = \lambda t^\beta$$

for some function λ. We multiply by $a_{\alpha\beta}\, t^\alpha$ and use (2.11) to get $\lambda = -\kappa_g$, and so

$$\frac{Dn_g^\alpha}{ds} = -\kappa_g\, t^\alpha. \tag{2.12}$$

Equations (2.10), (2.12) are sometimes called *the surface form of the Serret–Frenet equations of the curve*.

The definition of geodesic curvature given in this section shows that it is an intrinsic property, involving only the curve and the first fundamental form. We shall see shortly that this is compatible with the definition given in Chapter II.

EXERCISE 2.1. If G is the geodesic touching γ at P, and if Q, R are points on G, γ respectively at equal arc distance s from P, prove that as $s \to 0$ the vector $\overrightarrow{QR}$ becomes parallel to $\mathbf{n}_g$ and that $\lim_{s\to 0} 2QR/s^2 = \kappa_g$.

EXERCISE 2.2. Prove that a necessary and sufficient condition that a curve be a geodesic is that its geodesic curvature shall be zero.

EXERCISE 2.3. If γ is defined by the equation $\phi(u^1, u^2) = 0$, prove that the geodesic curvature is

$$\kappa_g = -\frac{\Delta_2 \phi}{\sqrt{(\Delta_1 \phi)}} - \Delta_1\left(\phi, \frac{1}{\sqrt{(\Delta_1 \phi)}}\right). \qquad (2.13)$$

(Equation (2.13) is known as *Beltrami's formula* for geodesic curvature. Since the equation is invariant, it is sufficient to verify that it is satisfied in some special coordinate system, for example taking as parametric curves the system $\phi =$ constant and the orthogonal trajectories.)

3. The fundamental formulae of surface theory

In this section we make use of the theory of connexions over sub-manifolds developed in Chapter VI, section 6, in the particular case when M' is a two-dimensional surface S, M is E_3, L' is the Riemannian connexion associated with $(a_{\alpha\beta})$, and L is the Riemannian connexion associated with (g_{ij}).

Take the covariant derivative of the double tensor $(x^i{}_\alpha)$ to get

$$x^i{}_{\alpha,\beta} = \frac{\partial^2 x^i}{\partial u^\alpha \partial u^\beta} + \begin{Bmatrix} i \\ j\,k \end{Bmatrix}_g x^j{}_\alpha x^k{}_\beta - \begin{Bmatrix} \delta \\ \alpha\beta \end{Bmatrix}_a x^i{}_\delta, \qquad (3.1)$$

where $\begin{Bmatrix} i \\ j\,k \end{Bmatrix}_g$ denotes the Christoffel symbols formed from (g_{ij}), and $\begin{Bmatrix} \delta \\ \alpha\beta \end{Bmatrix}_a$ denotes the symbols formed from $(a_{\alpha\beta})$. Evidently the tensor components $x^i{}_{\alpha,\beta}$ are symmetric in the suffixes α and β.

Take the covariant derivative of the relation

$$g_{ij} x^i{}_\alpha x^j{}_\beta = a_{\alpha\beta}$$

to get $\qquad g_{ij} x^i{}_{\alpha,\gamma} x^j{}_\beta + g_{ij} x^i{}_\alpha x^j{}_{\beta,\gamma} = a_{\alpha\beta,\gamma} = 0. \qquad (3.2)$

Cyclic interchange of α, β, and γ gives

$$g_{ij} x^i{}_{\beta,\alpha} x^j{}_\gamma + g_{ij} x^i{}_\beta x^j{}_{\gamma,\alpha} = 0, \qquad (3.2')$$

and again $\qquad g_{ij} x^i{}_{\gamma,\beta} x^j{}_\alpha + g_{ij} x^i{}_\gamma x^j{}_{\alpha,\beta} = 0. \qquad (3.2'')$

Add (3.2) to (3.2′) and subtract (3.2″) to get

$$g_{ij} x^i{}_{\alpha,\gamma} x^j{}_\beta = 0, \qquad (3.3)$$

after using the symmetry of $x^i{}_{\alpha,\beta}$.

Now the double tensor $(x^i{}_\alpha)$ can be interpreted as giving the components with respect to E_3 of the tangents to the parametric curves in S. For convenience $(x^i{}_\alpha)$ will be called a *surface vector* since for fixed α it is tangent to S. Equation (3.3) implies that (for fixed α, γ) the vector $(x^i{}_{\alpha,\gamma})$ is orthogonal to the surface S, i.e. parallel to the unit surface normal $\mathbf{N}$ with components N^i. Thus we may write

$$x^i{}_{\alpha,\beta} = \Omega_{\alpha\beta} N^i, \tag{3.4}$$

where $\Omega_{\alpha\beta}$ are components of some symmetric covariant tensor. Multiply (3.4) by $g_{ij} N^j$ to get

$$\Omega_{\alpha\beta} = g_{ij} x^i{}_{\alpha,\beta} N^j. \tag{3.5}$$

Comparison with III (1.3) shows that (3.5) is the tensor generalization of the formula for the coefficients of the second fundamental form of the surface. In fact, when (x^i) are rectangular Cartesian coordinates, $\Omega_{11} = L$, $\Omega_{12} = \Omega_{21} = M$, and $\Omega_{22} = N$. Moreover, equation (3.4) is the tensor generalization of III (9.1) and is thus *the tensor form of the Gauss equations*.

It has been seen that a vector field (λ^α) defined along a curve γ in S is mapped under the differential of the inclusion mapping ϕ into the vector field (λ^i). The question arises whether the intrinsic derivative $(D\lambda^\alpha/dt)$ along γ will map into the intrinsic derivative $(D\lambda^i/dt)$.

Since $\lambda^i = x^i{}_\alpha \lambda^\alpha$, we have

$$\frac{D\lambda^i}{dt} = x^i{}_{\alpha,\beta} \frac{du^\beta}{dt} \lambda^\alpha + x^i{}_\alpha \frac{D\lambda^\alpha}{dt}.$$

Using (3.4), this becomes

$$\frac{D\lambda^i}{dt} - x^i{}_\alpha \frac{D\lambda^\alpha}{dt} = \Omega_{\alpha\beta} N^i \frac{du^\beta}{dt} \lambda^\alpha,$$

i.e.

$$\frac{D}{dt}(x^i{}_\alpha \lambda^\alpha) - x^i{}_\alpha \frac{D\lambda^\alpha}{dt} = \Omega_{\alpha\beta} N^i \frac{du^\beta}{dt} \lambda^\alpha.$$

It follows that the two operators D/dt and $d\phi$ do not commute, and hence the vector with components $D\lambda^\alpha/dt$ does not map into the vector with components $D\lambda^i/dt$.

We next obtain the tensor form of Weingarten's formula for the derivatives of the normal vector $\mathbf{N}$. Since (N^i) is a unit vector it follows that

$$g_{ij} N^i N^j{}_{,\alpha} = 0$$

and hence $N^j{}_{,\alpha}$, being orthogonal to (N^i), must be expressible in the form

$$N^j{}_{,\alpha} = b^\gamma{}_\alpha x^j{}_\gamma \tag{3.6}$$

for some matrix $b^\gamma{}_\alpha$. By differentiating the relation $g_{ij} x^i{}_\alpha N^j = 0$ we get

$$g_{ij} x^i{}_{\alpha,\beta} N^j + g_{ij} x^i{}_\alpha N^j{}_{,\beta} = 0. \tag{3.7}$$

Substitute from (3.6) in (3.7) to get

$$g_{ij}x^i_{\alpha,\beta}N^j + g_{ij}x^i_\alpha b^\gamma_\beta x^j_\gamma = 0,$$

i.e.

$$\Omega_{\alpha\beta} + a_{\alpha\gamma}b^\gamma_\beta = 0. \tag{3.8}$$

Multiply by $a^{\alpha\epsilon}$ to get

$$b^\epsilon_\beta = -a^{\alpha\epsilon}\Omega_{\alpha\beta},$$

so substitution in (3.6) gives the required *Weingarten's formula*

$$N^i_{,\alpha} = -\Omega_{\alpha\beta}a^{\beta\gamma}x^i_\gamma. \tag{3.9}$$

Sometimes it is convenient to consider *the third fundamental quadratic form* associated with the surface, defined by

$$C \equiv c_{\alpha\beta}du^\alpha du^\beta, \tag{3.10}$$

where

$$c_{\alpha\beta} = g_{ij}N^i_{,\alpha}N^j_{,\beta}. \tag{3.11}$$

Using (3.9) and (3.11), we get

$$\begin{aligned} c_{\alpha\beta} &= g_{ij}\Omega_{\alpha\epsilon}a^{\epsilon\gamma}x^i_\gamma\Omega_{\beta\sigma}a^{\sigma\delta}x^j_\delta \\ &= a_{\gamma\delta}a^{\epsilon\gamma}a^{\sigma\delta}\Omega_{\alpha\epsilon}\Omega_{\beta\sigma} \quad \text{(using (2.6))} \\ &= \Omega_{\alpha\epsilon}a^{\epsilon\sigma}\Omega_{\sigma\beta}, \end{aligned} \tag{3.12}$$

which relates the coefficients of the three forms.

We now derive the tensor form of the equation of Gauss and the Mainardi–Codazzi relations. Differentiate (3.1) covariantly with respect to u^γ, interchange β and γ and subtract. After simplifying we get

$$x^i_{\alpha,\beta\gamma} - x^i_{\alpha,\gamma\beta} = R^\delta_{\alpha\beta\gamma}x^i_\delta. \tag{3.13}$$

But since $x^i_{\alpha,\beta} = \Omega_{\alpha\beta}N^i$, we have on differentiation

$$\begin{aligned} x^i_{\alpha,\beta\gamma} &= \Omega_{\alpha\beta,\gamma}N^i + \Omega_{\alpha\beta}N^i_{,\gamma} \\ &= \Omega_{\alpha\beta,\gamma}N^i - \Omega_{\alpha\beta}\Omega_{\gamma\delta}a^{\delta\epsilon}x^i_\epsilon \quad \text{(using (3.9))}. \end{aligned}$$

Substitute in (3.13) to get

$$x^i_\epsilon R^\epsilon_{\alpha\beta\gamma} = (\Omega_{\alpha\beta,\gamma} - \Omega_{\alpha\gamma,\beta})N^i - a^{\delta\epsilon}x^i_\epsilon(\Omega_{\alpha\beta}\Omega_{\gamma\delta} - \Omega_{\alpha\gamma}\Omega_{\beta\delta}). \tag{3.14}$$

Multiply both sides by $g_{ij}N^j$ to get

$$\Omega_{\alpha\beta,\gamma} = \Omega_{\alpha\gamma,\beta}. \tag{3.15}$$

These are the Mainardi–Codazzi equations in tensor form, and they reduce to III (9.19), previously obtained by vector calculus, when (x^i) are rectangular Cartesian coordinates.

Multiply (3.14) by $g_{ij}x^j_\delta$ to obtain

$$R_{\alpha\epsilon\beta\gamma} = (\Omega_{\alpha\beta}\Omega_{\gamma\epsilon} - \Omega_{\alpha\gamma}\Omega_{\beta\epsilon}), \tag{3.16}$$

which is the tensor form of the equation of Gauss, previously obtained in III (9.20).

From (3.16), the Gaussian curvature K satisfies

$$K\epsilon_{\rho\alpha}\epsilon_{\beta\gamma} = \Omega_{\rho\beta}\Omega_{\alpha\gamma} - \Omega_{\rho\gamma}\Omega_{\alpha\beta}. \tag{3.17}$$

Multiply by $a^{\rho\gamma}$ and use the relation

$$a_{\alpha\beta} = -a^{\rho\gamma}\epsilon_{\rho\alpha}\epsilon_{\beta\gamma} \tag{3.18}$$

to obtain $\qquad -Ka_{\alpha\beta} = a^{\rho\gamma}\Omega_{\rho\beta}\Omega_{\gamma\alpha} - a^{\rho\gamma}\Omega_{\rho\gamma}\Omega_{\alpha\beta}. \tag{3.19}$

Write $\qquad\qquad a^{\rho\gamma}\Omega_{\rho\gamma} = 2\mu, \tag{3.20}$

and substitute from (3.12) in (3.19) to get

$$c_{\alpha\beta} - 2\mu\Omega_{\alpha\beta} + Ka_{\alpha\beta} = 0, \tag{3.21}$$

giving another relation between the coefficients of the three forms.

4. Normal curvature and geodesic torsion

Let γ be a curve on the surface S, and let (t^α) be the tangent vector to γ at P. Then the corresponding space vector (t^i) satisfies

$$t^i = x^i{}_\alpha t^\alpha.$$

Differentiate intrinsically with respect to the arc length of γ to get

$$\frac{Dt^i}{ds} = x^i{}_\alpha\frac{Dt^\alpha}{ds} + t^\alpha x^i{}_{\alpha,\beta}\frac{du^\beta}{ds}. \tag{4.1}$$

Using (1.8), (2.10), (3.4), this becomes

$$\kappa n^i = x^i{}_\alpha\kappa_g n_g^\alpha + \Omega_{\alpha\beta}t^\alpha t^\beta N^i. \tag{4.2}$$

The curvature vector κn^i is decomposed into a normal component $\Omega_{\alpha\beta}t^\alpha t^\beta N^i$ and a tangential component $x^i{}_\alpha\kappa_g n_g^\alpha$. According to the non-intrinsic definition given in Chapter II, the geodesic curvature is the magnitude of the vector $x^i{}_\alpha\kappa_g n_g^\alpha$. Since

$$g_{ij}(x^i{}_\alpha\kappa_g n_g^\alpha)(x^j{}_\beta\kappa_g n_g^\beta) = a_{\alpha\beta}n_g^\alpha n_g^\beta\kappa_g^2$$
$$= \kappa_g^2,$$

it follows that *the two definitions are compatible.*

If θ is the angle between the curvature vector $\kappa\mathbf{n}$ and the surface normal $\mathbf{N}$, it follows on multiplying (4.2) by $g_{ij}N^j$ that

$$\kappa\cos\theta = \Omega_{\alpha\beta}t^\alpha t^\beta. \tag{4.3}$$

The right-hand member depends only upon the direction (t^α) and represents the normal curvature κ_n in this direction. Thus we have

$$\kappa_n = \Omega_{\alpha\beta}t^\alpha t^\beta = \kappa\cos\theta, \tag{4.4}$$

which is Meusnier's theorem on normal curvature.

Alternatively we have

$$\kappa_n = \frac{\Omega_{\alpha\beta}\,du^\alpha du^\beta}{a_{\alpha\beta}\,du^\alpha du^\beta}, \tag{4.5}$$

corresponding to equation III (1.2).

In order to find the principal curvatures and principal directions, it is necessary to find the stationary values of $\kappa_n = \Omega_{\alpha\beta}\,t^\alpha t^\beta$ subject to $a_{\alpha\beta}\,t^\alpha t^\beta = 1$. Write $\kappa = \Omega_{\alpha\beta}\,t^\alpha t^\beta - \lambda(a_{\alpha\beta}\,t^\alpha t^\beta - 1)$; then at a stationary value

$$\frac{1}{2}\frac{\partial\kappa}{\partial t^\alpha} = \Omega_{\alpha\beta}\,t^\beta - \lambda a_{\alpha\beta}\,t^\beta = 0. \tag{4.6}$$

Multiply through by t^α to get

$$\Omega_{\alpha\beta}\,t^\alpha t^\beta - \lambda a_{\alpha\beta}\,t^\alpha t^\beta = 0,$$

i.e.

$$\lambda = \kappa.$$

Thus the principal curvatures satisfy

$$\Omega_{\alpha\beta}\,t^\beta - \kappa a_{\alpha\beta}\,t^\beta = 0, \tag{4.7}$$

and are roots of the determinantal equation

$$|\Omega_{\alpha\beta} - \kappa a_{\alpha\beta}| = 0. \tag{4.8}$$

Writing $\Omega = |\Omega_{\alpha\beta}|$, $a = |a_{\alpha\beta}|$, this becomes

$$\kappa^2 - \kappa a^{\alpha\beta}\Omega_{\alpha\beta} + \Omega/a = 0.$$

Using (3.20) this becomes

$$\kappa^2 - 2\mu\kappa + \Omega/a = 0. \tag{4.9}$$

Thus the product of the principal curvatures satisfies

$$\kappa_1\kappa_2 = \Omega/a = K, \tag{4.10}$$

while

$$\tfrac{1}{2}(\kappa_1 + \kappa_2) = \mu. \tag{4.11}$$

Therefore the tensor formula for the mean curvature is

$$\mu = \tfrac{1}{2}a^{\alpha\beta}\Omega_{\alpha\beta}. \tag{4.12}$$

To find the principal directions we eliminate κ from equations (4.7). We have

$$\left.\begin{array}{l}\Omega_{1\beta}\,t^\beta = \kappa a_{1\beta}\,t^\beta \\ \Omega_{2\alpha}\,t^\alpha = \kappa a_{2\alpha}\,t^\alpha\end{array}\right\}. \tag{4.13}$$

Eliminate κ to get $(a_{2\beta}\Omega_{1\alpha} - a_{1\beta}\Omega_{2\alpha})t^\alpha t^\beta = 0$, which may be written

$$a_{\alpha\gamma}\,\epsilon^{\gamma\delta}\Omega_{\delta\beta}\,t^\alpha t^\beta = 0.$$

Define the tensor **h** with components $h_{\alpha\beta}$ by the equation

$$h_{\alpha\beta} = a_{\alpha\gamma}\,\epsilon^{\gamma\delta}\Omega_{\delta\beta}. \tag{4.14}$$

Then the principal directions are given by the equation

$$h_{\alpha\beta}\,du^{\alpha}du^{\beta} = 0, \tag{4.15}$$

which is the tensor form of equation III (2.8).

If at a point P the tensors $\mathbf{a}$, $\mathbf{\Omega}$ are proportional, i.e.

$$a_{\alpha\beta} = \lambda\Omega_{\alpha\beta},$$

then $h_{\alpha\beta} = 0$ and the principal directions are indeterminate. Thus the umbilics of S occur at points where the fundamental tensors are proportional.

The asymptotic lines on the surface have directions given by

$$\Omega_{\alpha\beta}\,du^{\alpha}du^{\beta} = 0. \tag{4.16}$$

From the general relation

$$\kappa n^{i} = \kappa_{g}\,n_{g}^{\alpha}\,x^{i}{}_{\alpha}+\Omega_{\alpha\beta}\,t^{\alpha}t^{\beta}N^{i}, \tag{4.17}$$

it follows that, at all points on an asymptotic line, the geodesic curvature is equal in magnitude to the ordinary curvature. Moreover, since the principal normal is a surface vector, the binormal must be parallel to the surface normal. From the Serret–Frenet formulae (1.10) we thus have

$$\frac{DN^{i}}{ds} = \pm\tau n^{i},$$

and hence

$$\tau^{2} = g_{ij}\frac{DN^{i}}{ds}\frac{DN^{j}}{ds} = g_{ij}\,N^{i}{}_{,\alpha}\,N^{j}{}_{,\beta}t^{\alpha}t^{\beta}$$

$$= c_{\alpha\beta}\,t^{\alpha}t^{\beta}, \quad \text{from (3.11).}$$

Using the relation (3.21) we get

$$\tau = \pm\sqrt{(-K)}, \tag{4.18}$$

which is known as *Enneper's formula*.

We now obtain a formula which expresses the relation between the torsion at P of a curve γ lying on a surface and the torsion of the (unique) geodesic which touches γ at P. The torsion of this geodesic is called the *geodesic torsion* of γ at P.

Using the formulae

$$\Omega_{\alpha\beta}\,t^{\alpha}t^{\beta} = \kappa\cos\theta, \tag{4.19}$$

$$\kappa_{g} = \kappa\sin\theta, \tag{4.20}$$

and equation (4.17), we get

$$n^{i} = \sin\theta(x^{i}{}_{\alpha}\,n_{g}^{\alpha})+\cos\theta\,N^{i}. \tag{4.21}$$

We differentiate this equation intrinsically with respect to the arc length of γ, and use (1.12) to get

$$\tau b^i - \kappa t^i = \sin\theta \frac{D}{ds}(x^i{}_\alpha n^\alpha_g) + \cos\theta \frac{d\theta}{ds}(x^i{}_\alpha n^\alpha_g) + \cos\theta N^i{}_{,\alpha} t^\alpha - \sin\theta \frac{d\theta}{ds} N^i.$$

Multiplying by N_i, and remembering that $N_i b^i = \sin\theta$, we get

$$\tau\sin\theta = N_i \sin\theta x^i{}_{\alpha,\beta} t^\beta n^\alpha_g - \sin\theta \frac{d\theta}{ds}$$

$$= \sin\theta \Omega_{\alpha\beta} t^\beta n^\alpha_g - \sin\theta \frac{d\theta}{ds}.$$

Thus, if $\sin\theta \neq 0$, we get

$$\tau + \frac{d\theta}{ds} = \Omega_{\alpha\beta} t^\beta n^\alpha_g. \tag{4.22}$$

Since $n^\alpha_g = \epsilon^{\beta\alpha} t_\beta$, this becomes

$$\tau + \frac{d\theta}{ds} = \epsilon^{\gamma\delta} a_{\gamma\alpha} \Omega_{\beta\delta} t^\alpha t^\beta,$$

i.e.

$$\tau + \frac{d\theta}{ds} = h_{\alpha\beta} t^\alpha t^\beta, \tag{4.23}$$

which gives a new interpretation to the tensor **h**.

The value of $\tau + d\theta/ds$ is the same for all curves of the surface which have the same direction (t^α) at P, and in particular for the geodesic at P in this direction. Since θ is constant for this geodesic we have

$$\tau_g = \tau + \frac{d\theta}{ds} = h_{\alpha\beta} t^\alpha t^\beta, \tag{4.24}$$

which is the required formula.

Differentiate intrinsically the identity

$$g_{ij}(x^i{}_\alpha n^\alpha_g) N^j = 0$$

to obtain

$$g_{ij} N^j \frac{D}{ds}(x^i{}_\alpha n^\alpha_g) + g_{ij} x^i{}_\alpha n^\alpha_g \frac{DN^j}{ds} = 0.$$

The second term of this equation is the scalar product of the two vectors $\mathbf{n}_g$ and $D\mathbf{N}/ds$, so we have

$$\mathbf{n}_g \cdot \frac{D\mathbf{N}}{ds} = -g_{ij} N^j \left(x^i{}_{\alpha,\beta} n^\alpha_g t^\beta + x^i{}_\alpha \frac{D}{ds} n^\alpha_g \right)$$

$$= -g_{ij} N^j (N^i \Omega_{\alpha\beta} t^\beta n^\alpha_g - x^i{}_\alpha \kappa_g t^\alpha), \text{ using (3.4) and (2.12)},$$

$$= -\Omega_{\alpha\beta} t^\beta n^\alpha_g.$$

From (4.22) it follows that an alternative formula for geodesic torsion is

$$\tau_g = -\mathbf{n}_g \cdot \frac{D\mathbf{N}}{ds} = \mathbf{N} \cdot \frac{D\mathbf{n}_g}{ds}. \tag{4.25}$$

EXERCISE 4.1. Prove that the geodesic torsion of a curve on a surface is zero if and only if the curve is a line of curvature.

5. The method of moving frames

The results obtained in sections 1 to 4 could equally well have been obtained by Cartan's method of moving frames. Little is gained by applying this method to the theory of space curves, but the power of the method begins to appear when applied to surfaces embedded in three-dimensional Euclidean space.

In the case of a space curve γ the vectors $(\mathbf{t}, \mathbf{n}, \mathbf{b})$ form an ortho-normal frame at each point P. Calculations similar to those of section 1 in the osculating Euclidean space at P lead to equations (1.13) and it follows immediately (as already mentioned in section 1) that the same equations hold for a curve γ embedded in a Riemannian space R_3. It is interesting to change notation from $(\mathbf{t}, \mathbf{n}, \mathbf{b})$ to $(\mathbf{e}_1, \mathbf{e}_2, \mathbf{e}_3)$ and write equations (1.13) in the form

$$d\mathbf{P} = ds\,\mathbf{e}_1 \tag{5.1}$$

$$\left.\begin{array}{l} D\mathbf{e}_1 = \kappa\,ds\,\mathbf{e}_2 \\ D\mathbf{e}_2 = -\kappa\,ds\,\mathbf{e}_1 + \tau\,ds\,\mathbf{e}_3 \\ D\mathbf{e}_3 = -\tau\,ds\,\mathbf{e}_2 \end{array}\right\}. \tag{5.2}$$

These are a particular case of the equations of structure

$$d\mathbf{P} = \sum \omega_j\,\mathbf{e}_j, \tag{5.3}$$

$$D\mathbf{e}_i = \sum_j \omega_{ji}\,\mathbf{e}_j, \tag{5.4}$$

when $\omega_1 = ds$, $\omega_2 = \omega_3 = 0$, and

$$\omega_{ji} = \begin{pmatrix} 0 & -\kappa\,ds & 0 \\ \kappa\,ds & 0 & -\tau\,ds \\ 0 & \tau\,ds & 0 \end{pmatrix}. \tag{5.5}$$

As an illustration of the use of moving frames we prove a theorem due to Dupin on triply orthogonal 1-parameter systems of surfaces in Euclidean 3-space, for example the quadrics of the confocal system

$$\frac{x^2}{a^2+\lambda} + \frac{y^2}{b^2+\lambda} + \frac{z^2}{c^2+\lambda} = 1.$$

Three surfaces of the system pass through an arbitrary point P, and the surfaces can be classified in three separate families. The theorem of Dupin states that the *curve of intersection of two surfaces belonging to two different families forms a line of curvature for each of these surfaces*.

To prove the theorem take as coordinates of P the parameters u^1, u^2, u^3 of the three surfaces which pass through P. Then the equations of structure referred to the natural frame are

$$d\mathbf{P} = du^1\mathbf{e}_1 + du^2\mathbf{e}_2 + du^3\mathbf{e}_3, \tag{5.6}$$

$$D\mathbf{e}_i = \omega^j{}_i\mathbf{e}_j. \tag{5.7}$$

The metric is given by

$$ds^2 = g_{11}(du^1)^2 + g_{22}(du^2)^2 + g_{33}(du^3)^2, \tag{5.8}$$

where $\qquad g_{ij} = \mathbf{e}_i.\mathbf{e}_j.$

Denote covariant differentiation with respect to u^1 by D_1. Then, since $\qquad \omega^j{}_i = \Gamma^j_{ik}\,du^k,$
from (5.7) we have

$$D_1\mathbf{e}_3 = \Gamma^1_{31}\mathbf{e}_1 + \Gamma^2_{31}\mathbf{e}_2 + \Gamma^3_{31}\mathbf{e}_3. \tag{5.9}$$

We recall that the parametric curve C_1 $(u^2 = u^2_0,\ u^3 = u^3_0)$ lying on the surfaces S_2 $(u^2 = u^2_0)$, S_3 $(u^3 = u^3_0)$ will be a line of curvature on surface S_3 if the normals to S_3 at neighbouring points P, P' on C_1 intersect. More precisely, we require that the vector $D_1\mathbf{e}_3\,du^1$ shall lie in the plane determined by $\mathbf{e}_1$, $\mathbf{e}_3$ when second-order terms are neglected. From (5.9) this will be the case if $\Gamma^2_{31} = 0$ and it is readily verified from the special form of (5.8) that this condition is satisfied. Hence C_1 is a line of curvature on S_3.

Similarly, C_1 will be a line of curvature on S_2 if $D_1\mathbf{e}_2\,du^1$ does not involve $\mathbf{e}_3$, i.e. if $\Gamma^3_{21} = 0$, which is certainly satisfied. This completes the proof of the theorem. Indeed we have proved a generalization of Dupin's theorem, since our proof is still valid when the system of surfaces is embedded in a three-dimensional *Riemannian* space which may not be Euclidean.

As a further illustration, consider a set of frames defined over a surface S, such that the vectors $\mathbf{e}_1$, $\mathbf{e}_2$ are orthogonal unit tangent vectors while $\mathbf{e}_3$ is the unit surface normal. Such a set of frames evidently exists, for $\mathbf{e}_1$, $\mathbf{e}_2$ can be along the directions of the principal curvature at each point P. Since the frames are orthonormal we may use the equations of structure in the covariant form VII

(19.10). If (ω_{ij}) is the connexion referred to this frame, then $\omega_{11} = \omega_{22} = \omega_{33} = 0$, $\omega_{12} = -\omega_{21}$, $\omega_{13} = -\omega_{31}$, and $\omega_{23} = -\omega_{32}$.

$$d\mathbf{P} = \omega_1 \mathbf{e}_1 + \omega_2 \mathbf{e}_2 + \omega_3 \mathbf{e}_3, \tag{5.10}$$

$$D\mathbf{e}_1 = \omega_{21} \mathbf{e}_2 + \omega_{31} \mathbf{e}_3, \tag{5.11}$$

$$D\mathbf{e}_2 = \omega_{12} \mathbf{e}_1 + \omega_{32} \mathbf{e}_3, \tag{5.12}$$

$$D\mathbf{e}_3 = \omega_{13} \mathbf{e}_1 + \omega_{23} \mathbf{e}_2. \tag{5.13}$$

The coefficients ω_{ij} are no longer given by Christoffel symbols since $\mathbf{e}_1$, $\mathbf{e}_2$, $\mathbf{e}_3$ are not necessarily specially related to a system of coordinates. Since the connexion has zero torsion,

$$d\omega_i + \omega_{ij} \wedge \omega_j = 0. \tag{5.14}$$

Write

$$d\omega_i = \tfrac{1}{2} c_{khi} \omega_k \wedge \omega_h, \tag{5.15}$$

where

$$c_{khi} + c_{hki} = 0. \tag{5.16}$$

Write

$$\omega_{ij} = \gamma_{ijk} \omega_k. \tag{5.17}$$

Then, from the equation $\omega_{ij} + \omega_{ji} = 0$, it follows that

$$\gamma_{ijk} + \gamma_{jik} = 0. \tag{5.18}$$

Thus we have from VII (19.18),

$$\gamma_{ijk} = \tfrac{1}{2}(c_{jki} + c_{kij} - c_{ijk}), \tag{5.19}$$

which expresses the γ's in terms of the c's.

The equations of structure of S are deduced from the previous equations by considering only displacements $d\mathbf{P}$ in the surface, i.e. displacements for which $\omega_3 = 0$. Equation (5.10) becomes

$$d\mathbf{P} = \omega_1 \mathbf{e}_1 + \omega_2 \mathbf{e}_2, \tag{5.20}$$

and the metric of S is

$$ds^2 = d\mathbf{P} . d\mathbf{P} = (\omega_1)^2 + (\omega_2)^2. \tag{5.21}$$

Equation (5.14) with $i = 3$, together with $\omega_3 = 0$, now gives

$$\omega_{31} \wedge \omega_1 + \omega_{32} \wedge \omega_2 = 0,$$

i.e.

$$\gamma_{312} \omega_2 \wedge \omega_1 + \gamma_{321} \omega_1 \wedge \omega_2 = 0,$$

from which

$$\gamma_{312} = \gamma_{321}. \tag{5.22}$$

In view of (5.18) this can also be written

$$\gamma_{132} = \gamma_{231}. \tag{5.23}$$

The second fundamental form Φ of the surface S is given by

$$\Phi = -D\mathbf{e}_3 . d\mathbf{P} = -(\omega_{13} \mathbf{e}_1 + \omega_{23} \mathbf{e}_2) . (\omega_1 \mathbf{e}_1 + \omega_2 \mathbf{e}_2) \quad \text{(from (5.13))}$$

$$= -\omega_1 \omega_{13} - \omega_2 \omega_{23}.$$

Using (5.17) we obtain
$$\Phi = -\gamma_{131}(\omega_1)^2 - \gamma_{132}\,\omega_1\,\omega_2 - \gamma_{231}\,\omega_1\,\omega_2 - \gamma_{232}(\omega_2)^2.$$
Writing
$$a_{11} = -\gamma_{131},\ a_{22} = -\gamma_{232},\ a_{12} = -\gamma_{132},\ a_{21} = -\gamma_{231}, \qquad (5.24)$$
we have from (5.23) that $a_{12} = a_{21}$, and hence Φ assumes the form
$$\Phi = a_{11}(\omega_1)^2 + 2a_{12}\,\omega_1\,\omega_2 + a_{22}(\omega_2)^2. \qquad (5.25)$$
Since ω_{13}, ω_{23} are differential forms representing the vectorial form $-D\mathbf{e}_3$, and ω_1, ω_2 represent the vectorial form $d\mathbf{P}$, it follows that the coefficients of (5.25) are components of a symmetric tensor. These components are uniquely determined by the choice of P and the vectors $\mathbf{e}_1$, $\mathbf{e}_2$. We now examine the geometrical significance of these coefficients.

Consider a curve C lying on S and having at P the same direction as $\mathbf{e}_1$. Then for displacements $d\mathbf{P}$ along this curve, $\omega_1 = ds$, $\omega_2 = 0$. From (5.11) we have
$$\frac{D\mathbf{e}_1}{ds} = \frac{\omega_{21}}{ds}\mathbf{e}_2 + \gamma_{311}\,\mathbf{e}_3$$

$$= \frac{\omega_{21}}{ds}\mathbf{e}_2 + a_{11}\,\mathbf{e}_3.$$

It follows that a_{11} is the normal curvature κ_n, so that
$$a_{11} = \kappa_n = \kappa\cos\theta, \qquad (5.26)$$
where κ is the curvature of C and θ is the angle between the principal normal to C and the normal to the surface.

Meusnier's theorem that the normal curvature is the same for all curves having the same direction at P is an immediate consequence. Moreover, the geodesic curvature $\kappa_g = \omega_{21}/ds = \kappa\sin\theta$ varies with the curve C because ω_{21} is not a tensorial form. This gives a new interpretation of the essential difference between the normal curvature and geodesic curvature of curves at P.

Using equation (4.25) we see that the geodesic torsion of C is
$$\tau_g = -\mathbf{e}_2.\frac{D\mathbf{e}_3}{ds}$$

$$= -\mathbf{e}_2.\left(\frac{\omega_{13}}{ds}\mathbf{e}_1 + \frac{\omega_{23}}{ds}\mathbf{e}_2\right)$$

$$= -\gamma_{231},$$
and hence, from (5.24), we have
$$\tau_g = a_{21}. \qquad (5.27)$$

We thus recover the result that the geodesic torsion of C is the same for all curves having the same direction as C at P.

The principal curvatures at P are given by finding the extremal values of
$$a_{11}(\omega_1)^2 + 2a_{12}\,\omega_1\,\omega_2 + a_{22}(\omega_2)^2,$$
subject to $\omega_1^2 + \omega_2^2 = ds^2$. By the usual method we find that
$$\frac{a_{11}\,\omega_1 + a_{12}\,\omega_2}{\omega_1} = \frac{a_{21}\,\omega_1 + a_{22}\,\omega_2}{\omega_2} = \kappa,$$
and hence κ satisfies $\kappa^2 - (a_{11} + a_{22})\kappa + a_{11}a_{22} - a_{12}^2 = 0$, so that
$$a_{11} + a_{22} = 2\mu, \tag{5.28}$$
where μ is the mean curvature of S at P. We have thus found a geometrical significance for each of the coefficients a_{11}, a_{12}, and a_{22}.

Consider the covariant derivative of the tensor $(a_{\alpha\beta})$ given by
$$Da_{\alpha\beta} = da_{\alpha\beta} - \sum_\gamma \omega_{\gamma\alpha}\,a_{\gamma\beta} - \sum_\gamma \omega_{\gamma\beta}\,a_{\alpha\gamma} = \sum_\gamma a_{\alpha\beta,\gamma}\,\omega_\gamma. \tag{5.29}$$

Writing out these equations in full we have
$$da_{11} - \omega_{21}a_{21} - \omega_{21}a_{12} = a_{11,1}\,\omega_1 + a_{11,2}\,\omega_2, \tag{5.30}$$
$$da_{12} - \omega_{21}a_{22} - \omega_{12}a_{11} = a_{12,1}\,\omega_1 + a_{12,2}\,\omega_2, \tag{5.31}$$
$$da_{22} - \omega_{12}a_{12} - \omega_{12}a_{21} = a_{22,1}\,\omega_1 + a_{22,2}\,\omega_2. \tag{5.32}$$

Consider now a displacement along the curve C tangent to $\mathbf{e}_1$ at P. Since $\omega_2 = 0$, we have from (5.30),
$$a_{11,1} = \frac{da_{11}}{ds} - 2a_{12}\frac{\omega_{21}}{ds}$$
$$= \frac{d\kappa_n}{ds} - 2\tau_g\,\kappa_g. \tag{5.33}$$

It follows that *the right-hand member of* (5.33) *has the same value for all curves which have the same direction at* P. This result was first obtained by Laguerre.

The second relation (5.31) gives
$$a_{12,1} = \frac{da_{12}}{ds} + (a_{11} - a_{22})\frac{\omega_{21}}{ds}$$
$$= \frac{d\tau_g}{ds} + 2\kappa_g(\kappa_n - \mu), \tag{5.34}$$
and the right-hand member of (5.34) again takes the same value for all curves with the same direction at P.

By considering second covariant derivatives of $(a_{\alpha\beta})$ similar expressions may be obtained which depend only upon the direction of the curves at P. It may be noted that the results obtained by these methods apply when the surface S is embedded in a three-dimensional Riemannian space whether Euclidean or not.

MISCELLANEOUS EXERCISES VIII

1. If a geodesic on a surface is a plane curve show that it is also a line of curvature, and prove that the converse holds.

2. Two surfaces S_1, S_2 intersect along a curve γ at a constant angle. Prove that the geodesic torsion of γ considered as a curve on S_1 is the same as its geodesic torsion considered as a curve on S_2. Prove also that if γ is a line of curvature for S_1, then it is also a line of curvature for S_2.

3. If two curves on a surface cut orthogonally, show that at the point of intersection the sum of their geodesic torsions is zero.

4. If the determinant $|\Omega_{\alpha\beta}|$ is non-zero, prove that a necessary and sufficient condition that the directions given by $p_{\alpha\beta}\lambda^\alpha\lambda^\beta = 0$ shall be conjugate is $\Omega^{\alpha\beta}p_{\alpha\beta} = 0$, where $(\Omega^{\alpha\beta})$ is the reciprocal tensor of $(\Omega_{\alpha\beta})$.

Show that the curves $\phi = $ constant, $\psi = $ constant form a conjugate system if $\Omega^{\alpha\beta}\phi_{,\alpha}\psi_{,\beta} = 0$.

5. Show that the expression

$$\frac{d\kappa}{ds}\cos\theta - \kappa\sin\theta\left(2\tau + 3\frac{d\theta}{ds}\right)$$

has the same value for all curves on a surface S which have the same direction at P, where θ is the angle between the principal normal and the normal to the surface.

EXERCISES

[Most of the following exercises are taken from examination papers of the University of Liverpool]

1. Prove that $\kappa/\tau = $ constant is necessary and sufficient for a helix. Prove that the curve

$$x = au, \qquad y = bu^2, \qquad z = cu^3$$

is a helix if and only if $3ac = \pm 2b^2$.

2. Two twisted curves C and C_1 are related in the following way: the principal normal at any point P of C meets C_1 at P_1, and PP_1 is the principal normal at P_1. Prove that (i) PP_1 is of constant length a, (ii) the osculating planes at P and P_1 cut at a constant angle α, and (iii) $\kappa \sin\alpha + \tau \cos\alpha = a^{-1}\sin\alpha$.

3. Prove that the centre of spherical curvature at the point $\mathbf{r}$ of a given curve is the point $\mathbf{r} + \rho\mathbf{n} + \rho'\sigma\mathbf{b}$.

Find the principal directions, curvature, and torsion of the locus of the centre of spherical curvature.

4. The parametric equations of a curve C are

$$x = 3au, \qquad y = 3bu^2, \qquad z = cu^3.$$

Prove that the osculating plane at the point $u = 1$ is

$$\frac{x}{a} - \frac{y}{b} + \frac{z}{c} = 1.$$

Prove that the tangents to C meet this plane at points which lie on the curve

$$x = a(2u+1), \qquad y = bu(u+2), \qquad z = cu^2.$$

Show that this curve is a parabola.

5. Write down the Serret–Frenet formulae for a space curve.

Calculate the curvature and torsion at the point u of the curve given by the parametric equations

$$x = a(3u - u^3), \qquad y = a(3u + u^3), \qquad z = 3au^2.$$

Show that the curve is a helix, and find the direction of the generators of the cylinder on which the curve is a helix.

6. Define the curvature κ and torsion τ of a twisted curve and establish the Serret–Frenet formulae

$$\mathbf{t}' = \kappa\mathbf{n}, \qquad \mathbf{n}' = \tau\mathbf{b} - \kappa\mathbf{t}, \qquad \mathbf{b}' = -\tau\mathbf{n}.$$

At a variable point P of a curve C a line l is drawn in the normal plane to make an angle θ with the principal normal at P. Prove that, if l is a tangent to another curve C_1 (an evolute of C), $d\theta/ds = -\tau$, where s, τ are the arc-length and torsion of C.

Show that C has infinitely many evolutes, and that, if C is a plane curve, its evolutes are helices.

Prove that, for any curve C and evolute C_1, the ratio of the curvature of C_1 to the torsion of C_1 is $\cot\theta$.

7. Define the intrinsic equations of a space curve. *Assuming* the existence of a curve with given intrinsic equations, prove that the curve is unique except for its position in space.

The intrinsic equations of a curve C are

$$\rho = a \sin(s/2a), \qquad \sigma = 2a.$$

Prove that C lies on a sphere of radius a.

Prove that the intrinsic equations of the locus of the centre of *circular* curvature of C are

$$\rho_1 = \tfrac{1}{2}a \sin(s_1/a), \qquad \sigma_1 = \tfrac{1}{3}a.$$

8. Define a helix of angle α, and prove that the condition $\kappa/\tau = \tan\alpha$ is necessary and sufficient for a curve to be a helix of angle α on some cylinder.

The generators of a cylinder are parallel to the z-axis, and its cross-section is the curve $r = ce^\theta$, where r, θ are plane polar coordinates and c is constant. Find parametric equations for the curve of intersection of this cylinder and the half-cone $x^2 + y^2 = a^2 z^2$, $z > 0$. Prove that this curve is a helix on the given cylinder, and find the angle of the helix. Prove also that the curve meets the generators of the given cone at a constant angle.

9. Write an account of geodesics on a surface.

10. Define direction coefficients on a surface, and obtain formulae for the sine and cosine of the angle between two given directions.

Prove that on a surface with metric

$$ds^2 = v^2\,du^2 + u^2\,dv^2,$$

the family of curves orthogonal to $uv = $ constant is given by $u/v = $ constant, and find the metric referred to $\log u \pm \log v$ as new parameters.

11. Define the geodesic curvature of a curve on a surface, and prove that

$$\kappa_g = \kappa \sin\theta = [\mathbf{N}, \mathbf{r}', \mathbf{r}''],$$

where κ_g is the geodesic curvature, κ the curvature of the curve, θ the angle between the principal normal to the curve and the unit surface normal $\mathbf{N}$, and $\mathbf{r}$ the position vector of a point on the curve.

Prove that the geodesic curvature of the curve $u = $ constant on the paraboloid $x = u\cos\theta$, $y = u\sin\theta$, $z = \tfrac{1}{2}au^2$ is $u^{-1}(1 + a^2u^2)^{-\frac{1}{2}}$.

12. Show that on the helicoid

$$x = u\cos v, \qquad y = u\sin v, \qquad z = \phi(u) + cv \quad (u \geqslant 0),$$

the orthogonal trajectories of the helices $u = $ constant are geodesics.

If $\phi(u) \equiv u^2$, prove that the geodesic through (u_0, v_0), in a direction orthogonal to $u = u_0$, meets the z-axis at a distance $c^2 \log(1 + c^{-2}u_0^2)$ from the point where $v = v_0$ meets this axis. Give a brief description of the surface and the family of geodesics considered in this question.

13. A surface of revolution is given by the equations

$$x = ak\cos u\cos v, \qquad y = ak\cos u\sin v,$$

$$z = a\int_0^u \sqrt{(1 - k^2\sin^2\theta)}\,d\theta.$$

Prove that the geodesic which passes through the point $(ak, 0, 0)$, and makes an angle α with the parallel through this point, is given by the equation

$$\tan u = \tan \alpha \sin kv.$$

Find the height above the plane $z = 0$ to which this geodesic rises.

14. A right circular cone has semi-vertical angle α. A curve is drawn on the cone to cut its generators at a constant angle β. Show that the coordinates of a point on the curve can be taken as $(r\cos\theta, r\sin\theta, r\cot\alpha)$, where

$$\frac{dr}{ds} = \cos\beta\sin\alpha, \qquad r\frac{d\theta}{ds} = \sin\beta.$$

Prove that the radius of curvature at a point on the curve is proportional to r.

15. A torus is determined by the parametric equations

$$x = (a+b\cos\phi)\cos\theta, \qquad y = (a+b\cos\phi)\sin\theta, \qquad z = b\sin\phi,$$

where $a > b$. Find the two differential equations which are the necessary condition that a curve on the surface is a geodesic. Show that only two geodesics lie completely in a plane parallel to the x, y plane.

A certain geodesic touches the curve $\phi = \frac{1}{2}\pi$. Show that it lies completely on the portion of the torus for which $-\frac{1}{2}\pi \leqslant \phi \leqslant \frac{1}{2}\pi$.

16. Defining geodesics on a surface by their variational property derive the Euler equations for geodesics and the single equation in terms of a general parameter. Prove that the single equation is sufficient.

Prove that the curves $u+v = $ constant are geodesics on a surface with metric
$$(1+u^2)\,du^2 - 2uv\,du\,dv + (1+v^2)\,dv^2.$$

17. A helicoid of pitch $2\pi a$ is generated by the screw motion of a straight line which meets the axis at an angle α. Find the orthogonal trajectories of the generators, and find the metric of the surface referred to the generators and their orthogonal trajectories as parametric curves.

Find a surface of revolution which is isometric with a region of the helicoid.

18. The metric on a surface is given by

$$ds^2 = E\,du^2 + 2F\,du\,dv + G\,dv^2.$$

Obtain a formula for the cosine of the angle between two directions on the surface at the point (u, v), and prove that the curves $\phi(u, v) = $ constant are orthogonal to the curves $\psi(u, v) = $ constant if

$$G\phi_1\psi_1 - F(\phi_1\psi_2 + \phi_2\psi_1) + E\phi_2\psi_2 = 0.$$

Given that $E = v^2$, $F = 0$, $G = u^2$, find the orthogonal trajectories of the curves $uv = $ constant, and find the metric of the surface referred to new parameters chosen so that the curves $uv = $ constant and their orthogonal trajectories become parametric curves. Deduce that the curves of *one* of these families are geodesics.

19. Write down the two equations for a geodesic, taking the arc length s as parameter. Show that, except for a parametric curve, either one of these equations implies the other.

An anchor ring is given by the equations

$$x = (b+a\cos u)\cos v, \qquad y = (b+a\cos u)\sin v, \qquad z = a\sin u,$$

where $b > a > 0$. Express du/dv as a function of u for a geodesic other than a parametric curve.

Prove that a geodesic which cuts the curve $u = 0$ at an angle $\alpha\,(0 < \alpha < \tfrac{1}{2}\pi)$ will also meet the curve $u = \pi$ provided that

$$\cos \alpha \leqslant (b-a)/(b+a).$$

20. On a surface with metric $ds^2 = e^2\,du^2 + g^2\,dv^2$, where e, g are functions of u and v, the geodesic curvature of a curve which cuts the curves $v =$ constant at an angle θ is

$$\kappa_g = \frac{d\theta}{ds} + \frac{1}{eg}\,(g_1 \sin \theta - e_2 \cos \theta),$$

where $g_1 = \partial g/\partial u$, etc. Assuming this formula, prove that, if C is the boundary of a simply connected region R of the surface,

$$\int_C \kappa_g\,ds = 2\pi - \int_R K\,dS,$$

where K is a function of position independent of C. Give an expression for K in terms of e, g and their derivatives.

Prove that $K = 0$ at every point of a plane, and that a surface at every point of which $K = 0$ is locally applicable to a plane.

21. Defining the geodesic curvature of a curve on a surface as the tangential component of the principal curvature prove that, on a surface with metric

$$\lambda^2\,du^2 + \mu^2\,dv^2,$$

the geodesic curvature of the curve $u = c$ is

$$\frac{1}{\lambda\mu}\,\frac{\partial \mu}{\partial u}.$$

A surface S is such that every geodesic circle with centre at a given point O on S has constant geodesic curvature (which may vary from one circle to another). Prove that S is applicable to a surface of revolution.

22. Obtain the equation for geodesics (excluding the parametric curves $u =$ constant) in the form

$$\frac{d}{ds}\left(\frac{\partial T}{\partial v'}\right) - \frac{\partial T}{\partial v} = 0, \qquad T = \tfrac{1}{2}(Eu'^2 + 2Fu'v' + Gv'^2).$$

(The Euler equations in Calculus of Variations may be assumed.)

A surface of revolution is obtained by rotating the curve

$$x = a \sin u, \qquad y = 0, \qquad z = a(\cos u + \log \tan \tfrac{1}{2}u) \quad (\tfrac{1}{2}\pi < u < \pi)$$

about the z-axis. Show that the geodesics are given by the equation

$$A(v^2 + \operatorname{cosec}^2 u) + Bv + C = 0,$$

where A, B, C are arbitrary constants, and deduce that the surface can be mapped on a plane in such a way that the geodesics correspond to straight lines.

23. (i) Explain what is meant by the statement that a region of a surface is simple and convex with respect to the geodesics. State an existence theorem concerning such regions.

On a circular cylinder of radius a find the maximum radius of a geodesic circle which is the boundary of a simple convex region, and prove that a geodesic circle of greater radius is convex but not simple.

(ii) On the paraboloid of revolution

$$x = 2au\cos v, \qquad y = 2au\sin v, \qquad z = au^2,$$

a geodesic C cuts a meridian orthogonally at the point $(2ab, 0, ab^2)$. Show that C is given by the equation

$$\frac{dv}{du} = \pm\frac{b}{u}\left(\frac{u^2+1}{u^2-b^2}\right)^{\frac{1}{2}},$$

and explain the alternative sign.

Without integrating this equation, deduce that C intersects itself infinitely often.

24. Give an intrinsic definition of the geodesics of a two-dimensional surface $\mathbf{r} = \mathbf{r}(u,v)$, and prove that a necessary and sufficient condition for the curve $u = u(t)$, $v = v(t)$ to be a geodesic is

$$U\frac{\partial T}{\partial \dot{v}} - V\frac{\partial T}{\partial \dot{u}} = 0,$$

where

$$U = \frac{d}{dt}\left(\frac{\partial T}{\partial \dot{u}}\right) - \frac{\partial T}{\partial u}, \qquad V = \frac{d}{dt}\left(\frac{\partial T}{\partial \dot{v}}\right) - \frac{\partial T}{\partial v},$$

$$T = \tfrac{1}{2}(E\dot{u}^2 + 2F\dot{u}\dot{v} + G\dot{v}^2),$$

and the other symbols have their usual meaning.

(i) Find the condition that the curves $v = $ constant shall be geodesics.

(ii) Show that the curves of the family $u = ct^2$, $v = ct^3$ are geodesics on the surface with metric

$$2v^2\,du^2 - 2uv\,du\,dv + u^2\,dv^2 \quad (u > 0, \ v > 0).$$

25. A system S of curves on a surface with parameters u, v is given by the differential equation $P\,du + Q\,dv = 0$, where P, Q are continuous functions of u and v which do not vanish together. Show that there is a system S' of curves orthogonal to S, given by a differential equation similar to the equation for S.

A system of curves on the cylinder $x^2 + (y-a)^2 = a^2$ is given by the intersection of the cylinder and the paraboloids $xz = \lambda y$, where λ is a parameter. Show that every orthogonal trajectory of the system lies on a sphere with fixed centre.

26. Defining a geodesic on a surface as a curve whose principal normal is normal to the surface, derive the equation for geodesics

$$\frac{\partial T}{\partial \dot{u}}V_t - \frac{\partial T}{\partial \dot{v}}U_t = 0$$

in terms of a general parameter t, and explain the notation.

Prove that if $F = 0$ and the curves $v = $ constant are geodesics, then E is independent of v.

The metric of a surface is $du^2 + \mu^2\,dv^2$, and the curves $v/u = $ constant are geodesics. Find a partial differential equation for μ as a function of u and v. Show that $\mu = $ constant is a solution, and find another solution of the form $\mu = \theta(u)/v$.

27. Given $E = 1$, $F = 0$, $G = \lambda^2$, express $\mathbf{r}_{11}$, $\mathbf{r}_{12}$, $\mathbf{r}_{22}$ in terms of λ and L, M, N, the symbols having their usual meaning. Deduce that
$$LN - M^2 = -\lambda\lambda_{11},$$
and explain the significance of this equation.

Define geodesic polar coordinates (describing carefully any theorems on geodesic parallels that you use), and prove that, if A is the area of a geodesic circle of centre P and radius r, the Gaussian curvature at P is
$$\lim_{r \to 0} \frac{\pi r^2 - A}{\frac{1}{12}\pi r^4}.$$

28. Define the principal curvatures at a point of a surface, and show that they are given by the equation
$$(EG - F^2)\kappa^2 - (EN - 2FM + GL)\kappa + (LN - M^2) = 0.$$
Prove that the roots of this equation are real.

Obtain the quadratic equation for the principal directions and discuss the case when these directions are indeterminate.

Find the Gaussian curvature and the mean curvature at a general point of the surface
$$x = a(u+v), \qquad y = b(u-v), \qquad z = uv,$$
and obtain parametric equations of the surface referred to parameters θ and ϕ such that the lines of curvature are given by $\theta = $ constant and $\phi = $ constant.

29. The metric of a surface is $du^2 + G\,dv^2$. By considering the identity $\mathbf{N}_{12} = \mathbf{N}_{21}$, derive the Codazzi equations
$$L_2 - M_1 = \tfrac{1}{2}M\frac{G_1}{G}, \qquad N_1 - M_2 = \tfrac{1}{2}\left(LG_1 - M\frac{G_2}{G} + N\frac{G_1}{G}\right).$$

Prove that, if the curves $v = $ constant are asymptotic, G is of the form $Au^2 + 2Bu + C$, where A, B, C are functions of v only. (The Gauss characteristic equation $LN - M^2 = -\tfrac{1}{2}G_{11} + \tfrac{1}{4}G_1^2/G$ may be assumed.)

Deduce, or prove otherwise, that, if a surface S and a system of geodesics on S are given, it is not in general possible to find a ruled surface S' such that S corresponds to S' isometrically with the given geodesics on S corresponding to generators on S'.

30. Define principal directions and lines of curvature on a surface, and prove that a necessary and sufficient condition for a curve to be a line of curvature is
$$\frac{d\mathbf{N}}{ds} \propto \frac{d\mathbf{r}}{ds},$$
where derivatives are taken along the curve.

(i) At a variable point of a line of curvature C a line is taken in the tangent plane and perpendicular to C. Prove that this line generates a developable surface.

(ii) Surfaces S, S' are in one-one correspondence so that, if P and P' are corresponding points, PP' is normal to S at P and normal to S' at P'. Prove that the lines of curvature on S correspond to lines of curvature on S'.

31. Write an account of **either** the conformal mapping of one surface on another, **or** ruled surfaces.

32. A ruled surface is defined by the equation
$$\mathbf{R}(u, v) \equiv \mathbf{r}(u) + v\mathbf{d}(u),$$
where $|\mathbf{d}(u)| = 1$. Show that the curve $\mathbf{R} \equiv \mathbf{r}(u)$ is the line of striction on the surface if $\frac{d}{du}\mathbf{d}(u)$ is either zero or in the direction of the normal to the surface at $(u, 0)$.

If the surface is not cylindrical and $\mathbf{R} = \mathbf{r}(u)$ is the line of striction, prove that the line of striction is an asymptotic line if $\mathbf{d}$ is parallel to $\kappa\mathbf{b} + \tau\mathbf{t}$.

[Along an asymptotic line $\mathbf{n} \cdot \mathbf{N} = 0$.]

33. A surface $\mathbf{r} \equiv \mathbf{r}(u, v)$ has a metric
$$ds^2 = \phi(u, v)\,du^2 + \psi(u, v)\,dv^2.$$

Find expressions for $\mathbf{r}_{uu}$, $\mathbf{r}_{uv}$, $\mathbf{r}_{vv}$ in terms of ϕ, ψ, $\mathbf{r}_u$, $\mathbf{r}_v$, $\mathbf{N}$, where $\mathbf{N}$ is the unit normal to the surface.

Indicate how the formula
$$LN - M^2 = -\tfrac{1}{2}\sqrt{(\phi\psi)}\left\{\frac{\partial}{\partial u}\left(\frac{1}{\sqrt{(\phi\psi)}}\frac{\partial\psi}{\partial u}\right) + \frac{\partial}{\partial v}\left(\frac{1}{\sqrt{(\phi\psi)}}\frac{\partial\psi}{\partial v}\right)\right\}$$

can be obtained. State its significance as regards 'intrinsic' properties of the surface.

Investigate for what value or values of n the surface with metric
$$ds^2 = v^n(du)^2 + u^n(dv)^2$$
is a developable.

34. Prove that the metric of a surface S in the neighbourhood of a curve C can be put in the form
$$ds^2 = E(u, v)\,du^2 + dv^2,$$
where C is the curve $v = 0$, and $E(u, 0) = 1$ for all u.

At each point of a curve C, lying on a surface S, that tangent line of S is taken which is orthogonal to C; the set of tangent lines generates a ruled surface S'. Prove that if C is the line of striction of S', it is a geodesic of S and also of S'.

35. A surface $\mathbf{r} \equiv \mathbf{r}(u, v)$ is such that, with the usual notation, $F = M = 0$. Show that, if $\mathbf{N}$ is the normal to the surface, then along any parametric curve $\mathbf{N}'$ is parallel to $\mathbf{r}'$.

A second surface is defined by
$$\mathbf{r}^*(u, v) = \mathbf{r}(u, v) + a\mathbf{N}(u, v),$$
where a is a constant. Show that the normals at corresponding points of the two surfaces (points with the same values of the parameters u, v) are parallel. Prove that the lines of curvature are the parametric curves.

36. Define an asymptotic direction and an asymptotic curve on a surface, and prove that there are in general two families of asymptotic curves on a surface.

A surface is obtained by rotating a parabola about the tangent at its vertex. Prove that the orthogonal projections of the asymptotic curves on a plane perpendicular to the axis of revolution are equiangular spirals.

37. A ruled surface is defined by the equation

$$\mathbf{r} \equiv \mathbf{r}_0(u) + v\mathbf{d}(u),$$

where $\mathbf{d}$ is a unit vector and u is the arc length of the curve $\mathbf{r} \equiv \mathbf{r}_0(u)$. Find the condition that $v = 0$ is the line of striction of the surface.

Show that if the surface is developable then it is either a cylinder or a surface generated by the tangents to the line of striction.

38. Prove that any curve is a geodesic on the surface generated by its binormals.

The curve C, $\mathbf{r} = \mathbf{r}(u)$, where u is the arc length, has constant torsion equal to 1. The surface S is generated by the binormals to C. Show that the geodesic on S which passes through $\mathbf{r}(u_0)$ and makes an angle α with C, in the direction of u increasing, lies between two fixed generators of S. Show also that the distance between these generators increases indefinitely as $\alpha \to 0$.

39. Find the asymptotic lines on the surface

$$z = y \sin x.$$

Give a sketch showing the projection of these curves on the plane $z = 0$ for $-\frac{1}{2}\pi \leqslant x \leqslant \frac{1}{2}\pi$.

State what is the section of the surface by a plane parallel to and near the tangent plane at a point where $x = \frac{1}{2}\pi$. What are the asymptotic lines when $x = \frac{1}{2}\pi$?

40. Prove that the Gauss curvature of the surface of revolution $r = \{u\cos v,\ u\sin v,\ f(u)\}$ is

$$K = \frac{f_1 f_{11}}{u(1+f_1^2)^2}.$$

If S is a surface of revolution of constant negative curvature $-k^2$ which cuts the xy-plane orthogonally along the circle $x^2+y^2 = a^2/k^2$, prove that S has no points outside the cylinder $x^2+y^2 = (a^2+1)/k^2$.

41. Define the lines of curvature on a surface and prove that the normals to the surface at the points of a curve C form a developable surface if and only if C is a line of curvature.

Prove that the developable is a cone if and only if the corresponding principal radius of curvature is constant along C, and show that if it is a cylinder the Gaussian curvature of the surface is zero at all points of C.

42. Prove that the normal curvature at any point of the surface $\mathbf{r} \equiv \mathbf{r}(u, v)$ in the direction (du, dv) is given by

$$\kappa_n = \frac{L\,du^2 + 2M\,du\,dv + N\,dv^2}{E\,du^2 + 2F\,du\,dv + G\,dv^2},$$

where L, M, N, E, F, G have their usual meaning. Show also that the curvature of a section of the surface by a plane through the same tangent line making an angle θ with the normal plane is $(\kappa_n \sec \theta)$.

Prove that the curvature κ at any point P of the curve of intersection of two surfaces is given by

$$\kappa^2 \sin^2\alpha = \kappa_1^2 + \kappa_2^2 - 2\kappa_1\kappa_2 \cos\alpha,$$

where κ_1, κ_2 are the normal curvatures of the surfaces in the direction of the curve at P, and α is the angle between their normals at that point.

43. Define an asymptotic line on a surface.

Prove that the generators of a ruled surface are asymptotic lines.

Prove that a surface cannot have a one-parameter family of plane asymptotic curves other than straight lines, unless it is a plane.

If all asymptotic lines are straight lines, then the surface is a developable or a quadric.

44. Write down an expression for Gaussian curvature in terms of the six fundamental coefficients of a surface.

Show that the surface generated by the helicoidal motion of a circle about a tangent is given by

$$x = a(1+\cos u)\cos v, \qquad y = a(1+\cos u)\sin v, \qquad z = a(v+\sin u),$$

the radius of the circle being a, and the pitch of the helicoid $2\pi a$. Prove that the Gaussian curvature at a distance c from the axis is $(3c-4a)/4a^3$.

45. Explain briefly and without details how the Gauss and Codazzi equations are derived.

By using the identity

$$\mathbf{r}_{11}\cdot\mathbf{r}_{22}-\mathbf{r}_{12}^2 = (\mathbf{r}_1\cdot\mathbf{r}_{22})_1-(\mathbf{r}_1\cdot\mathbf{r}_{12})_2$$

or otherwise, prove that any surface with metric $ds^2 = v^2\,du^2+u^2\,dv^2$ is developable.

46. Prove that a surface with constant Gaussian curvature $1/a^2$ is applicable to a sphere of radius a. State (without proof) a similar theorem for a surface of constant negative Gaussian curvature.

A surface is given by the parametric equations

$$x = \int_0^{u+v} (\tfrac{1}{2}\cos t)^{\frac{1}{2}}\,dt, \qquad y = \cos u+\cos v, \qquad z = \sin u-\sin v.$$

Calculate the second fundamental coefficients, and hence, or otherwise, prove that the surface is applicable to a sphere. Find the radius of this sphere.

[You may assume the formulae $K = (LN-M^2)/H^2$ for any parameters, and

$$K = -\frac{1}{2H}\left\{\frac{\partial}{\partial u}\left(\frac{G_1}{H}\right)+\frac{\partial}{\partial v}\left(\frac{E_2}{H}\right)\right\}$$

for orthogonal parameters.]

47. State (without proof) the Gauss–Bonnet theorem for a simply connected region of a surface. Deduce (i) an intrinsic definition of the Gaussian curvature at a point, (ii) the total curvature of an oval closed surface, and (iii) the total curvature of a torus.

Prove that the total curvature of a paraboloid of revolution is 2π. Is this also the total curvature of one sheet of a two-sheeted hyperboloid of revolution?

48. State the Gauss–Bonnet theorem for a simply connected region of a surface, and indicate (without giving all details) how it is proved. Deduce a formula for the total curvature of a closed surface of genus p.

The vertex of a right circular semi-cone of semi-vertical angle α is smoothed so that it is no longer a singularity. Find the total curvature of the surface.

49. State (without proof) the Gauss–Bonnet theorem for a simply connected region of a surface and deduce that the total curvature of a regular closed surface of connectivity c is $2\pi(3-c)$.

On a regular closed connected surface are n given points. Prove that, if n is sufficiently large, these points can be joined by arcs which do not intersect (except at the given points) in such a way that these arcs divide the surface into simply connected quadrangular regions. (Each such region has four of the given points as vertices and is bounded by four of the arcs.)

Prove that, if the connectivity of the surface is c, the number of quadrangular regions is $n+c-3$.

50. State the Gauss–Bonnet theorem on the total curvature of a simply connected region of a surface. Describe briefly, without giving all details, a proof of this theorem.

Show how the Gauss–Bonnet theorem may be used to find the total curvature of a region which is not simply connected.

The vertex of a right circular half-cone of semi-vertical angle α is smoothed so that there is no longer a singularity on the surface. Prove that the total curvature of the surface is increased by $2\pi(1-\sin\alpha)$.

51. Define the covariant derivative $\lambda^i{}_{,j}$ of a contravariant vector field λ^i, with respect to a connexion L^i_{jk} and prove that $\lambda^i{}_{,kj}-\lambda^i{}_{,jk}$ is of the form

$$\lambda^h L^i{}_{hjk}+2\lambda^i{}_{,h}\Omega^h_{jk},$$

where L^i_{hjk} and Ω^h_{jk} depend upon the connexion coefficients but not upon the components λ^i or their derivatives.

If a connexion with coefficients $\bar{L}^i_{jk}$ is given by the relations

$$\bar{L}^i_{jk} = L^i_{jk}+2\delta^i_j\mu_k,$$

where μ_k are the components of a vector, prove that

$$\bar{L}^i_{hjk}-L^i_{hjk} = 2\delta^i_h\left(\frac{\partial\mu_k}{\partial x^j}-\frac{\partial\mu_j}{\partial x^k}\right),$$

where $\bar{L}^i_{hjk}$ are the components of the curvature tensor corresponding to $\bar{L}^i_{jk}$.

52. The components of a symmetric affine connexion are Γ^k_{st}, and the components of an arbitrary contravariant vector are ξ^i. Prove that

$$\xi^i_{,st}-\xi^i_{,ts} = \xi^a B^i_{ast},$$

where the comma denotes covariant differentiation with respect to the connexion, and

$$B^i_{rst} = \partial_t\Gamma^i_{rs}-\partial_s\Gamma^i_{rt}+\Gamma^a_{rs}\Gamma^i_{at}-\Gamma^a_{rt}\Gamma^i_{as}.$$

The components of another symmetric affine connexion are $\bar{\Gamma}^k_{st}$. Writing $b^k_{st} = \bar{\Gamma}^k_{st}-\Gamma^k_{st}$, prove that b^k_{st} are the components of a tensor of type (1.2).

If B^i_{jkl}, $\bar{B}^i_{jkl}$ denote the curvature tensors corresponding to Γ^k_s and $\bar{\Gamma}^k_{st}$, prove that

$$B^i_{jkl}-\bar{B}^i_{jkl} = b^i_{jl,k}-b^i_{jk,l}+b^h_{jl}b^i_{hk}-b^h_{jk}b^i_{hl},$$

where covariant derivation is with respect to the connexion Γ^k_{st}.

53. Explain what is meant by

(i) an n-dimensional differentiable manifold of class ∞;

(ii) parallel transport;

(iii) the absolute differential of a tensor of type $(2, 1)$.

If $\boldsymbol{\lambda}_\alpha$ $(\alpha = 1, 2, ..., n)$ are n linearly independent tangent vectors defined over an n-dimensional differentiable manifold, obtain an affine connexion with respect to which the vectors are parallel. Show that if this connexion is symmetric, the vectors $[\boldsymbol{\lambda}_\alpha, \boldsymbol{\lambda}_\beta]$ are all zero vectors.

54. The components of the curvature tensor associated with a Riemannian metric tensor of components g_{ij} are $R^h{}_{ijk}$. Establish the Bianchi identity

$$R^h{}_{ijk,l} + R^h{}_{ikl,j} + R^h{}_{ilj,k} = 0.$$

[*If normal coordinates are used, their relevant properties should be stated clearly.*]

If $\qquad R_{ij} = R^h{}_{ijh}, \qquad R^l_j = g^{li} R_{ij}, \qquad R = R^i_i,$

show that $\qquad R^l_{j,l} = \tfrac{1}{2} \partial_j R.$

Hence, or otherwise, show that if $n > 2$, the scalar curvature R of an Einstein space is constant.

Show also that if the relation

$$R_{hijk} = b(g_{hj} g_{ik} - g_{hk} g_{ij})$$

holds at every point of a neighbourhood, then b is constant.

55. Define the Beltrami differential parameters $\Delta_1(u, v)$, $\Delta_2(u)$ on a Riemannian manifold M.

If M is orientable and closed, show that for any two functions u, v which are C^2 everywhere on M

$$\int_M u\, \Delta_2(v)\, d\tau = -\int_M \Delta_1(u, v)\, d\tau = \int_M v \Delta_2(u)\, d\tau,$$

where $d\tau$ is the element of volume.

Prove that u satisfies the equation

$$\Delta_2(u) = 0$$

at all points of M if and only if u is a constant.

56. (i) Given that E_n, F_p are vectors spaces of dimension n and p respectively, explain what is meant by

(1) the tensor product of E_n and F_p;

(2) an affine tensor attached to E_n.

Prove that t_{ij} are the components of a covariant tensor attached to E_n if and only if, for arbitrary vectors with components x^i, y^i, the expression $t_{ij} x^i y^j$ is invariant under any change of base.

(ii) If $a_{\alpha\beta}$, b_α are numbers satisfying

$$a_{\alpha\beta} = a_{\beta\alpha}, \qquad a_{\alpha\beta} b_\gamma + a_{\gamma\alpha} b_\beta + a_{\beta\gamma} b_\alpha = 0,$$

prove that either all the $a_{\alpha\beta}$ are zero or all the b_α are zero.

A Riemannian space with curvature tensor R_{hijk} admits a non-zero tensor field K_{lm} such that

$$R_{hijk}K_{lm} + R_{lmhi}K_{jk} + R_{jklm}K_{hi} = 0.$$

Prove that the space is flat.

57. Define the Christoffel symbols $[k, ij]$, Γ^i_{jk} in terms of the fundamental tensor g_{ij} of a Riemannian space and prove that

$$\Gamma^i_{ij} = \frac{1}{\sqrt{g}} \frac{\partial \sqrt{g}}{\partial x^j},$$

where $g = |g_{ij}|$.

Define the operator ∇^2 for a Riemannian space and prove that

$$\nabla^2 V = \frac{1}{\sqrt{g}} \frac{\partial}{\partial x^i}\left(\sqrt{g}\, g^{ij} \frac{\partial V}{\partial x^j}\right).$$

The metric of a 3-space is $u^{-2}(dx^2 + dy^2 + dz^2)$, where $u = 1 + \frac{1}{4}Kr^2$, K is a constant, and $r^2 = x^2 + y^2 + z^2$. Find the general solution of $\nabla^2 V = 0$, where V is a function of r alone.

58. Let g_{ij}, R_{hijk} denote the fundamental and curvature tensors on a Riemannian space V_n ($n > 2$). Assuming Bianchi's identity

$$R_{hijk,l} + R_{hilj,k} + R_{hikl,j} = 0,$$

prove that, if $\qquad R_{hijk} = K(g_{hj}g_{ik} - g_{hk}g_{ij})$

at every point, then K is constant.

On a space V_n ($n > 2$) of non-zero constant curvature, the covariant derivative of a tensor (a_{ij}) is zero. Prove that $a_{ij} = a_{ji}$ and that $a_{ij} = \rho g_{ij}$, where ρ is constant.

59. Obtain the Gauss and Codazzi equations for a surface with the metric

$$ds^2 = du^2 + \lambda^2 dv^2.$$

Prove that the Gaussian curvature is $-\dfrac{1}{\lambda} \dfrac{\partial^2 \lambda}{\partial u^2}$.

60. Define an n-dimensional manifold of class r.

Prove that a compact n-dimensional Riemannian manifold M with positive definite metric cannot have a scalar ϕ defined on it such that $\nabla^2 \phi > 0$ at every point.

Hence prove that the vanishing of the rth covariant derivatives of a contravariant vector field on M implies the vanishing of the first.

Give an example showing that this may be false for a non-compact Riemannian manifold.

61. Give two definitions of a geodesic line on a surface. Deduce from one of these the equations

$$\frac{d}{ds}(Eu' + Fv') = \frac{1}{2}(E_1 u'^2 + 2F_1 u'v' + G_1 v'^2),$$

$$\frac{d}{ds}(Fu' + Gv') = \frac{1}{2}(E_2 u'^2 + 2F_2 u'v' + G_2 v'^2).$$

Two surfaces touch along a curve. If the curve is a geodesic on one of the surfaces, prove that it is a geodesic on the other.

62. Prove that a necessary and sufficient condition that the tangents to the curves $x^2 = $ constant on a V_2 be parallel with respect to a curve C, is that C is an integral curve of

$$\begin{Bmatrix} 2 \\ 1\,i \end{Bmatrix} dx^i = 0 \quad (i = 1, 2).$$

Show that, if the tangents to the parametric curves of either family are parallel with respect to the curves of the other family, then g_{11} is a function of x^1 only and g_{22} is a function of x^2 only.

$$[ds^2 = g_{ij}\,dx^i dx^j.]$$

63. Assuming Euler's equations, prove that the differential equations of geodesics in a V_n are

$$\frac{d^2x^i}{ds^2} + \begin{Bmatrix} i \\ j\,k \end{Bmatrix}\frac{dx^j}{ds}\frac{dx^k}{ds} = 0 \quad (i = 1, 2,...,n),$$

where s is the arc length of the curve.

If the metric of the V_n is

$$ds^2 = g_{ij}\,dx^i dx^j + g_{nn}(dx^n)^2 \quad (i,j = 1, 2,...,n-1),$$

where all the g_{ij} are independent of x^n, prove that the hypersurface $x^n = $ constant is totally geodesic in V_n (i.e. every geodesic in the hypersurface is a geodesic in V_n).

64. Give two definitions of the lines of curvature of a surface. Use one of these to show that the lines of curvature satisfy the differential equation

$$(EM - FL)\,du^2 + (EN - GL)\,du\,dv + (FN - GM)\,dv^2 = 0.$$

If the parametric curves of $S: \mathbf{r} = \mathbf{r}(u, v)$ are lines of curvature of S and if $\mathbf{N}$ denotes the unit normal to S at (u, v), prove that the lines $u = $ constant, $v = $ constant are lines of curvature of $S': \mathbf{r} = \mathbf{r}(u, v) + a\mathbf{N}$, where a is a constant.

Prove also that the Gauss curvatures of S and S', at points with the same parameters, are in constant ratio if and only if the principal curvatures of S satisfy a certain bilinear relation.

65. Prove that the element of length of the surface generated by the binormals to a curve $\mathbf{r} = \mathbf{r}(u)$, where u is the arc length of the curve, is given by

$$ds^2 = (1 + v^2\tau^2)\,du^2 + dv^2,$$

where v is the distance measured along a binormal from the base curve.

Prove also that the curved asymptotic lines of the surface satisfy

$$\frac{dv}{du} = v^2\kappa\tau - \frac{v\tau'}{\tau} + \frac{\kappa}{\tau},$$

and solve this equation when the curve is a right circular helix.

66. Show that if $a_{hijk}\lambda^h\mu^i\lambda^j\mu^k = 0$, where (λ^i), (μ^i) are any two arbitrary vectors in V_n, then

$$a_{hijk} + a_{hkji} + a_{jihk} + a_{jkhi} = 0.$$

If in addition

$$a_{hijk} = -a_{ihjk}, \qquad a_{hijk} = a_{jkhi},$$
and
$$a_{hijk} + a_{hjki} + a_{hkij} = 0,$$
show that
$$a_{hijk} = 0.$$

67. Prove that, in a V_3, the components R_{hijk} of the curvature tensor can be expressed in terms of the Ricci tensor and the fundamental metric tensor by the relation

$$R_{hijk} = g_{hk} R_{ij} + g_{ij} R_{hk} - g_{hj} R_{ik} - g_{ik} R_{hj} - \tfrac{1}{2} R(g_{hk} g_{ij} - g_{hj} g_{ik}),$$

where $R = g^{ij} R_{ij}$.

Deduce that if a V_3 is an Einstein space, then it is a space of constant curvature.

SUGGESTIONS FOR FURTHER READING

BIANCHI, L., *Lezioni di Geometria Differenziale*, I, II, Zanichelli, Bologna (1923).

BLASCHKE, W., *Einführung in die Differentialgeometrie*, Springer (1950).

CARTAN, É., *Leçons sur la géométrie des espaces de Riemann*, Gauthier–Villars (1946).

DARBOUX, G., *Théorie générale des surfaces*, I, II, III, IV, Gauthier–Villars (1888–96).

EISENHART, L. P., *A Treatise on the Differential Geometry of Curves and Surfaces*, Ginn and Co. (1909).

—— *Riemannian Geometry*, Princeton University Press (1926; 2nd edition, 1949).

FLANDERS, H., *Differential Forms*, Academic Press (1963).

GUGGENHEIMER, H., *Differential Geometry*, McGraw-Hill (1963).

HELGASON, S., *Differential Geometry and Symmetric Spaces*, Academic Press (1962).

HODGE, W. V. D., *The Theory and Applications of Harmonic Integrals*, Cambridge (1941; 2nd edition, 1952).

KOBAYASHI, S., and NOMIZU, K., *Foundations of Differential Geometry*, Interscience Publishers (1963).

LANG, S., *Introduction to Differentiable Manifolds*, Interscience Publishers (1962).

LICHNEROWICZ, A., *Théorie globale des connexions et des groupes d'holonomie*, C.N.D.R. Monografie Matematiche, Rome (1955).

—— *Géométrie des groupes de transformations*, Dunod, Paris (1958).

NOMIZU, K., *Lie Groups and Differential Geometry*, Tokyo (1956).

DE RHAM, G., *Variétés différentiables*, Hermann (1955).

SCHOUTEN, J. A., *Ricci-Calculus*, Springer (1954).

STRUIK, D. T., *Lectures on Classical Differential Geometry*, Addison-Wesley Press (1950).

YANO, K., and BOCHNER, S., *Curvature and Betti Numbers*, Princeton (1953).

INDEX

PRINTED IN GREAT BRITAIN
AT THE UNIVERSITY PRESS, OXFORD
BY VIVIAN RIDLER
PRINTER TO THE UNIVERSITY